Use R!

Series Editors:
Robert Gentleman   Kurt Hornik   Giovanni Parmigiani

For other titles published in this series, go to
http://www.springer.com/series/6991

M. Henry H. Stevens

A Primer of Ecology with R

M. Henry H. Stevens
Department of Botany
Miami University
Oxford, OH 45056, USA
hstevens@muohio.edu

ISBN 978-0-387-89881-0 e-ISBN 978-0-387-89882-7
DOI 10.1007/978-0-387-89882-7
Springer Dordrecht Heidelberg London New York

Library of Congress Control Number: 2009927709

Printed on acid-free paper

9 8 7 6 5 4 3 2 (Corrected at 2nd printing 2010)

Springer is part of Springer Science+Business Media (www.springer.com)

To my perfect parents, Martin and Ann,
to my loving wife, Julyan,
and to my wonderfully precocious kids, Tessa and Jack.

Preface

Goals and audience

In spite of the presumptuous title, my goals for this book are modest. I wrote it as

- the manual I wish I had in graduate school, and
- a primer for our graduate course in Population and Community Ecology at Miami University[1]

It is my hope that readers can enjoy the *ecological content* and ignore the R code, if they care to. Toward this end, I tried to make the code easy to ignore, by either putting boxes around it, or simply concentrating code in some sections and keeping it out of other sections.

It is also my hope that ecologists interested in learning R will have a rich yet gentle introduction to this amazing programming language. Toward that end, I have included some useful functions in an R package called `primer`. Like nearly all R packages, it is available through the R projects repositories, the CRAN mirrors. See the Appendix for an introduction to the R language.

I have a hard time learning something on my own, unless I can *do* something with the material. Learning ecology is no different, and I find that my students and I learn theory best when we write down formulae, manipulate them, and explore consequences of rearrangement. This typically starts with copying down, verbatim, an expression in a book or paper. Therefore, I encourage readers to take pencil to paper, and fingers to keyboard, and copy expressions they see in this book. After that, make sure that what I have done is correct by trying some of the same rearrangements and manipulations I have done. In addition, try things that aren't in the book — have fun.

A pedagogical suggestion

For centuries, musicians and composers have learned their craft in part by *copying by hand* to works of others. Physical embodiment of the musical notes

[1] Miami University is located in the Miami River valley in Oxford, Ohio, USA; the region is home to the Myaamia tribe that dwelled here prior to European occupation.

and their sequences helped them learn composition. I have it on great authority that most theoreticians (and other mathematicians) do the same thing — they start by copying down mathematical expressions. This physical process helps get the content under their skin and through their skull. I encourage you to do the same. Whether otherwise indicated or not, let the first assigned problem at the end of each chapter be to copy down, with a pencil and paper, the mathematical expression presented in that chapter. In my own self-guided learning, I have often taken this simple activity for granted and have discounted its value — to my own detriment. I am not surprised how often students also take this activity for granted, and similarly suffer the consequences. *Seeing* the logic of something is not always enough — sometimes we have to actually *recreate* the logic for ourselves.

Comparison to other texts

It may be useful to compare this book to others of a similar ilk. This book bears its closest similarities to two other wonderful primers: Gotelli's *A Primer of Ecology*, and Roughgarden's *Primer of Theoretical Ecology*. I am more familiar with these books than any other introductory texts, and I am greatly indebted to these authors for their contributions to my education and the discipline as a whole.

My book, geared toward graduate students, includes more advanced material than Gotelli's primer, but most of the ecological topics are similar. I attempt to start in the same place (e.g., "What is geometric growth?"), but I develop many of the ideas much further. Unlike Gotelli, I do not cover life tables at all, but rather, I devote an entire chapter to *demographic matrix models*. I include a chapter on community structure and diversity, including *multivariate distances*, *species-abundance distributions*, *species-area relations*, and *island biogeography*, as well as *diversity partitioning*. My book also includes code to implement most of the ideas, whereas Gotelli's primer does not.

This book also differs from Roughgarden's primer, in that I use the Open Source R programming language, rather than Matlab®, and I do not cover physiology or evolution. My philosphical approach is similar, however, as I tend to "talk" to the reader, and we fall down the rabbit hole together[2].

Aside from Gotelli and Roughgarden's books, this book bears similarity in content to several other wonderful introductions to mathematical ecology or biology. I could have cited repeatedly (and in some places did so) the following: Ellner and Guckenheimer (2006), Gurney and Nisbet (1998), Kingsland (1985), MacArthur (1972), Magurran (2004), May (2001), Morin (1999), Otto and Day (2006), and Vandermeer and Goldberg (2007). Still others exist, but I have not yet had the good fortune to dig too deeply into them.

Acknowledgements

I am indebted to Scott Meiners and his colleagues for their generous sharing of data, metadata, and statistical summaries from the Buell-Small Succession

[2] From *Alice's Adventures in Wonderland* (1865), L. Carroll (C. L. Dodgson).

Study (http://www.ecostudies.org/bss/), a 50+ year study of secondary succession (supported in part by NSF grant DEB-0424605) in the North American temperate deciduous forest biome. I would like to thank Stephen Ellner for Ross's Bombay death data and for R code and insight over the past few years. I am also indebted to Tom Crist and his colleagues for sharing some of their moth data (work supported by The Nature Conservancy Ecosystem Research Program NSF DEB-0235369).

I am grateful for the generosity of early reviewers and readers, each of whom has contributed much to the quality of this work: Jeremy Ash, Tom Crist, David Gorchov, Raphael Herrera-Herrera, Thomas Petzoldt, James Vonesh, as well as several anonymous reviewers, and the students of our Population and Community Ecology class. I am also grateful for the many conversations and emails shared with four wonderful mathematicians and theoreticians: Jayanth Banavar, Ben bolker, Stephen Ellner, Amit Shukla, and Steve Wright — I never have a conversation with these people without learning something. I have been particularly fortunate to have team-taught Population and Community Ecology at Miami University with two wonderful scientists and educators, Davd Gorchov and Thomas Crist. Only with this experience, of working closely with these colleagues, have I been able to attempt this book. It should go without saying, but I will emphasis, that the mistakes in this book are mine, and there would be many more but for the sharp eyes and insightful minds of many other people.

I am also deeply indebted to the R Core Development Team for creating, maintaining and pushing forward the R programming language and environment [173]. Like the air I breathe, I cannot imagine my (professional) life without it. I would especially like to thank Friedrich Leisch for the development of `Sweave`, which makes literate programming easy [106]. Because I rely on Aquamacs, ESS, LaTeX, and a host of other Open Source programs, I am deeply grateful to those who create and distribute these amazing tools.

A few other R packages bear special mention. First, Ben Bolker's text [13] and packages for modeling ecological data (`bbmle` and `emdbook`) are broadly applicable. Second, Thomas Petzoldt's and Karsten Rinke's `simecol` package provides a general computational architecture for ecological models, and implements many wonderful examples [158]. Much of what is done in this primer (especially in chapters 1, 3–6, 8) can be done with `simecol`, and sometimes done better. Third, Robin Hankin's `untb` package is an excellent resource for exploring ecological neutral theory (chapter 10) [69]. Last, I relied heavily on the `deSolve` [190] and `vegan` packages [151].

Last, and most importantly, I would like to thank those to whom this book is dedicated, whose love and senses of humor make it all worthwhile.

Martin Henry Hoffman Stevens
Oxford, OH, USA, Earth
February, 2009

Contents

Part I Single Species Populations

Part III Special Topics

Part I

Single Species Populations

1

Simple Density-independent Growth

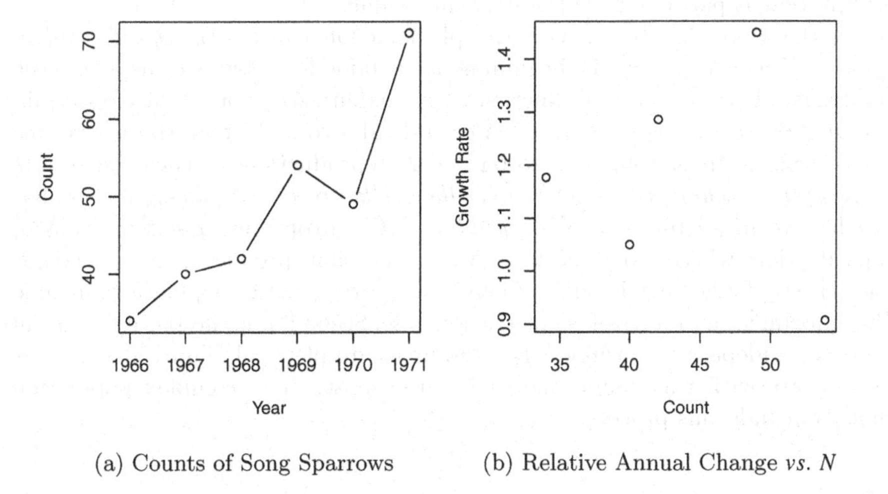

(a) Counts of Song Sparrows

(b) Relative Annual Change *vs.* N

Fig. 1.1: Song Sparrow (*Melospiza melodia*) counts in Darrtown, OH, USA. From Sauer, J. R., J. E. Hines, and J. Fallon. 2005. The North American Breeding Bird Survey, Results and Analysis 1966–2004. Version 2005.2. USGS Patuxent Wildlife Research Center, Laurel, MD.

Between 1966 and 1971, Song Sparrow (*Melospiza melodia*) abundance in Darrtown, OH, USA, seemed to increase very quickly, seemingly unimpeded by any particular factor (Fig. 1.1a). In an effort to manage this population, we may want to predict its future population size. We may also want to describe its growth rate and population size in terms of mechanisms that could influence its growth rate. We may want to compare its growth and relevant mechanisms to those of other Song Sparrow populations or even to other passerine populations. These goals, *prediction*, *explanation*, and *generalization*, are frequently the goals toward which we strive in modeling anything, including populations, communities, and ecosystems. In this book, we start with simple models of populations and slowly add complexity to both the form of the model, and the analysis of its behavior. As we move along, we also practice applying these models to real populations.

What is a model, and why are they important in ecology? First, a model is an abstraction of reality. A road map, for instance, that you might use to find your way from Mumbai to Silvasaa is a model of the road network that allows you to predict which roads will get you to Silvasaa. As such, it *excludes* far more information about western India than it includes. Partly as a result of excluding this information, it is eminently useful for planning a trip. Models in ecology are similar. Ecological systems are potentially vastly more complex than just about any other system one can imagine for the simple reason that ecosystems are composed of vast numbers of genetically distinct individuals, each of which is composed of at least one cell (e.g., a bacterium), and all of these individuals may interact, at least indirectly. Ecological models are designed to capture particular key features of these potentially complex systems. The goal is to capture a key feature that is particularly interesting and useful.

In this book, we begin with the phenomenon called *density-independent growth.* We consider it at the beginning of the book for a few reasons. First, the fundamental process of reproduction (e.g., making seeds or babies) results in a *geometric series.*[1] For instance, one cell divides to make two, those two cells each divide to make four, and so on, where reproduction for each cell results in two cells, *regardless of how many other cells are in the population* — that is what we mean by *density-independent.* This myopically observed event of reproduction, whether one cell into two, or one plant producing many seeds, is the genesis of a geometric series. Therefore, most models of populations include this fundamental process of geometric increase. Second, populations can grow in a density-independent fashion when resources are plentiful. Third, it behooves us to start with this simple model because most, more complex population models include this process.

1.1 A Very Specific Definition

Density-independence in a real population is perhaps best defined quite specifically and operationally as a lack of a statistical relation between the density of a population, and its *per capita* growth rate. The power to detect a significant relation depends, in part, on those factors which govern power in statistical relations between any two continuous variables: the number of observations, and the range of the predictor variable. Therefore, our conclusion, that a particular population exhibits density-independent growth, may be trivial if our sample size is small (i.e., few generations sampled), or if we sampled the population over a very narrow range of densities. Nonetheless, it behooves us to come back

[1] A mathematical series is typically a list of numbers that follow a rule, and that you sum together.

to this definition if, or when, we get caught up in the biology of a particular organism.

We could examine directly the relation between the growth rate and population size of our Song Sparrow population (Fig. 1.1b). We see that there is no apparent relation between the growth rate and the density of the population.[2] That is what we mean by "density-independent growth."

1.2 A Simple Example

Let's pretend you own a small piece of property and on that property is a pond. Way back in June 2000, as a present for Mother's Day, you were given a water lily (*Nymphaea odorata*), and you promptly planted it, with its single leaf or frond, in the shallows of your pond. The summer passes, and your lily blossoms, producing a beautiful white flower. The following June (2001) you notice that the lily grew back, and that there were three leaves, not just one. Perhaps you cannot discern whether the three leaves are separate plants. Regardless, the pond now seems to contain three times the amount of lily pad that it had last year.

The following June (2002) you are pleased to find that instead of three leaves, you now have nine. In June 2003, you have 27 leaves, and in 2004 you have 81 leaves (Fig. 1.3). How do we describe this pattern of growth? How do we predict the size of the population in the future? Can we take lessons learned from our water lily and apply it to white-tailed deer in suburbia, or to bacteria in the kitchen sink?

We rely on theory to understand and describe the growth of our water lily in such a way as to apply it to other populations. The role of theory, and theoretical ecology, is basically three-fold. We might like theory to allow us to describe the pattern in sufficient detail (1) to provide a *mechanistic explanation* for how the lily grew as fast or as slowly as it did, (2) allow us to make *predictions* about the population size in the future, and (3) allow us to *generalize* among other lily populations or other species. These goals typically compete with each other, so real models are mathematical descriptions that result from tradeoffs among these goals which depend precisely on our particular needs [109].

1.3 Exploring Population Growth

So, how fast are the lilies of the example growing? Between years 1 and 2, it increased by 2 fronds; between years 2 and 3, it increased by 6. In subsequent years it increased by 18, and 54 fronds. The number changes each year (Fig. 1.3), so how do we predict the future, or even explain the present? Can we find a general rule that works for any year?

---

[2] Consider that if area is fixed, "count" or population size differs from density by a fixed multiplier

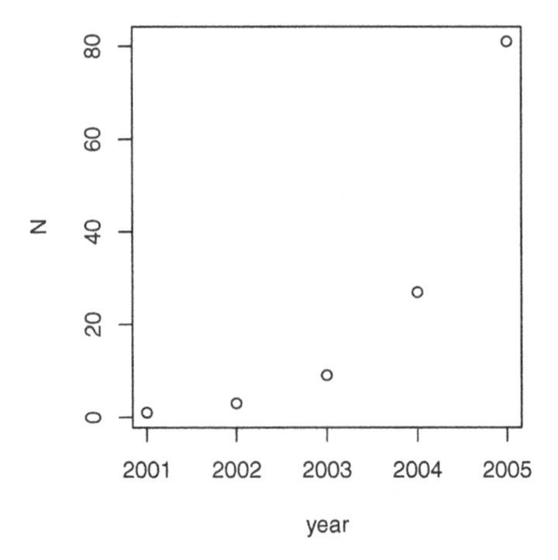

Fig. 1.2: Hypothetical water lily population size through time.

Simple Graphing of Population Size (Fig. 1.3)

Here we create two vectors: population size, N, and years. Using `c()` allows us to create an arbitrary vector, and the colon, :, provides a sequence of consecutive integers.

```
> N <- c(1, 3, 9, 27, 81)
> year <- 2001:2005
> plot(year, N)
```

The lily population (Fig. 1.3) increases by a different amount each year. What about proportions — does it increase by a different proportion each year? Let's divide each year's population size by the previous year's size, that is, perform N_{t+1}/N_t for all t, where t is any particular point in time, and $t+1$ is the next point in time. For N, that amounts to $3/1$, $9/3$, ... What do we notice?

Vectorized math

Here we divide each element of one vector (the second through fifth element of N) by each element of another vector (the first through fourth elements of N).

```
> rates = N[2:5]/N[1:4]
> rates

[1] 3 3 3 3
```

Lo, and behold! all of these proportions are the same: 3. Every year, the lilies increase by the same proportion — they triple in abundance, increasing by 200%. That is the general rule that is specific to our population. It is general because it appears to apply to each year, and could potentially describe other populations; it is not, for instance, based on the photosynthetic rate in lily pads. It is specific because it describes a specific rate of increase for our water lily population. We can represent this as

$$N_{2002} = 3 \times N_{2001}$$

where N_{2002} is the size of the population in 2002. If we rearrange this, dividing both sides by N_{2001}, we get

$$\frac{N_{2002}}{N_{2001}} = 3$$

where 3 is our rate of increase.

Generalizing this principle, we can state

$$N_{t+1} = 3N_t$$

$$\frac{N_{t+1}}{N_t} = 3.$$

1.3.1 Projecting population into the future

The above equations let us describe the rate of change population size N from one year to the next, but how do we predict the size 10 or 20 years hence? Let's start with one year and go from there.

$$\begin{aligned} N_{2002} &= 3N_{2001} \\ N_{2003} &= 3N_{2002} = 3\,(3N_{2001}) \\ N_{2004} &= 3N_{2003} = 3\,(3N_{2002}) = 3\,(3\,(2N_{2001})) \end{aligned}$$

So, ... what is the general rule that is emerging for predicting water lily N, some years hence? Recall that $3 \times 3 \times 3 = 3^3$ or $a \times a \times a = a^3$, so more generally, we like to state

$$N_t = \lambda^t N_0 \qquad (1.1)$$

where t is the number of time units (years in our example), N_0 is the size of our population at the beginning, λ is the per capita rate of increase over the specified time interval and N_t is the predicted size of the population after t time units.

Note that lambda, λ, is the *finite rate of increase*. It is the per capita rate of growth of a population *if the population is growing geometrically*. We discuss some of the finer points of λ in Chapter 2. We can also define a related term, the *discrete growth factor*, r_d, where $\lambda = (1 + r_d)$.

Note that "time" is not in calendar years but rather in years since the initial time period. It is also the number of time steps. Imagine that someone sampled a population for five years, 1963–1967, then we have four time steps, and $t = 4$.

Projecting population size

Here we calculate population sizes for 10 time points beyond the initial. First we assign values for N_0, λ, and time.

```
> N0 <- 1
> lambda <- 2
> time <- 0:10
```

Next we calculate N_t directly using our general formula.

```
> Nt <- N0 * lambda^time
> Nt

 [1]    1    2    4    8   16   32   64  128  256  512 1024
```

1.3.2 Effects of initial population size

Let's explore the effects of initial population size. First, if we just examine equation 1.1, we will note that $N_t = N_0 \times \text{stuff}$. Therefore, if one population starts out twice as big as another, then it will always be twice as big, given geometric growth (Fig. 1.3a). We see that small initial differences diverge wildly over time (Fig. 1.3a), because "twice as big" just *looks* a lot bigger as the magnitude increases.

Effects of Initial Population Size

We first set up several different initial values, provide a fixed λ, and set times from zero to 4.

```
> N0 <- c(10, 20, 30)
> lambda <- 2
> time <- 0:4
```

We calculate population sizes at once using **sapply** to *apply* a function (**n*lambda^time**) to each element of the first argument (each element of N0).

```
> Nt.s <- sapply(N0, function(n) n * lambda^time)
> Nt.s

     [,1] [,2] [,3]
[1,]   10   20   30
[2,]   20   40   60
[3,]   40   80  120
[4,]   80  160  240
[5,]  160  320  480
```

The result is a matrix, and we see N_0 in the first row, and each population is in its own column. Note that population 2 is always twice as big as population 1.

If we change the y-axis scale to logarithms, however, we see that the lines are parallel! Logarithms are a little weird, but they allow us to look at, and think about, many processes where rates are involved, or where we are especially interested in the relative magnitudes of variables. Consider the old rule we get when we take the logarithm of both sides of an equation, where the right hand side is a ratio.

$$y = \frac{a}{b} \tag{1.2}$$

$$\log y = \log\left(\frac{a}{b}\right) = \log a - \log b \tag{1.3}$$

Thus, when we change everything into logarithms, ratios (like λ) become *differences*, which result in straight lines in graphs (Fig. 1.3b). On a linear scale, populations that are changing at the same *rates* can look very different (Fig. 1.3a), whereas on a logarithmic scale, the populations will have *parallel* trajectories (Fig. 1.3b).

Graphing a Matrix (Figs. 1.3a, 1.3b)

We can use `matplot` to plot a matrix *vs.* a single vector on the X-axis. By default it labels the points according to the number of the column

```
> matplot(time, Nt.s, pch = 1:3)
```

We can also plot it with a log scale on the y-axis.

```
> matplot(time, Nt.s, log = "y", pch = 1:3)
```

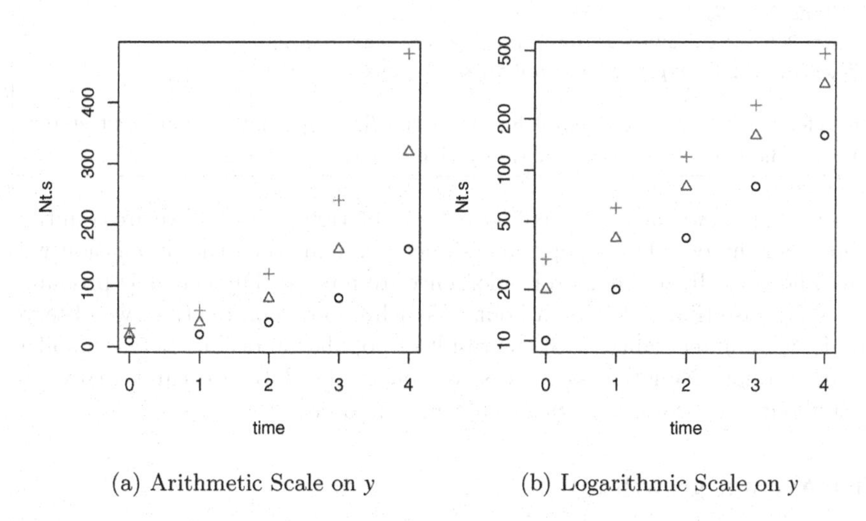

(a) Arithmetic Scale on y (b) Logarithmic Scale on y

Fig. 1.3: Effects of variation in initial N on population size, through time. Different symbols indicate different populations.

Note that changing the initial population size changes the intercept. It also changes the slope in linear space, but not in log-linear space. It changes the absolute rate of increase $(N_2 - N_1)$, but not the relative rate of increase (N_2/N_1).

1.3.3 Effects of different per capita growth rates

Perhaps the most important single thing we can say about λ is that when $\lambda < 1$, the population shrinks, and when $\lambda > 1$ the population grows. If we examine eq 1.1, $N_t = \lambda^t N_0$, we will note that λ is exponentiated, that is, raised to a power.[3] It will always be true that when $\lambda > 1$ and $t > 1$, $\lambda^t > \lambda$. It will also be true that when $\lambda < 1$ and $t > 1$, $\lambda^t < \lambda$ (Fig. 1.4).

Thus we see the basis of a very simple but important truism. *When $\lambda > 1$, the population grows, and when $\lambda < 1$ the population shrinks* (Fig. 1.4). When $\lambda = 1$, the population size does not change, because $1^t = 1$ for all t.

Effects of Different λ (Fig. 1.4)

Here we demonstrate the effects on growth of $\lambda > 1$ and $\lambda < 1$. We set $N_0 = 100$, and time, and then pick three different λ.

```
> N0 <- 100
> time <- 0:3
> lambdas <- c(0.5, 1, 1.5)
```

We use `sapply` again to apply the geometric growth function to each λ. This time, `x` stands for each λ, which our function then uses to calculate population size. We then plot it, and add a reference line and a little text.

```
> N.all <- sapply(lambdas, function(x) N0 * x^time)

> matplot(time, N.all, xlab = "Years", ylab = "N", pch = 1:3)
> abline(h = N0, lty = 3)
> text(0.5, 250, expression(lambda > 1), cex = 1.2)
> text(0.5, 20, expression(lambda < 1), cex = 1.2)
```

The reference line is a *h*orizontal line with the *l*ine *ty*pe dotted. Our text simply indicates the regions of positive and negative growth.

We note that we have graphed *discrete* population growth. If we are counting bodies, and the population reproduces once per year, then the population will jump following all the births (or emergence from eggs). Further, it is probably always the case that following a bout of synchronous reproduction, we observe chronic ongoing mortality, with the result of population decline between spikes of reproduction. Nonetheless, unless we collect the data, we can't really say much about what goes on in between census periods.

1.3.4 Average growth rate

In any real data set, such as from a real population of *Nymphaea*, N_{t+1}/N_t will vary from year to year. Let's examine this with a new data set in which annual growth rate varies from year to year.

[3] What happens to y^x as x increases, if $y > 1$ — does y^x increase? What happens if $y < 1$ — does y^x decrease? The answer to both these questions is yes.

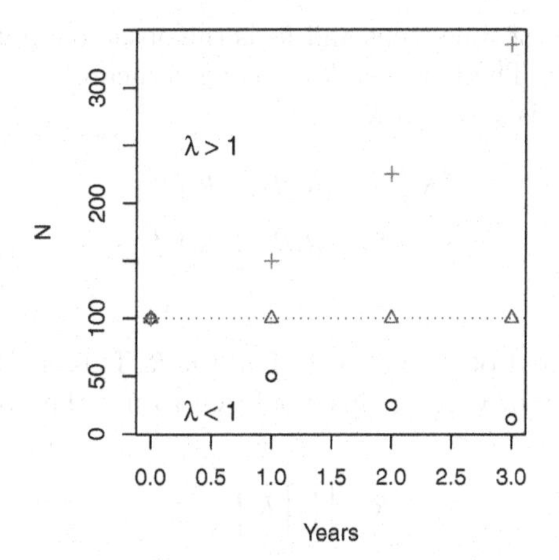

Fig. 1.4: Effects of variation in λ on population size through time. The dotted line indicates no change ($N_t = N_0$; $\lambda = 1$). Different symbols (circles, triangles, crosses) indicate populations resulting from $\lambda = (0.5, 1.0, 1.5)$, respectively. Any λ greater than 1 results in positive geometric growth; any $\lambda < 1$ results in negative geometric growth, or population decline.

Since growth rate varies from year to year, we may want to calculate average annual growth rate over several years. As we see below, however, the arithmetic averages are not really appropriate.

Consider that N_{t+1}/N_t may be a random variable which we will call R.[4] That is, this ratio from any one particular year could take on a wide variety of values, from close to zero, up to some (unknown) large number. Let's pick two out of a hat, where $R = 0.5, 1.5$. The arithmetic average of these is 1.0, so this might seem to predict that, on average, the population does not change. Let's project the population for two years using each R once.

$$
\begin{aligned}
N_0 &= 100 \\
N_1 &= N_0\,(0.5) = 50 \\
N_2 &= N_1\,(1.5) = 75
\end{aligned}
$$

We started with 100 individuals, but the population shrank! Why did that happen? It happens because, in essence, we *multiply* the λ together, where $N_2 = N0\,R_1\,R_2$. In this case, then, what is a sensible "average"?

How do we calculate an average for things that we multiply together? We would like a value for R which would provide the solution to

$$\bar{R}^t = R_1 R_2 \ldots R_t \tag{1.4}$$

[4] Some authors use R for very specific purposes, much as one might use λ; here we just use it for a convenient letter to represent observed per capita change.

where t is the number of time steps and R_1 is the observed finite rate of increase from year 1 to year 2. The bar over R indicates a mean.

All we have to do is solve for R.

$$\left(\bar{R}^t\right)^{1/t} = (R_1 R_2 \ldots R_t)^{1/t} \tag{1.5}$$

$$\bar{R} = (R_1 R_2 \ldots R_t)^{1/t} \tag{1.6}$$

$$\tag{1.7}$$

We take the t-th root of the product of all the R. This is called the *geometric average*. Another way of writing this would be to use the product symbol, Π, as in

$$\bar{R} = \left(\prod_{i=1}^{t} R_i\right)^{1/t} \tag{1.8}$$

If we examine the Song Sparrow data (Fig. 1.5), we see that projections based on the geometric average R are less than when based on the arithmetic average; this is always the case.

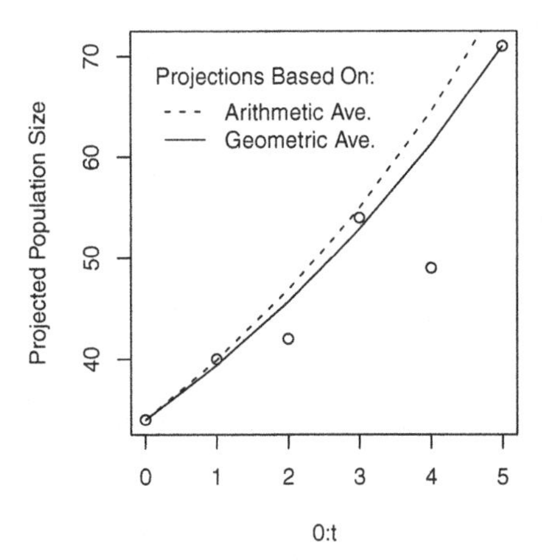

Fig. 1.5: Song Sparrow population sizes, and projections based on arithmetic and geometric mean R.

Comparing arithmetic and geometric averages (Fig. 1.5)

First we select the number of observed R ($t = 5$); this will require that we use six years of Song Sparrow data.

```
> t <- 5
> data(sparrows)
> SS6 <- sparrows[1:(t + 1), ]
```

Next we calculate λ for each generation, from t to $t+1$, and calculate the arithmetic and geometric means.

```
> SSgr <- SS6$Count[2:(t + 1)]/SS6$Count[1:t]
> lam.A <- sum(SSgr)/t
> lam.G <- prod(SSgr)^(1/t)
```

Now we can plot the data, and the projections based on the two averages (Fig. 1.5).

```
> N0 <- SS6$Count[1]
> plot(0:t, SS6$Count, ylab = "Projected Population Size")
> lines(0:t, N0 * lam.A^(0:t), lty = 2)
> lines(0:t, N0 * lam.G^(0:t), lty = 1)
> legend(0,70,c("Arithmetic Ave.", "Geometric Ave."),title = "Projections Based On:",
+      lty = 2:1, bty = "n", xjust = 0)
```

1.4 Continuous Exponential Growth

Although many, many organisms are modeled well with discrete growth models (e.g., insects, plants, seasonally reproducing large mammals), many populations are poorly represented by discrete growth models. These populations (e.g., bacteria, humans) are often modeled as continuously growing populations. Such models take advantage of simple calculus, the mathematics of rates.

Whereas geometric growth is proportional change in a population over a specified finite time interval, exponential growth is proportional *instantaneous* change over, well, an instant.

Imagine a population of *Escherichia coli* started by inoculating fresh Luria-Bertania medium with a stab of *E. coli* culture. We start at time zero with about 1000 cells or CFUs (colony forming units), and wind up the next day with 10^{10} cells. If we used (incorrectly) a discrete growth model, we could calculate N_{t+1}/N_t and use this as an estimate for λ, where $\lambda = 10^{10}/10^3 = 10^7$ cells per cell per day. We know, however, that this is a pretty coarse level of understanding about the dynamics of this system. Each cell cycle is largely asynchronous with the others, and so many cells are dividing each second. We could simply define our time scale closer to the average generation time of a cell, for example $\lambda = 2$ cells cell^{-1} 0.5 h^{-1}, but the resulting discrete steps in population growth would still be a poor representation of what is going on. Rather, we see population size changing very smoothly from hour to hour, minute to minute. Can we come up with a better description? Of course.

1.4.1 Motivating continuous exponential growth

If we assume that our *E. coli* cells are dividing asynchronously, then many cells are dividing each fraction of a second — we would like to make that fraction of a second an *infinitely* small time step. Unfortunately, that would mean that we have an infinitely large number of time steps between $t = 0$ and $t = 1$ day, and we couldn't solve anything.

A long time ago, a very smart person[5] realized that geometric growth depends on how often you think a step of increase occurs. Imagine you think a population increases at an annual growth rate $\lambda = 1.5$. This represents a 50% increase or

$$N_1 = N_0\,(1 + 0.5)$$

so the *discrete growth increment* is $r_d = 0.5$. You could reason that twice-annual reproduction would result in half of the annual r_d. You could then do growth over two time steps, and so we would then raise λ^2, because the population is taking two, albeit smaller, time steps. Thus we would have

$$N_1 = N_0\,(1 + 0.5/2)^2 = N_0\,(1 + 0.25)^2$$

What if we kept increasing the number of time steps, and decreasing the growth increment? We could represent this as

$$N_1 = N_0\left(1 + \frac{r_d}{n}\right)^n$$

$$\frac{N_1}{N_0} = \left(1 + \frac{r_d}{n}\right)^n$$

Our question then becomes, what is the value of $\left(1 + \frac{r_d}{n}\right)^n$ as n goes to infinity? In mathematics, we might state that we would like the solution to

$$\lim_{n \to \infty}\left(1 + \frac{r_d}{n}\right)^n. \tag{1.9}$$

To begin with, we simply try larger and larger values of n, graph eq. 1.9 *vs.* n, and look for a limit (Fig. 1.6).

[5] Jacob Bernoulli (1654–1705)

Numerical approximation of e

Here we use brute force to try to get an approximate solution to eq. 1.9. We'll let n be the number of divisions within one year. This implies that the finite rate of increase during each of these fractional time steps is r_d/n. Let the $\lambda = 2$ and therefore $r_d = 1$. Note that because $N_0 = 1$, we could ignore it, but let's keep it in for completeness.

```
> n <- 0:100
> N0 <- 1
> rd <- 1
```

Next, we calculate $\left(1 + \frac{r_d}{n}\right)^n$ for ever larger values of n.

```
> N1 <- N0 * (1 + rd/n)^n
```

Last, we plot the ratio and add some fancy math text to the plot (see `?plotmath` for details on mathematical typesetting in R).

```
> plot(n, N1/N0, type = "l")
> text(50, 2, "For n = 100,")
> text(50, 1.6, bquote((1 + frac("r"["d"], "n"))^"n" == .(round(N1[101]/N0,
+      3))))
```

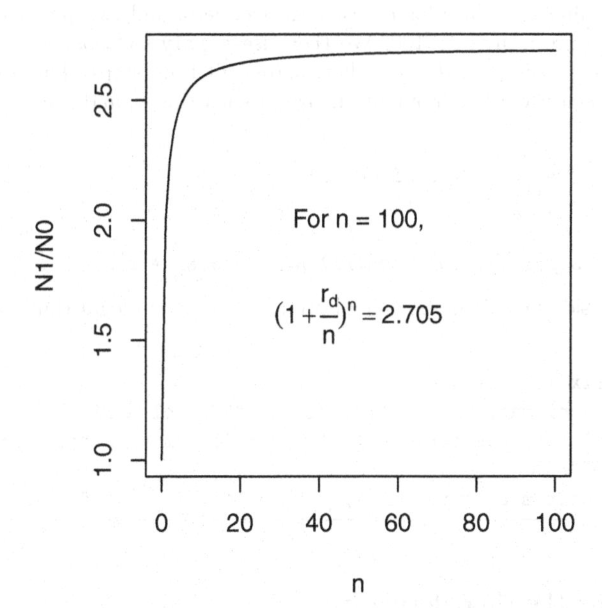

Fig. 1.6: The limit to subdividing reproduction into smaller steps. We can compare this numerical approximation to the true value, $e^1 = 2.718$.

Thus, when reproduction occurs continuously, the population can begin to add to itself right away. Indeed, if a population grew in a discrete annual step

$N_{t+1} = N_t(1 + r_d)$, the same r_d, divided up into many small increments, would result in a much larger increase.

It turns out that the increase has a simple mathematical expression, and we call it the *exponential*, e. As you probably recall, e is one of the magic numbers in mathematics that keeps popping up everywhere. In this context, we find that

$$\lim_{n\to\infty}\left(1 + \frac{r}{n}\right)^n = e^r \tag{1.10}$$

where e is the *exponential*.

This means that when a population grows geometrically, with infinitely small time steps, we say the population grows *exponentially*, and we represent that as,

$$N_t = N_0 e^{rt}. \tag{1.11}$$

We call r the *instantaneous* per capita growth rate, or the *intrinsic rate of increase*.

Projection of population size with continuous exponential growth is thus no more difficult than with discrete growth (Fig. 1.7).

Projecting a continuous population

We select five different values for r: two negative, zero, and two positive. We let t include the integers from 1 to 100. We then use `sapply` to *apply* our function of continuous exponential growth to each r, across all time steps. This results in a matrix where each row is the population size at each time t, and each column uses a different r.

```
> r <- c(-0.03, -0.02, 0, 0.02, 0.03)
> N0 <- 2
> t <- 1:100
> cont.mat <- sapply(r, function(ri) N0 * exp(ri * t))
```

Next we create side-by-side plots, using both arithmetic and logarithmic scales, and add a legend.

```
> layout(matrix(1:2, nrow = 1))
> matplot(t, cont.mat, type = "l", ylab = "N", col = 1)
> legend("topleft", paste(rev(r)), lty = 5:1, col = 1, bty = "n",
+     title = "r")
> matplot(t, cont.mat, type = "l", ylab = "N", log = "y", col = 1)
```

1.4.2 Deriving the time derivative

We can also differentiate eq. 1.11 with respect to time to get the differential equation for instantaneous population growth rate. Recall that the chain rule tells us that the derivative of a product of two terms is the sum of the products of the derivative of one times the other original term.

$$\frac{\mathrm{d}}{\mathrm{d}t}(XY) = \frac{\mathrm{d}X}{\mathrm{d}t}Y + \frac{\mathrm{d}Y}{\mathrm{d}t}X$$

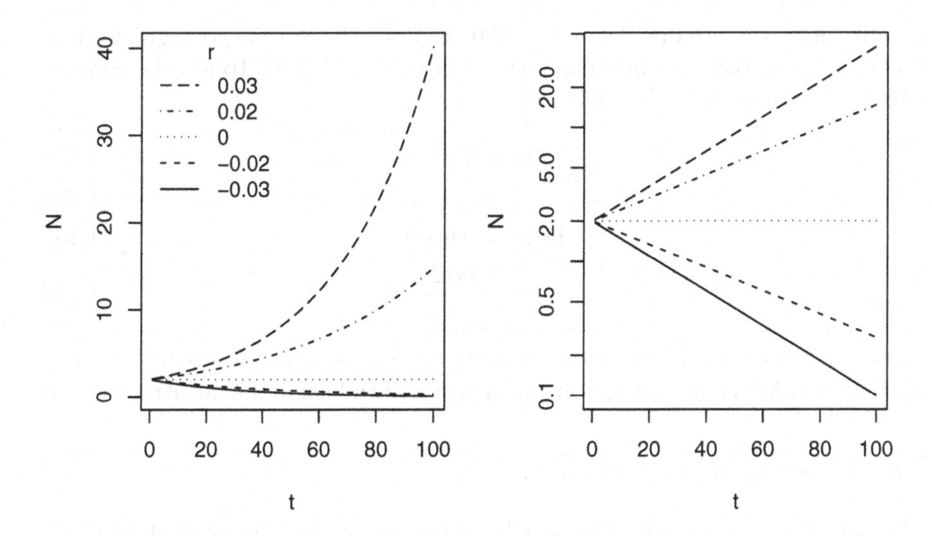

Fig. 1.7: Projecting continuous populations with different r.

Therefore to begin to differentiate eq. 1.11, with respect to t, we have,

$$\frac{\mathrm{d}}{\mathrm{d}t}N_0e^{rt} = \frac{\mathrm{d}}{\mathrm{d}t}N_0 \cdot (e^r)^t + \frac{\mathrm{d}}{\mathrm{d}t}(e^r)^t \cdot N_0$$

Recall also that the derivative of a constant is zero, and the derivative of a^t is $\ln a\,(a^t)$, resulting in,

$$\frac{\mathrm{d}}{\mathrm{d}t}N_0e^{rt} = 0 \cdot (e^r)^t + \ln e^r \cdot (e^r)^t \cdot N_0$$

Given that $\ln e = 1$, and that $N_0e^{rt} = N$ for any time t, this reduces to eq. 1.12. The time derivative, or differential equation, for exponential growth is

$$\frac{\mathrm{d}N}{\mathrm{d}t} = rN. \tag{1.12}$$

1.4.3 Doubling (and tripling) time

For heuristic purposes, it is frequently nice to express differences in growth rates in terms of something more tangible than a dimensionless number like r. It is relatively concrete to say population X increases by about 10% each year ($\lambda = 1.10$), but another way to describe the rate of change of a population is to state the amount of *time* associated with a dramatic but comprehensible change. The *doubling time* of a population is the time required for a population to double in size. Shorter doubling times therefore mean more rapid growth.

We could determine doubling time graphically. If we examine the expanding population in Fig. 1.4, we see that it takes about one and half years for the population size to change from $N = 100$ to $N = 200$. Not surprisingly, we can do better than that. By doubling, we mean that $N_t = 2N_0$. To get the time at which this occurs, we solve eq. (1.11) for t,

$$2N_0 = N_0 e^{rt} \tag{1.13}$$
$$2 = e^{rt} \tag{1.14}$$
$$\ln(2) = rt \ln(e) \tag{1.15}$$
$$t = \frac{\ln(2)}{r}. \tag{1.16}$$

Thus, eq. 1.16 gives us the time required for a population to double, given a particular r. We could also get any arbitrary multiple m of any arbitrary initial N_0.

Creating a function for doubling time

We can create a *function* for this formula, and then evaluate it for different values of m and r. For $m = 2$, we refer to this as "doubling time." When we define the function and include arguments `r` and `m`, we also set a default value for `m=2`. This way, we do not always have to type a value for m; be default the function will return the doubling time.

```
> m.time <- function(r, m = 2) {
+     log(m)/r
+ }
```

Now we create a vector of r, and then use `m.time` to generate a vector of doubling times.

```
> rs <- c(0, 1, 2)
> m.time(rs)
```

```
[1]    Inf 0.6931 0.3466
```

Note that R tells us that when $r = 0$, it takes an infinite (`Inf`) amount of time to double. This is what we get when we try to divide by zero!

1.4.4 Relating λ and r

If we assume constant exponential and geometric growth, we can calculate r from data as easily as λ. Note that, so rearranging, we see that

$$N_t = N_0 e^{rt}$$
$$\ln(N_t) = \ln(N_0) + rt.$$

In other words, r is the slope of the linear relation between $\ln(N_t)$ and time (Fig. 1.7), and $\ln(N_0)$ is the y-intercept. If the data make sense to fit a straight regression line to log-transformed data, the slope of that line would be r.

It also is plain that,

$$\lambda = e^r \tag{1.17}$$

$$\ln \lambda = r. \tag{1.18}$$

Summarizing some of what we know about how λ and r relate to population growth:

No Change $\lambda = 1$, $r = 0$
Population Growth $\lambda > 1$, $r > 0$
Population Decline $\lambda < 1$, $r < 0$

Remember — λ is for populations with discrete generations, and r is for continuously reproducing populations.

Units

What are the units for λ and r? As λ is a ratio of two population sizes, the units could be individuals/individual, thus rendering λ dimensionless. Similarly, we can view λ as the net number of individuals produced by individuals in the population such that the units are net new individuals per individual per time step, or inds $\text{ind}^{-1}\,t^{-1}$. The intrinsic rate of increase, r, is also in units of inds $\text{ind}^{-1}\,t^{-1}$

Converting between time units

A nice feature of r as opposed to λ is that r can be used to scale easily among time units. Thus, $r = 0.1$ inds $\text{ind}^{-1}\,\text{year}^{-1}$ becomes $r = 0.1/365 = 0.00027$ inds $\text{ind}^{-1}\,\text{day}^{-1}$. *You cannot do this with* λ. If you would like to scale λ from one time unit to another, first convert it to r using logarithms, make the conversion, then convert back to λ.

1.5 Comments on Simple Density-independent Growth Models

It almost goes without saying that if we are considering density-independent growth models to be representative of real populations, we might feel as though we are making a very long list of unrealistic assumptions. These might include no immigration or emigration, no population structure (i.e. all individuals are identical), and you can probably come up with many others [58]. However, I would argue vociferously that we are making only one assumption:

> *N increases by a constant per capita rate over the time interval(s) of interest.*

Think about that. I am *not* saying that competition is not occurring, or that no death is occurring, or that all individuals are reproductively viable, or there is no abiotic disturbance, or that there is no population genetic structure. I am just saying that for the time period of interest, all things balance out, or are of no substantive consequence, and the population chugs along at a particular pace.

If the per capita rate is constant, then there can be no statistical relation between the size of the population and its per capita growth rate. *In the absence of such a relation*, we say that the growth rate is density-independent.

Other ecologists will disagree with my sentiments regarding an absence of assumptions. That's OK — still others may agree with these sentiments. Take it upon yourself to acquire multiple perspectives and evaluate them yourself.

Both λ and r obviously depend on birth rates and death rates. For instance, we could view geometric growth as

$$N_{t+1} = N_t + BN_t - DN_t \tag{1.19}$$

where B is the number of births per individual and D is the probability of an individual dying during the specified time interval. Lambda, in this case, is $1 + (B - D)$ and $r_d = B - D$. This form would be nice if we had data on births and deaths, because, after all, one goal of Science is to explain complex phenomena (e.g., λ) in terms of their determinants (e.g., B and D). Similarly, we can state $r = b - d$ where b and d are per capita instanteous rates. Such an advance in understanding the determinants would be great.

Perhaps now is a good time to think about all the assumptions others might tell us we are making when we present the above formulation. Are all individuals in the population equally likely to produce progeny and/or die? Will birth and death rates vary over time or as the size of the population changes more? How will resource supply rate influence these rates? Is there really no immigration or emigration? These are very interesting questions.

Simple density-independent growth provides, in some sense, a null hypothesis for the dynamic behavior of a population. Malthus warned us that organisms produce more progeny than merely replacement value, and population growth is an exponential (or geometric) process [125]. The question then becomes "What causes population growth to differ from a constant rate of change?" That, in a nutshell, is what the rest of the book is about.

1.6 Modeling with Data: Simulated Dynamics

The main purpose of this section[6] is to begin to understand the mechanics of simulation. The preceding sections (the bulk of the chapter) emphasized understanding the deterministic underpinnings of simple forms of density independent growth: geometric and exponential growth. This section explores the simulation of density independent growth.

When we model populations, perhaps to predict the size of a population in the future, we can take a variety of approaches. One type of approach emphasizes deterministic prediction, using, for instance, $\bar{R}$. Another approach is to *simulate* population dynamics and we take this up in this next section.

[6] This section emphasizes work in R.

To project population growth into the future should include some quantification of the uncertainty with our guess. Simulation is one way we can project populations and quantify the uncertainty. The way one often does that is to use the original data and sample it randomly to calculate model parameters. In this fashion, the simulations are random, but based on our best available knowldge, i.e., the real data. The re-use of observed data occurs in many guises, and it is known generally as bootstrapping or resampling.

1.6.1 Data-based approaches

In using our data to predict population sizes, let us think about three levels of biological organization and mechanism: population counts, changes in population counts, and individual birth and death probabilities. First, our count data alone provide a sample of a very large number of different possible counts. If we assume that there will be no trend over time, then a simple description of our observed counts (e.g., mean and confidence intervals) provide all we need. We can say "Song Sparrow counts in the Breeding Bird Survey in Darrtown, OH, are about 51."

Second, we could use the observed *changes* in population counts $R_t = N_{t+1}/N_t$ as our data. We would then draw an R_t at random from among the many observed values, and project the population one year forward. We then repeat this into the future, say, for ten years. Each simulation of a ten year period will result in a different ten year trajectory because we draw R_t at random from among the observed R_t. However, if we do many such simulations, we will have a *distribution* of outcomes that we can describe with simple statistics (e.g., median, mean, quantiles).

Third, we might be able to estimate the individual probabilities of births and deaths in the entire Darrtown population, and use those probabilities and birth rates to simulate the entire population into the future. In such an *individual-based* simulation, we would simulate the fates of individuals, keeping track of all individual births and deaths.

There are myriad others approaches, but these give you a taste of what might be possible. In this section we focus on the second of these alternatives, in which we use observed R_t to simulate the dynamics of Song Sparrow counts.

Here we investigate Song Sparrow (*Melospize melodia*) dynamics using data from the annual U.S. Breeding Bird Survey (http://www.mbr-pwrc.usgs.gov/bbs/). Below we will

1. look at and collecting the data (annual R's),
2. simulate one projection,
3. scale up to multiple simulations,
4. simplify simulations and perform 1000's, and
5. analyze the output.

1.6.2 Looking at and collecting the data

Let's start by looking at the data. *Looking at the data is always a good idea — it is a principle of working with data.* We first load the data from the `primer`

R package, and look at the names of the data frame. We then choose to `attach` the data frame, because it makes the code easier to read.[7]

```
> names(sparrows)

[1] "Year"              "Count"             "ObserverNumber"

> attach(sparrows)
```

Now we plot these counts through time (Fig. 1.8).

```
> plot(Count ~ Year, type = "b")
```

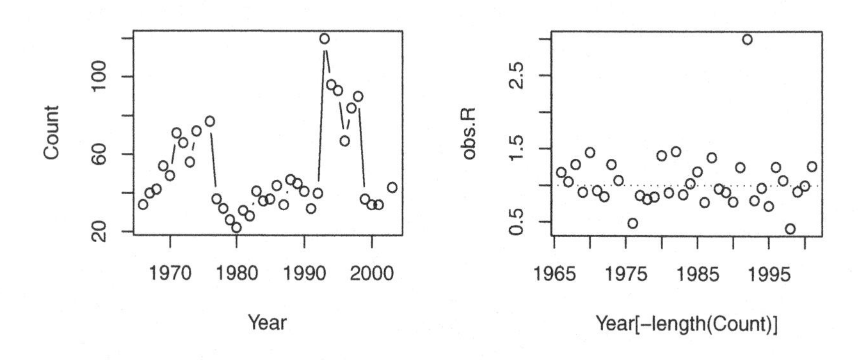

Fig. 1.8: Observations of Song Sparrows in Darrtown, OH (http://www.mbr-pwrc.usgs.gov/bbs/).

We see that Song Sparrow counts[8] at this site (the DARRTOWN transect, OH, USA) fluctuated a fair bit between 1966 and 2003. They never were completely absent and never exceeded ~ 120 individuals.

Next we calculate annual $R_t = N_{t+1}/N_t$, that is, the observed growth rate for each year t.[9]

```
> obs.R <- Count[-1]/Count[-length(Count)]
```

Thus our *data* are the observed R_t, not the counts *per se*. These R form the basis of everything else we do. Because they are so important, let's plot these as well. Let's also indicate $R = 1$ with a horizontal dotted line as a visual cue

[7] I typically do not use `attach` but rather choose to always define explicitly the parent data frame I am using. It helps me reduce silly mistakes.

[8] Recall that these are *samples* or observations of sparrows. These are *not* population sizes. Therefore, we will be simulating sparrows counts, not sparrow population sizes.

[9] The use of "`-`" in the index tells R to exclude that element (e.g., `-1` means "exclude the first element of the vector").

for zero population growth. Note that we exclude the last year because each R_t is associated with N_t rather than N_{t+1}.

```
> plot(obs.R ~ Year[-length(Count)])
> abline(h = 1, lty = 3)
```

One thing that emerges in our graphic data display (Fig. 1.8) is we have an unusually high growth rate in the early 1990's, with the rest of the data clustered around 0.5–1.5. We may want to remember that.

1.6.3 One simulation

Now that we have our randomly drawn Rs, we are ready to simulate dynamics. A key assumption we will make is that *these R are representative of R in the future, and that each is equally likely to occur.* We then *resample* these observed R *with replacement* for each year of the simulation. This random draw of observed R's then determines one random realization of a possible population trajectory. Let's begin.

First, we decide how many years we want to simulate growth.

```
> years <- 50
```

This will result in 51 population sizes, because we have the starting year, year 0, and the last year.

Next we draw 50 R at random with replacement. This is just like having all 35 observed R written down on slips of paper and dropped into a paper bag. We then draw one slip of paper out of the bag, write the number down, and put the slip of paper back in the bag, and then repeat this 49 more times. This is *resampling with replacement*[10]. The R function `sample` will do this. Because this is a pseudorandom[11] process, we use `set.seed` to make your process the same as mine, i.e., repeatable.

```
> set.seed(3)
> sim.Rs <- sample(x = obs.R, size = years, replace = TRUE)
```

Now that we have these 50 R, all we have to do is use them to determine the population size through time. For this, we need to use what programmers call a *for-loop* (see B.6 for further details). In brief, a for-loop repeatedly *loops* through a particular process, with one loop *for* each value of some indicator variable. Here we calculate each sparrow count in the next year, N_{t+1}, using the count in the current year N_t and the randomly drawn R *for* each year t.

[10] We could resample *without* replacement. In that case, we would be assuming that all of these R_t are important and *will* occur at some point, but we just don't know when — they constitute the entire universe of possiblities. Sampling *with* replacement, as we do above, assumes that the observed R_t are all equally likely, but none is particularly important — they are just a sample of what is possible, and they might be observed again, or they might not.

[11] A *pseudorandom* process is the best computers can do — it is a complex deterministic process that generates results that are indistinguishable from random.

We begin by creating an empty output vector that is the right length to hold our projection, which will be the number of Rs plus one.[12]

```
> output <- numeric(years + 1)
```

We want to start the projection with the sparrow count we had in the last year (the "maximum," or biggest, year) of our census data.

```
> output[1] <- Count[Year == max(Year)]
```

Now the fun really starts to take off, as we finally use the for-loop. For each year t, we multiply N_t by the randomly selected R_t to get N_{t+1} and put it into the $t + 1$ element of `output`.

```
> for (t in 1:years) output[t + 1] <- {
+     output[t] * sim.Rs[t]
+ }
```

Let's graph the result.

```
> plot(0:years, output, type = "l")
```

It appears to work (Fig. 1.9a) — at least it did something! Let's review what we have done. We

- had a bird count each year for 36 years. From this we calculated 35 R (for all years except the very last).
- decided how many years we wanted to project the population (50 y).
- drew *at random and with replacement* the observed R — one R for each year we want to project.
- got ready to do a simulation with a for-loop — we created an empty vector and put in an initial value (the last year's real data).
- performed each year's calculation, and put it into the vector we made.

So what does Fig. 1.9a represent? It represents one possible outcome of a trajectory, if we assume that R has an equal probability of being any of the observed R_t. This *particular* trajectory is very unlikely, because it would require one particular sequence of Rs. However, our simulation assumes that it is *no less likely* than any other particular trajectory.

As only one realization of a set of randomly selected R, Fig. 1.9a tells us very little. What we need to do now is to replicate this process a very large number of times, and examine the *distribution* of outcomes, including moments of the distribution such as the mean, median, and confidence interval of eventual outcomes.

1.6.4 Multiple simulations

Now we create a way to perform the above simulation several times. There are a couple tricks we use to do this. We still want to start small so we can figure out the steps as we go. Here is what we would do next.

[12] Remember that we always have one more population count than we do R_t.

- We start by specifying that we want to do 10 simulations, where one simulation is what we did above.
- We will need to use $50 \times 10 = 500$ randomly drawn Rs and store those in a matrix.
- To do the ten separate, independent simulations, we will use `sapply`, to "apply" our simulations ten times. We have to use a for-loop for each population simulation, because each N_t depends on the previous N_{t-1}. We use `sapply` and related functions for when we want to do more than one independent operation.

Here we specify 10 simulations, create a matrix of the 10×50 randomly drawn R.

```
> sims = 10
> sim.RM <- matrix(sample(obs.R, sims * years, replace = TRUE),
+     nrow = years, ncol = sims)
```

Next we get ready to do the simulations. First, to hold each projection temporarily, we will reuse `output` as many times as required. We then *apply* our for-loop projection as many times as desired, for each value of `1:sims`.

```
> output[1] <- Count[Year == max(Year)]
> outmat <- sapply(1:sims, function(i) {
+     for (t in 1:years) output[t + 1] <- output[t] * sim.RM[t,
+         i]
+     output
+ })
```

Now let's peek at the results (Fig. 1.9b). This is fun, but also makes sure we are not making a heinous mistake in our code. Note we use log scale to help us see the small populations.

```
> matplot(0:years, outmat, type = "l", log = "y")
```

What does it mean that the simulation has an approximately even distribution of final population sizes *on the log scale* (Fig. 1.9b)? If we plotted it on a linear scale, what would it look like?[13]

Rerunning this simulation, with new R each time, will show different dynamics every time, and that is the point of simulations. Simulations are a way to make a few key assumptions, and then leave the rest to chance. In that sense it is a null model of population dynamics.

[13] Plotting it on the log scale reveals that the relative change is independent of population size; this is true because the rate of change is geometric. If we plotted it on a linear scale, we would see that many trajectories result in small counts, and only a few get really big. That is, the median size is pretty small, but a few populations get huge.

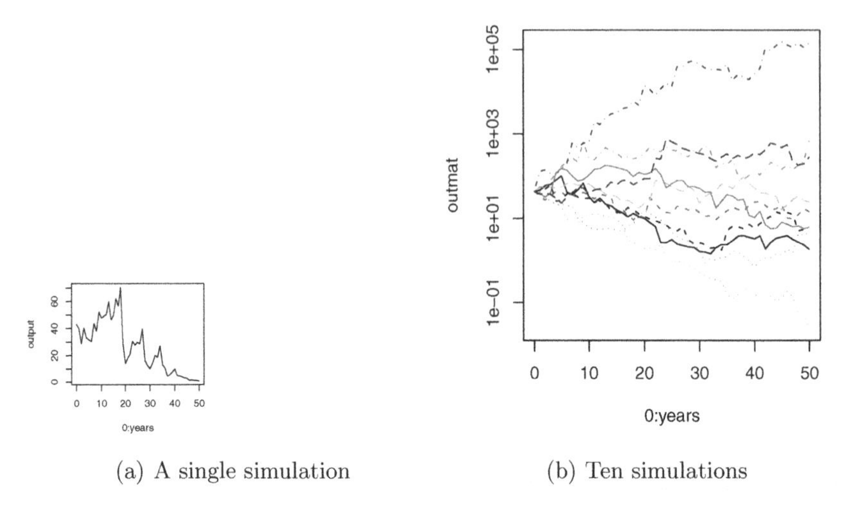

(a) A single simulation (b) Ten simulations

Fig. 1.9: Simulated population dynamics with *R* drawn randomly from observed Song Sparrow counts.

1.6.5 Many simulations, with a function

Let's turn our simulation into a user-defined function[14] that simplifies our lives. We also add another assumption: individuals are irreducible. Therefore, let us use `round(,0)` to round to zero decimal places, i.e., the nearest integer.[15]

Our user defined function, `PopSim`, simply wraps the previous steps up in a single function.[16] The output is a matrix, like the one we plotted above.

```
> PopSim <- function(Rs, N0, years = 50, sims = 10) {
+     sim.RM = matrix(sample(Rs, size = sims * years, replace = TRUE),
+         nrow = years, ncol = sims)
+     output <- numeric(years + 1)
+     output[1] <- N0
+     outmat <- sapply(1:sims, function(i) {
+         for (t in 1:years) output[t + 1] <- round(output[t] *
+             sim.RM[t, i], 0)
+         output
+     })
+     return(outmat)
+ }
```

If you like, try to figure out what each step of the simulation is doing. Consider it one of the end-of-chapter problems. Rely on on the code above to help you decipher the function.

[14] For user-defined functions, see sec. B.4.1.

[15] We could use also use `floor` to round down to the lowest integer, or `ceiling` to round up.

[16] This process, of working through the steps one at a time, and then wrapping up the steps into a function, is a useful work flow.

Now we have the pleasure of using this population simulator to examine a number of things, including the sizes of the populations after 50 years. I first simulate 1000 populations,[17] and use `system.time` to tell me how long it takes on my computer.

```
> system.time(output <- PopSim(Rs = obs.R, N0 = 43, sims = 1000))

   user  system elapsed
  0.404   0.004   0.407
```

This tells me that it took less than half a second to complete 1000 simulations. That helps me understand how long 100 000 simulations might take. We also check the dimensions of the output, and they make sense.

```
> dim(output)

[1]   51 1000
```

We see that we have an object that is the size we think it should be. We shall assume that everything worked way we think it should.

1.6.6 Analyzing results

We extract the last year of the simulations (last row), and summarize it.

```
> N.2053 <- output[51, ]
> summary(N.2053, digits = 6)

    Min.  1st Qu.   Median     Mean  3rd Qu.     Max.
     0.0     14.0     66.0   1124.6    291.8 332236.0
```

We see from this summary that the median final population size, among the 1000 simulations, is 66 individuals (median=50% quantile). While at least one of the populations has become extinct (min. = 0), the maximum is huge (max. = 332236). The `quantile` function allows us to find a form of empirical confidence intervals, including, approximately, the central 95% of the observations.[18]

```
> quantile(N.2053, prob = c(0.0275, 0.975))

2.75% 97.5%
    0  5967
```

These quantiles suggest that in 2053, we might observe sparrow counts anywhere from 0 to 5967, where zero and ~ 6000 are equally likely.

Notice the huge difference between the mean, $N = 1125$, and the median, N=66. Try to picture a histogram for which this would be true. It would be skewed right (long right hand tail), as with the lognormal distribution; this is common in ecological data.

[17] If we were doing this in a serious manner, we might do 10–100 000 times.

[18] Note that there are many ways to estimate quantiles (R has nine ways), but they are approximately similar to percentiles.

Let's make a histogram of these data. Exponentially growing populations, like this one, frequently show a lognormal distribution of abundances. Indeed, some say the "natural" unit of a population is *log*(N), rather than N. We will plot two frequency distributions of the final populations, one on the orignal scale, one using the logarithms of the final population sizes plus 1 (we use $N+1$ so that we can include 0's — what is *log*(0)? *log*(1)?).

```
> hist(N.2053, main = "N")
> hist(log10(N.2053 + 1), main = "log(N+1)")
> abline(v = log10(quantile(N.2053, prob = c(0.0275, 0.975)) +
+     1), lty = 3)
```

We added some reference lines on the second histogram, showing the 2.5 and 97.5% quantiles (Fig. 1.10). You can see that the logarithms of the population sizes are much more well-behaved, more symmetrical.

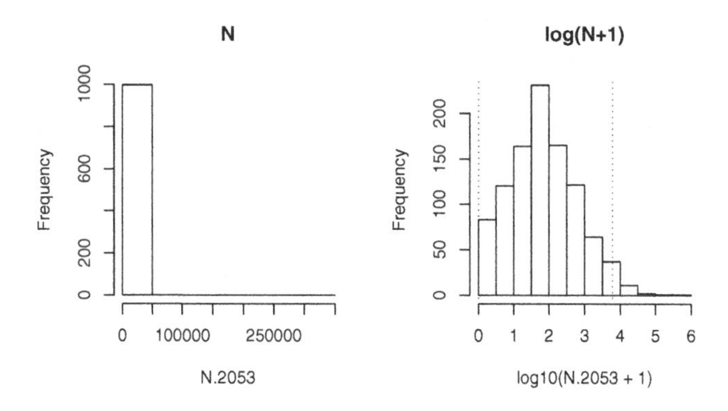

Fig. 1.10: Exploratory graphs of the distributions of the final simulated population sizes.

Can we really believe this output? To what can we compare our output? One thing that occurs to me is to compare it to the lower and upper bounds that we might contrive from *deterministic* projections.

To compare the simulation to deterministic projections, we could find the 95% t-distribution based confidence limits for the geometric mean of R. If we use our rules regarding the geometric mean, we would find that the logarithm of the *geometric* mean of R is the arthmetic mean of the $\log R$. So, one thing we could do is calculate the t-based confidence limits[19] on $\log R$, backtransform these to project the population out to 2053 with lower and upper bounds.

Here we take find the logarithms, caculate degrees of freedom and the relevant quantiles for the t distribution.

[19] Remember: the t-distribution needs the degrees of freedom, and a 95% confidence region goes from the 2.5% and the 97.5% percentiles.

```
> logOR <- log(obs.R)
> n <- length(logOR)
> t.quantiles <- qt(c(0.025, 0.975), df = n - 1)
```

Next we calculate the standard error of the mean, and the 95% confidence limits for $\log R$.

```
> se <- sqrt(var(logOR)/n)
> CLs95 <- mean(logOR) + t.quantiles * se
```

We backtransform to get R, and get a vector of length 2.

```
> R.limits <- exp(CLs95)
> R.limits

[1] 0.8968 1.1302
```

What do we see immediately about these values? One is less than 0, and one is greater than 0. This means that for the lower limit, the population will shrink (geometrically), while for the upper limit, the population will increase (geometrically). Let's go ahead and make the 50 y projection.

```
> N.Final.95 <- Count[Year == max(Year)] * R.limits^50
> round(N.Final.95)

[1]     0 19528
```

Here we see that the lower bound for the deterministic projection is the same (extinction) as the simulation, while the upper bound is much greater than that for the simulation. Why would that be? Perhaps we should examine the assumptions of our deteministic approach.

We started by assuming that the $\log R$ could be approximated with the t distribution, one of the most pervasive distributions in statistics and life. Let's check that assumption. We will compare the $\log R$ to the theoretical values for a t distribution. We scale `logOR` to make the comparison more clear.

```
> qqplot(qt(ppoints(n), df = n - 1), scale(logOR))
> qqline(scale(logOR))
```

How do we interpret these results? If the distribution of an observed variable is consistent with a particular theoretical distribution, the ordered quantiles of data will be a linear (straight line) function of the theoretical quantiles of the theoretical distribution. Deviations from that straight line illustrate how the data deviate. Here we see that the data have three outliers that are much more extreme (greater and smaller) than expected in the t-distribution, and more data are cluster around the center of the distribution than we would expect. We should ask whether those extreme values are mistakes in data collection or recording or whether they are every bit as accurate as all the other measurements.

Compare our two different confidence limits. These provide two different answers to our original question, "what might be the Song Sparrow count at this site in 2053?" Both of these assume a similar underlying model, density

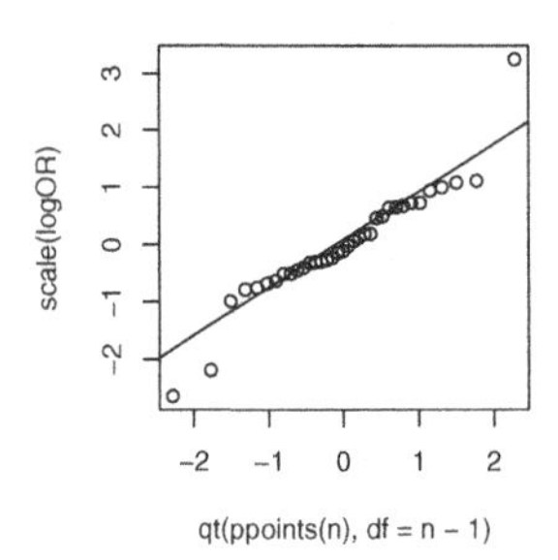

Fig. 1.11: Quantile-quantile plot used to compare $\log R$ to a t-distribution. Scaling `logOR` in this case means that we subtracted the mean and divided by the standard deviation. A histogram performs a similar service but is generally less discriminating and informative.

independent growth, but give different answers. Of which approach are we more confident? Why? What assumptions does each make?

We can be quite sure that our assumption regarding the t-distribution of our R is unsupported — our data have outliers, relative to a t-distribution. What would this do? It would increase the variance of our presumed distribution, and lead to wider confidence intervals, even though most of the data conform to a narrower distribution. Our simulation procedure, on the other hand, rarely samples those extreme points and, by chance, samples observed R that fall much closer to the median. This can occasionally be a problem in simulations based on too little data — the data themselves do not contain enough variability. Imagine the absurdity of a data-based simulation that relies on one observation — it would be *very* precise (but wrong)!

Our conclusions are based on a model of discrete density-independent population growth — what assumptions are we making? are they valid? Are our unrealistic assumptions perhaps nonetheless a good approximation of reality? We will revisit these data later in the book (Chapter 3) to examine one of these assumptions. We do not need to answer these questions now, but it is essential, and fun, to speculate.

1.7 Summary

In this chapter, we have explored the meaning of density-independent population growth. It is a statistically demonstrable phenomenon, wherein the per captia growth rate exhibits no relation with population density. It is a useful starting point for conceptualizing population growth. We have derived discrete geometric and continuous exponential growth and seen how they are related. We have caculated doubling times. We have discussed the assumptions that different people might make regarding these growth models. Last, we have used simulation to explore prediction and inference in a density-independent context.

Problems

1.1. Geometric growth Analyze the following data, relying on selected snippets of previous code.
(a) In the years 1996 through 2005, lily population sizes are $N = 150, 100, 125, 200, 225, 150, 100, 175, 100, 150$. Make a graph of population size versus time.
(b) Calculate R for each year; graph R *vs.* time.
(c) Calculate arithmetic and geometric average growth rates of this population.
(d) Based on the appropriate average growth rate, what would be the expected population size in 2025? What would the estimated population size be if you used the inappropriate mean? Do not use simulation for this.
(d*) Given these data, develop simulations as above with the user-defined function, `PopSim`. Describe the distribution of projected population sizes for 2010.

1.2. Doubling Time
(a) Derive the formula for doubling time in a population with contiunous exponential growth.
(b) What is the formula for tripling time?
(c) If we are modeling humans or *E. coli*, would a model of geometric, or exponential growth be better? Why?
(d) If an *E. coli* population grew from 1000 cells to 2×10^9 cells in 6 h, what would its intrinsic rate of increase be? Its doubling time?

1.3. Human Population Growth
(a) There were about 630 million people on the planet in 1700, and 6.3 billion in 2003 [33]. What was the intrinsic rate of increase, r?
(b) Graph the model of human population size population size from 1700 to 2020.
(c) Add points on the graph indicating the population doublings from 1700 onward.
(d*) What will prevent humans from outweighing the planet by the end of this century? What controls human population growth? Do these controls vary spatially across the planet? See Cohen [33] to get going.

1.4. R functions
Find the R functions in Chapter 1. Demonstrate their uses.

2

Density-independent Demography

Different populations have different numbers of individuals of different ages. Consider the human populations of Mexico and Sweden in 1990. Mexico had more individuals in total than Sweden, and a larger fraction of their population was of child bearing age or younger (Figs. 2.1a, 2.1b).

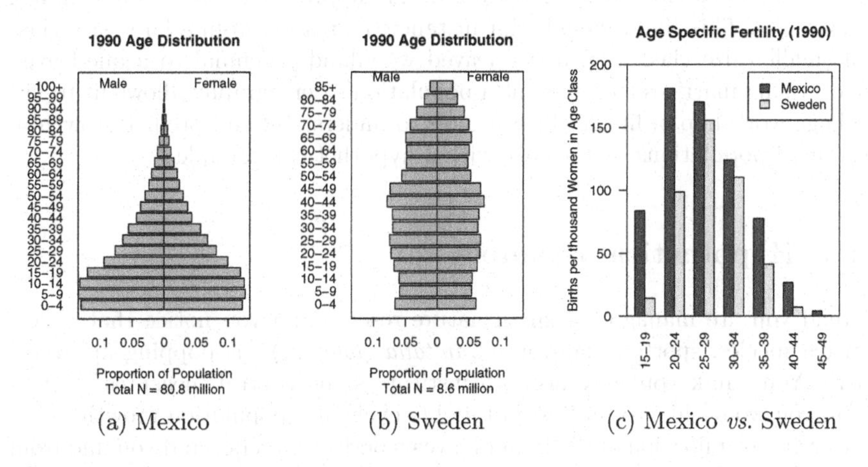

(a) Mexico (b) Sweden (c) Mexico *vs.* Sweden

Fig. 2.1: Demography of human populations of Mexico and Sweden. Based on 1990 data from US Census Bureau, Population Division, International Programs Center.

In addition, the age-specific fertility rate is higher in Mexico, especially for younger women (Fig. 2.1c). How did this happen, and why does Mexico have so many young people? What are the consequences of this for their culture, their use of resources, their domestic and foreign policies, and their future population growth? How about Sweden?

Demography is the study of populations with special attention to age or stage structure [113]. Originally, age-based human demography was the provenance of actuaries who helped governments keep track of the number citizens of different ages and thus, for instance, know how many would be available for conscription into the military.[1]The demography of a population is the age (or stage) structure and the survival, fertility, and other demographic rates associated with those ages or life history stages. *Age structure* is the number or relative abundance of individuals of different ages or age classes. *Stage structure* is the number or relative abundance of individuals of different stages. Stages are merely useful categories of individuals, such as size classes (e.g. diameters of tropical trees) or life history stages (e.g. egg, larvae, and adult anurans). Stages are particularly useful when (i) age is difficult to determine, and/or (ii) when stage is a better predictor of demographic rates (e.g. birth, death, survival) than is age. Demography is, in part, the study of how demographic rates vary among ages or stages, and the consequences of those differences.

There are a few ways to study a population's demography, and all ecology text books can provide examples. *Life tables* are lists of important demographic parameters such as survivorship, birth and death rates each age or age class.

Commonly, both age and stage based demography now take advantage of matrix algebra to simplify and synthesize age and stage specific demography [23]. This approach is essential when individuals don't proceed through stages in a simple sequential manner, perhaps reverting to an "earlier" stage. When used with age-based demography, these matrices are referred to as Leslie matrices [107]. L. P. Lefkovitch [100] generalized this approach to allow for complex demography. This could include, for instance, regression from a large size class to a smaller size class (e.g. a two-leaved woodland perennial to a one-leaved stage). Using matrices to represent a population's demography allows us to use the huge workshop of linear algebra tools to understand and predict growth in structured populations. Let's start with a hypothetical example.

2.1 A Hypothetical Example

Pretend you are managing a small nature reserve and you notice that a new invasive species, spotted mallwort (*Capitalia globifera*),[2] is popping up everywhere. You think you may need to institute some control measures, and to begin, you would like to understand its life cycle and population growth.

Careful examination of the flowers reveals perfect flowers,[3] and you find from the literature that mallwort is a perennial capable of self-fertilizing. The seeds germinate in early fall, grow the following spring and early summer to a small adult that has 2–3 leaves and which sometimes produce seeds. In the second year and beyond, if the plants survive, they grow to large adults which have four or more leaves and only then do they produce the majority of their seeds.

[1] In his chapter entitled "Interesting Ways to Think about Death" G.E. Hutchinson [84] cites C. F. Trenerry, E. L. Gover and A. S. Paul (*The Origins and Early History of Insurance*, London, P. S. King & Sons, Ltd.) for description of early Roman actuarial tables.

[2] Not a real species.

[3] Individual flowers possess both female and male reproductive structures.

The seeds do not seem to survive more than one year, so there is probably no seed bank.

You summarize what you have learned in the form of a *life cycle graph* (Fig. 2.1). Demographers use a life cycle graph to summarize stages that may be observed at a single repeated point in time (e.g., when you go out to explore in June). It also can include the probabilities that an individual makes the transition from one stage to another over one time step (e.g. one year), as well as the fecundities.

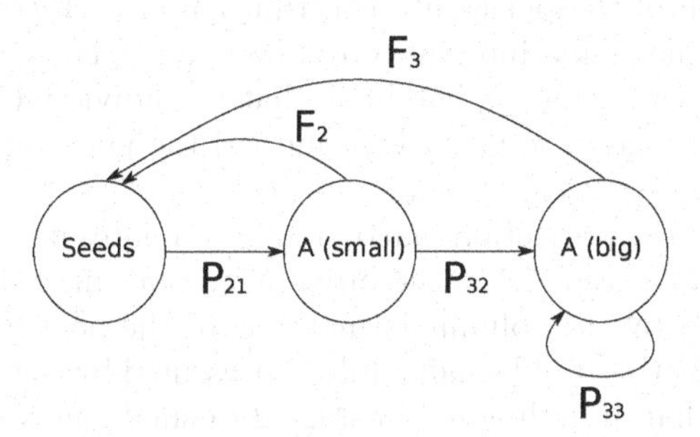

Fig. 2.2: Life cycle graph of the imaginary spotted mallwort (*Capitalia globifera*). P_{ij} is the probability that an individual in stage j transitions to stage i over a single fixed time interval between samples. F_i the number of progeny (transitioning into stage 1) produced by an individual in stage j. Thus, for mallwort, P_{21} is the probability that a seed (Seeds) makes it into the small adult stage (A-Small). P_{32} is the probability that a small adult shows up as a large adult the next year. F_2 is the average fertility for individuals in the small adult stage and F_3 is the average fertility for individuals in the large adult stage.

As the manager responsible for this small reserve, you decide to keep track of this new exotic species. After identifying a general area where the plant seems to have obtained a foothold, you established 50 permanent 1 m^2 sample plots, located randomly throughout the invasion area. Each year, for two years, you sample in early summer when the fruits are on the plants (when the weather is pleasant and you can find interns and volunteers to help). In all plots you tag and count all first year plants (2–3 leaves), and all older plants (4+ leaves). You also are able to count fruits and have determined that there is one seed per fruit.

Now that you have your data for two years, you would like to figure out how quickly the population growing. You could simply keep track of the total population size, N, or just the large adults. You realize, however, that different stages may contribute very differently to growth, and different stages may be better for focused control efforts. A description, or model, of the population that includes different stages will provide this. We call such a model a *demographic model*, and it consists of a *population projection matrix*. The population projection matrix is a matrix that represents the life cycle graph.

We use the projection matrix to calculate all kinds of fun and useful stuff including

- The finite rate of increase, λ (the asymptotic population growth rate).
- The stable stage distribution (the population structure that would emerge if the demographic rates (P, F) do not change).
- Elasticty, the relative importance of each transition to λ.

2.1.1 The population projection matrix

The population projection matrix (a.k.a. the transition matrix) is simply an organized collection of the per capita contribution of each stage j to the next stage i in the specified time interval (often one year). These contributions, or transitions, consist of (i) the probabilities that an individual in stage j in one year makes it into stage i the next year, and (ii) the per capita fecundities for reproductive stages (eq. 2.1).

Each element of the projection matrix (eq. 2.1) relates its column to its row. Thus P_{21} in our matrix, eq. 2.1 is the probability that an individual in stage 1 (seeds; respresented by the column 1) makes it to the next census period and shows up in stage 2 (1 year old small adult; represented by row 2). Similarly, P_{32} is the probability that an individual in stage 2 (a small one year old adult) has made it to the large adult stage at the next census period. The fecundities are not probabilities, of course, but are the per capita contribution of propagules from the adult stage to the seed stage. The population projection matrix allows us to multiply all of these transition elements by the abundances in each age class in one year to predict, or *project*, the abundances of all age classes in the following year.

$$\begin{pmatrix} 0 & F_2 & F_3 \\ P_{21} & 0 & 0 \\ 0 & P_{32} & P_{33} \end{pmatrix} \tag{2.1}$$

2.1.2 A brief primer on matrices

We refer to matrices by their rows and columns. A matrix with three rows and one column is a 3×1 matrix (a "three by one" matrix); we *always* state the number of rows first. Matrices are composed of *elements*; an element of a matrix is signified by its row and column. The element in the second row and first column is a_{21}.

To multiply matrices, we multiply and then sum each row by each column (eq. B.3). More specifically, we multiply each row element of matrix **A** times each column element of matrix **B**, sum them, and place this sum in the respective element of the final matrix. Consider the matrix multiplication in eq. B.3. We first multiply each element of row 1 of **A** ($a \quad b$), times the corresponding elements of column 1 of **B** ($m \quad n$), sum these products and place the sum in the first row, first column of the resulting matrix. We then repeat this for each row of **A** and each column of **B**

$$\mathbf{A} = \begin{pmatrix} a\, b \\ c\, d \end{pmatrix};\ \mathbf{B} = \begin{pmatrix} m\ o \\ n\ p \end{pmatrix} \tag{2.2}$$

$$\mathbf{AB} = \begin{pmatrix} (am + bn)\ (ao + bp) \\ (cm + dn)\ (co + dp) \end{pmatrix} \tag{2.3}$$

This requires that the number of columns in the first matrix must be the same as the number of rows in the second matrix. It also means that the resulting matrix will have the same number of rows as the first matrix, and the same number of columns as the second matrix.

Matrices in R

Let's define two 2 × 2 matrices, filling in one by rows, and the other by columns.

```
> M <- matrix(1:4, nr = 2, byrow = T)
> M

     [,1] [,2]
[1,]    1    2
[2,]    3    4
```

```
> N <- matrix(c(10, 20, 30, 40), nr = 2)
> N

     [,1] [,2]
[1,]   10   30
[2,]   20   40
```

Following our rules above, we would multiply and then sum the first row of M by the first column of N, and make this element a_{11} of the resulting matrix product.

```
> 1 * 10 + 2 * 20

[1] 50
```

We multiply matrices using %*% to signify that we mean *matrix* multiplication.

```
> M %*% N

     [,1] [,2]
[1,]   50  110
[2,]  110  250
```

2.1.3 Population projection

With our spotted mallwort we could multiply our projection matrix by the observed abundances (seeds=Sd, small adults - SA, large adults - LA) to *project* the abundances of all age classes in subsequent years.

$$\begin{pmatrix} 0 & F_2 & F_3 \\ P_{21} & 0 & 0 \\ 0 & P_{32} & P_{33} \end{pmatrix} \begin{pmatrix} N_{Sd} \\ N_{SA} \\ N_{LA} \end{pmatrix} = \begin{pmatrix} (0 \times N_{Sd} + F_2 \times N_{SA} + F_3 \times N_{LA}) \\ (P_{21} \times N_{Sd} + 0 \times N_{SA} + 0 \times N_{LA}) \\ (0 \times N_{Sd} + P_{32} \times N_{SA} + 0 \times N_{LA}) \end{pmatrix} \tag{2.4}$$

The next step is to create the projection matrix. Let's pretend that over the two years of collecting these data, you found that of the small adults we tagged, about half (50%) survived to become large adults the following year. This means that the transition from stage 2 (small adults) to stage 3 (large adults) is $P_{32} = 0.50$. Of the large adults that we tagged, about 90% of those survived to the next year, thus $P_{33} = 0.90$. We estimated that, on average, each small adult produces 0.5 seeds (i.e. $F_2 = 0.50$) and each large adult produces 20 seeds (i.e. $F_3 = 20$). Last, we found that, on average, for every 100 seeds (fruits) we counted, we found about 30 small adults (one year olds), meaning that $P_{21} = 0.30$. Note that this requires that seeds survive until germination, germinate, and then survive until we census them the following summer. We can now fill in our population projection matrix, **A**.

$$\mathbf{A} = \begin{pmatrix} 0 & F_2 & F_3 \\ P_{21} & 0 & 0 \\ 0 & P_{32} & P_{33} \end{pmatrix} = \begin{pmatrix} 0 & 0.5 & 20 \\ 0.30 & 0 & 0 \\ 0 & 0.50 & 0.90 \end{pmatrix} \tag{2.5}$$

Next we can multiply it the projection matrix, **A**, by the last year for which we have data.

$$\begin{pmatrix} 0 & 0.5 & 20 \\ 0.3 & 0 & 0 \\ 0 & 0.5 & 0.9 \end{pmatrix} \begin{pmatrix} 100 \\ 250 \\ 50 \end{pmatrix} = \begin{pmatrix} (0 \times 100 + 0.5 \times 250 + 20 \times 50) \\ (0.3 \times 100 + 0 \times 250 + 0 \times 50) \\ (0 \times 100 + 0.5 \times 250 + 0.9 \times 50) \end{pmatrix} = \begin{pmatrix} 1125 \\ 30 \\ 170 \end{pmatrix} \tag{2.6}$$

If we wanted more years, we could continue to multiply the projection matrix by each year's projected population. We will observe that, at first, each stage increases or decreases in its own fashion (Fig. 2.3a), and that over time, they tend to increase in a more similar fashion. This is typical for demographic models. It is one reason why it is important to examine *stage-structured growth* rather than trying to lump all the stages together — we have a much richer description of how the population is changing.

Stage structured growth - one step

First, we create a population projection matrix, and a vector of stage class abundances for year zero.

```
> A <- matrix(c(0, 0.5, 20, 0.3, 0, 0, 0, 0.5, 0.9), nr = 3,
+     byrow = TRUE)
> N0 <- matrix(c(100, 250, 50), ncol = 1)
```

Now we perform matrix multiplication between the projection matrix and N_0.

```
> N1 <- A %*% N0
> N1

     [,1]
[1,] 1125
[2,]   30
[3,]  170
```

Note that the first stage declined, while the second and third stages increased.

Stage structured growth - multiple steps

Now we project our population over six years, using a for-loop. We use a for-loop, rather than `sapply`, because each year depends on the previous year (see the Appendix, sec. B.6). First, we set the number of years we want to project, and then create a matrix to hold the results. We put N_0 in the first column.

```
> years <- 6
> N.projections <- matrix(0, nrow = nrow(A), ncol = years +
+     1)
> N.projections[, 1] <- N0
```

Now we perform the iteration with the for-loop.

```
> for (i in 1:years) N.projections[, i + 1] <- A %*% N.projections[,
+     i]
```

Last, we graph the results for each stage (Fig. 2.3a). To graph a matrix, R is expecting that the data will be in columns, not rows, and so we need to transpose the projection matrix.

```
> matplot(0:years, t(N.projections), type = "l", lty = 1:3,
+     col = 1, ylab = "Stage Abundance", xlab = "Year")
> legend("topleft", legend = c("Seeds", "Small Adult", "Large Adult"),
+     lty = 1:3, col = 1, bty = "n")
```

2.1.4 Population growth

We have projected the stages for six years — what is its observed rate of increase, $R_t = N_{t+1}/N_t$? How do we even think about R and N in stage structured growth? The way we think about and calculate these is to add all the individuals in all stages to get a total N, and calculate R with that, as we did in Chapter 1.

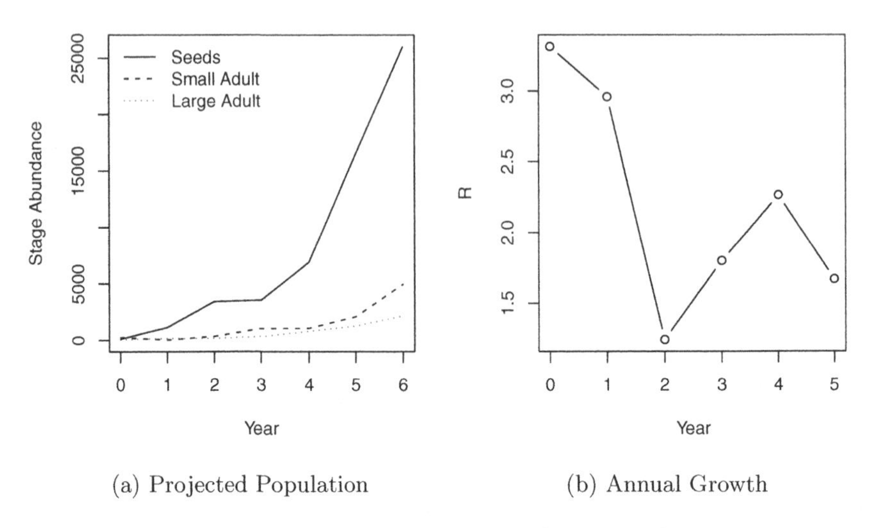

(a) Projected Population (b) Annual Growth

Fig. 2.3: Population dynamics and annual growth ($R = N_{t+1}/N_t$) of spotted mallwort. Note that stage abundance is on a log-scale.

$$R_t = N_{t+1}/N_t. \tag{2.7}$$

If we do that for our mallwort, we can see that R_t changes with time (Fig. 2.3b). We can summarize the projection as $\mathbf{n}(t) = \mathbf{A}^t\mathbf{n}_0$, where $\mathbf{A}^t$ is $\mathbf{A}$ multiplied by itself t times.

Annual growth rate

Now let's calculate $R_t = N_{t+1}/N_t$ for each year t. We first need to sum all the stages, by *applying* the `sum` function to each column.

```
> N.totals <- apply(N.projections, 2, sum)
```

Now we get each R_t by dividing all the N_{t+1} by each N_t. Using negative indices cause R to drop that element.

```
> Rs <- N.totals[-1]/N.totals[-(years + 1)]
```

We have one fewer Rs than we do years, so let's plot each R in each year t, rather than each year $t+1$ (Fig. 2.3b).

```
> plot(0:(years - 1), Rs, type = "b", xlab = "Year", ylab = "R")
```

2.2 Analyzing the Projection Matrix

You seem to have a problem on your hands (Fig. 2.3a). Being a well-trained scientist and resource manager, several questions come to mind: What the heck do I do now? What is this population likely to do in the future? Can these data provide insight into a control strategy? How confident can I be in these projections?

After you get over the shock, you do a little more research on demographic models; Caswell [23] is the definitive treatise. You find that, indeed, there is a lot you can do get more information about this population that might be helpful.

Once you have obtained the projection matrix, **A**, you can analysis it using *eigenanalysis* to estimate

- λ, the finite rate of increase,
- stable stage structure,
- reproductive value, and
- sensitivities and elasticities.

Below, we explain each of these quantities. These quantities will help you determine which stages of spotted mallwort on which to focus eradication efforts.

2.2.1 Eigenanalysis

Eigenanalysis is a mathematical technique that summarizes multivariate data. Ecologists use eigenanalysis frequently, for (i) multivariate statistics such as ordination, (ii) stability analyses with two or more species, and (iii) analyzing population projection matrices. Eigenanalysis is simply a method to transform a square matrix into independent, orthogonal, and useful chunks — the eigenvectors and their eigenvalues. In demography, the most useful piece is the dominant eigenvalue and its corresponding vector.

Eigenanalysis is a technique that finds all the solutions for λ and **w** of

$$\mathbf{Aw} = \lambda\mathbf{w}, \tag{2.8}$$

where **A** is a particular summary of our data. [4] With projection matrix analysis, **A** is the projection matrix. λ is an *eigenvalue* and **w** is an *eigenvector*. If we write out eq. 2.8 for a 3×3 matrix, we would have

$$\begin{pmatrix} a_{11} & a_{12} & a_{13} \\ a_{21} & a_{22} & a_{33} \\ a_{31} & a_{32} & a_{33} \end{pmatrix} \begin{pmatrix} w_{11} \\ w_{21} \\ w_{31} \end{pmatrix} = \lambda \begin{pmatrix} w_{11} \\ w_{21} \\ w_{31} \end{pmatrix} \tag{2.9}$$

There are typically an infinite number of solutions to this equation, and what eigenanalysis does is find set of solutions that are all independent of each other, and which capture all of the information in **A** in a particularly useful way.[5] Typically, the first solution captures the most important features of the

[4] For ordination, we analyze a correlation or covariance matrix, and for stability analyses, we use the matrix of pairwise partial differential equations between each pair of species. In these eigenanalyses of a square $i \times j$ matrix **A**, we can think of the elements of **A** describing the "effect" of stage (or species) j on stage (or species) i, where j is a column and i is a row.

[5] The number of solutions is infinite because they are just simple multiples of the set found with eigenanalysis.

projection matrix. We call this the dominant eigenvalue, λ_1 and its corresponding eigenvector, w_1. The first solution does not capture all of the information; the second solution captures much of the important remaining information. To capture all of the information in **A** requires as many solutions as there are columns of **A**. Nonetheless, the first solution is usually the most useful.

Eigenanalysis in R

Here we perform eigenanalysis on **A**.

```
> eigs.A <- eigen(A)
> eigs.A

$values
[1]  1.834+0.000i -0.467+1.159i -0.467-1.159i

$vectors
               [,1]                [,2]                [,3]
[1,] 0.98321+0i  0.97033+0.00000i  0.97033+0.00000i
[2,] 0.16085+0i -0.08699-0.21603i -0.08699+0.21603i
[3,] 0.08613+0i -0.02048+0.06165i -0.02048-0.06165i
```

Each eigenvalue and its corresponding eigenvector provides a solution to eq. 2.8.

The first, or dominant, eigenvalue is the long term asymptotic finite rate of increase λ. Its corresponding eigenvector provides the *stable stage distribution.*

We can also use eigenanalysis get the *reproductive values* of each stage out of **A**. To be a little more specific, **w** we described above are *right* eigenvectors, so-called because we solve for them with **w** on the right side of **A**. We will also generate *left* eigenvectors **v** (and their corresponding eigenvalues), where $\mathbf{vA} = \lambda(v)$. The dominant left eigenvector provides the reproductive values (section 2.2.4).

2.2.2 Finite rate of increase – λ

The asymptotic annual growth rate finite rate of increase is the dominant *eigenvalue* of the projection matrix. Eigenvalues are always referred to with the Greek symbol λ, and provides a solution to eq. (2.8). The dominant eigenvalue of any matrix, λ_1, is the eigenvalue with the largest absolute value, and it is frequently a complex number.[6] With projection matrices, λ_1 will always be positive and real.

We use eigenanalysis to solve eq. 2.8 and give us the answers — like magic. Another way to find λ_1 is to simply iterate population growth a very large number of times, that is, let t be very large. As t grows, the annual growth rate, N_{t+1}/N_t, approaches λ_1 (Fig. 2.4).

[6] When you perform eigenanalysis, it is common to get complex numbers, with real and imaginary parts. Eigenanalysis is, essentially, solving for the roots of the matrix, and, just like when you solved quadratic equations by hand in high school, it is possible to get complex numbers.

Finding λ

Next we explicitly find the index position of the largest absolute value of the eigenvalues. In most cases, it is the first eigenvalue.

```
> dom.pos <- which.max(eigs.A[["values"]])
```

We use that index to extract the largest eigenvalue. We keep the real part, using `Re`, dropping the imaginary part. (Note that although the dominant eigenvalue will be real, R will include an imaginary part equal to zero ($0i$) as a place holder if *any* of the eigenvalues have a non-zero imaginary part).

```
> L1 <- Re(eigs.A[["values"]][dom.pos])
> L1
```

```
[1] 1.834
```

`L1` is λ_1, the aysmptotic finite rate of increase.

Power iteration method of eigenanalysis

Because growth is an exponential process, we can figure out what is most important in a projection matrix by multiplying it by the stage structure many times. This is actually one way of performing eigenanalysis, and it is called the *power iteration method.* It is not terribly efficient, but it works well in some specific applications. (This method is *not* used by modern computational languages such as R.) The population size will grow toward infinity, or shrink toward zero, so we keep rescaling N, dividing the stages by the total N, just to keep things manageable.
Let t be big, and rescale N.

```
> t <- 20
> Nt <- N0/sum(N0)
```

We then create a for-loop that re-uses N_t for each time step, making sure we have an empty numeric vector to hold the output.

```
> R.t <- numeric(t)
> for (i in 1:t) R.t[i] <- {
+     Nt1 <- A %*% Nt
+     R <- sum(Nt1)/sum(Nt)
+     Nt <- Nt1/sum(Nt1)
+     R
+ }
```

Let's compare the result directly to the point estimate of λ_1 (Fig. 2.4).

```
> par(mar = c(5, 4, 3, 2))
> plot(1:t, R.t, type = "b", main = quote("Convergence Toward  " *
+     lambda))
> points(t, L1, pch = 19, cex = 1.5)
```

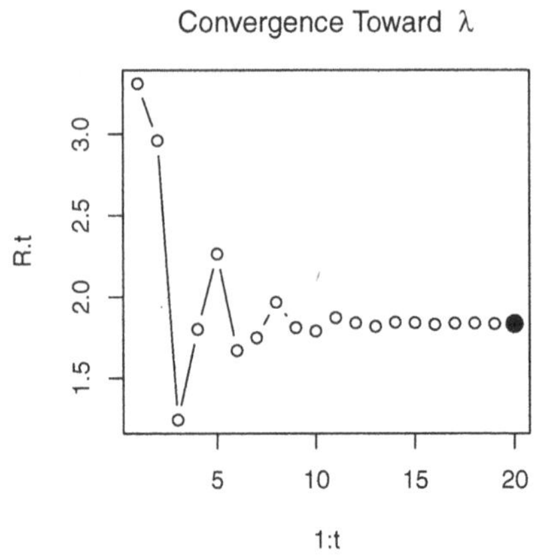

Fig. 2.4: Iterating the population, and recalculating $R_t = N_{t+1}/N_t$ at each time step converges eventually at the dominant eigenvalue, indicated as a solid point. It is possible to use the same power iteration method to get the other eigenvalues, but it is not worth the trouble.

2.2.3 Stable stage distribution

The relative abundance of the different life history stages is called the *stage distribution*, that is, the distribution of individuals among the stages. A property of a stage structured population is that, if all the demographic rates (elements of the population projection matrix) remain constant, its stage structure will approach a *stable stage distribution*, a stage distribution in which the **relative** number of individuals in each stage is constant. Note that a population can grow, so that the absolute number of individuals increases, but the relative abundances of the stages is constant; this is the stable stage distribution. If the population is not actually growing ($\lambda = 1$) and demographic parameters remain constant, then the population is *stationary* and will achieve a *stationary stage distribution*, where neither absolute nor relative abundances change.

How do we find the stable stage distribution? It also turns out that w_1, which is the corresponding eigenvector of λ_1 (eq. (2.8)), provides the necessary information. We scale the eigenvector w_1 by the sum of its elements because we are interested in the *distribution*, where all the stages should sum to one. [7] Therefore the stable stage distribution is

$$SSD = \frac{w_1}{\sum_{i=1}^{S} w_1} \tag{2.10}$$

where S is the number of stages.

Once a population reaches its stable stage distribution it grows exponentially,

[7] Eigenvectors can only be specified up to a constant, arbitrary multiplier.

$$\mathbf{N_t} = \mathbf{A^t N_0}$$
$$N_t = \lambda^t N_0$$

represented either in the matrix notation (for all stages), or simple scalar notation (for total N only).

Calculating the stable stage distribution

The dominant eigenvector, **w**, is in the same position as the dominant eigenvalue. We extract **w**, keeping just the real part, and divide it by its sum to get the stable stage distribution.

```
> w <- Re(eigs.A[["vectors"]][, dom.pos])
> ssd <- w/sum(w)
> round(ssd, 3)
```

```
[1] 0.799 0.131 0.070
```

This shows us that if the projection matrix does not change over time, the population will eventually be composed of 80% seeds, 13% small adults, and 7% large adults. Iterating the population projection will also eventually provide the stable stage distribution (e.g., Fig. 2.3a).

2.2.4 Reproductive value

If the stage structure gives us one measure of the importance of a stage (its abundance), then the *reproductive value* gives us one measure of the importance of an *individual* in each stage. Reproductive value is the expected contribution of each individual to present and future reproduction. We characterize all individuals in a stage using the same expected reproductive value.

We find each stage's reproductive value by solving for the dominant *left* eigenvector **v**, where

$$\mathbf{vA} = \lambda\mathbf{v} \tag{2.11}$$

Like the relation between the dominant right eigenvector and the stable stage distribution, this vector is actually *proportional* to the reproductive values. We typically scale it for $v_0 = 1$, so that all reproductive values are relative to that of the first stage class (e.g. newborns or seeds).

$$RV = \frac{v_1}{\sum_{i=1}^{S} v_1} \tag{2.12}$$

Calculating reproductive value

We get the left eigenvalues and -vectors by performing eigenanalysis on the transpose of the projection matrix. The positions of the dominant right and left eigenvalues are the same, and typically they are the first. We perform eigenanalysis, extracting just the the dominant left eigenvector; we then scale it, so the stage 1 has a reproductive value of 1.0.

```
> M <- eigen(t(A))
> v <- Re(M$vectors[, which.max(Re(M$values))])
> RV <- v/v[1]
> RV
```

```
[1]  1.000  6.113 21.418
```

Here we see a common pattern, that reproductive value, v, increases with age. In general, reproductive value of individuals in a stage increases with increasing probability of reaching fecund stages.

2.2.5 Sensitivity and elasticity

Sensitivity and elasticity tell us the relative importance of each transition (i.e. each arrow of the life cycle graph or element of the matrix) in determining λ. They do so by combining information on the stable stage structure and reproductive values.

The stage structure and reproductive values each in their own way contribute to the importance of each stage in determining λ. The stable stage distribution provides the relative abundance of individuals in each stage. Reproductive value provides the contribution to future population growth of individuals in each stage. Sensitivity and elasticity combine these to tell us the relative importance of each transition in determining λ.

Sensitivities of a population projection matrix are the direct contributions of each transition to determining λ. We would say, speaking in more mathematical terms, that the sensitivities for the elements a_{ij} of a projection matrix are the changes in λ, given small changes in each element, or $\delta\lambda/\delta a_{ij}$. Not surprisingly, then, these are derived from the stable stage distribution and the reproductive values. Specifically, the sensitivities are calculated as

$$\frac{\delta\lambda}{\delta a_{ij}} = \frac{v_{ij}w_{ij}}{\mathbf{v}\cdot\mathbf{w}} \tag{2.13}$$

where $v_i w_j$ is the product of each pairwise combination of elements of the dominant left and right eigenvectors, $\mathbf{v}$ and $\mathbf{w}$. The *dot product*, $\mathbf{v}\cdot\mathbf{w}$, is the sum of the pairwise products of each vector element. Dividing by this causes the sensitivities to be relative to the magnitudes of $\mathbf{v}$ and $\mathbf{w}$.

Sensitivity of projection matrices

Let's calculate sensitivities now. First we calculate the numerator for eq. 2.13.

```
> vw.s <- v %*% t(w)
```

Now we sum these to get the denominator, and then divide to get the sensitivities. (The dot product $\mathbf{v} \cdot \mathbf{w}$ yields a 1×1 matrix; in order to divide by this quantity, the simplest thing is to cause the dot product to be a simple scalar rather than a matrix (using `as.numeric`), and then R will multiply each element.)

```
> (S <- vw.s/as.numeric(v %*% w))

      [,1]    [,2]   [,3]
[1,] 0.258 0.04221 0.0226
[2,] 1.577 0.25798 0.1381
[3,] 5.526 0.90396 0.4840
```

We see from this that the most important transition exhibited by the plant is s_{21}, surviving from the seed stage to the second stage (the element s_{31} is larger, but is not a transition that the plant undergoes).

Elasticities are sensitivities, weighted by the transition probabilities. Sensitivities are large when reproductive value and or the stable age distribution are high, and this makes sense biologically because these factors contribute a lot to λ. We may, however, be interested in how a *proportional* change in a transition element influences lambda—how does a 10% increase in seed production, or a 25% decline in juvenile survival influence λ? For these answers, we need to adjust sensitivities to account for the relative magnitudes of the transition elements, and this provides the elasticities, e_{ij}, where

$$e_{ij} = \frac{a_{ij}}{\lambda} \frac{\delta\lambda}{\delta a_{ij}}. \tag{2.14}$$

Elasticity of projection matrices

In R, this is also easy.

```
> elas <- (A/L1) * S
> round(elas, 3)

      [,1]  [,2]  [,3]
[1,] 0.000 0.012 0.246
[2,] 0.258 0.000 0.000
[3,] 0.000 0.246 0.238
```

Note that all the elasticities except the seed production by small adults appear equally important. Specifically, the same proportional change in any of these elements will result in approximately the same change in λ.

There are two nice features of elasticities. First, *impossible transitions have elasticities equal to zero*, because we multiply by the projection matrix itself.

Second, the *elasticities sum to zero*, and so it is easier to compare elasticities among differ matrices and different organisms.

Once we have the sensitivities and elasticities, we can really begin to see what is controlling the growth rate of a stage (or age) structured population. Although these values do not tell us which stages and transitions will, in reality, be influenced by natural phenomona or management practices, they provide us with the predicted *effects* on λ of a proportional change in a demographic rate, P or F. This is particularly important in the management of invasive (or endangered) species where we seek to have the maximum impact for the minimum amount of effort and resources [23, 48].

2.2.6 More demographic model details

Births

For demographic models, a "birth" is merely the appearance in the first stage. If we census birds, a "birth" might be a fledging, if this is the youngest age class we sampled. If we census plants, we might choose to count seeds as the first age class, or we might use seedling, or some size threshold as the first stage. Regardless of the first stage- or age-class we use, a birth is the first appearance of an individual in the first stage.

Pre- vs. post-breeding census

Note that you are sampling the population of mallwort at a particular time of year. This sampling happens to be a *postbreeding census* because you captured everything right after breeding, when progeny were observed directly. The projection matrix would look different, and the interpretation of the matrix elements would differ, if we had used a *prebreeding census*, sampling the population before breeding. In particular, the projection matrix would have only two stages (small and large adults), because no seeds would be present at the time of sampling. The contribution of adults to the youngest stage, therefore, would represent both fertility and survival to the juvenile stage in late spring. Nonetheless, both models would be equivalent, generating the same λ.

Birth pulse vs. birth flow model

Another assumption we are making is that individuals set seed, or give birth, all at once. We refer to the relevant model as a *birth-pulse model.* On the other hand, if we assume that we have continuous reproduction, we do things quite differently, and would refer to this as a *birth-flow model.* Whether a population is breeding continuously over a year, or whether reproduction is seasonal, will influence how we estimate fecundities. Even for synchronously breeding populations, many models pool years into a single age class or stage. As result, we need to be careful about how we approximate probabilities that will differ among individuals within the age- or stage-class.

These details can get very confusing, and smart people don't always get it right. Therefore, consult an expert [23, 48], and remember that the stages of life cycle graph and matrix are the stages that you collect at one point in time.

2.3 Confronting Demographic Models with Data

This section uses R extensively throughout.

It is common to create a demographic matrix model with real data, and then use that model for an applied purpose (e.g., [44, 50]). A central question, however, is just how confident we can be in our model, and the values we derive from it. It turns out that we can use our data to derive confidence intervals on important parameters.

In Chapter 1, we used resampling to draw observed annual changes in bird counts at random to generate growth trajectories and confidence intervals on population size. Here we resample raw data to find confidence limits on λ. The method used here, *bootstrapping*, and related data-based inference techniques have a large literature [126]. Davison and Hinkley [46] have an comprehensive R-based text. Such randomization methods are very useful for a wide range of models in ecology, where the data do not conform clearly to parametric distributions or to situations like demographic models [140] or null models [60] for which analytical approximations are difficult or not possible.

The basic idea of bootstrapping is to

1. calculate the *observed parameter(s)* of interest (e.g., a mean, or λ) with your model and data,
2. resample your data *with replacement* to create a large number of datasets and recalculate your parameter(s) for each resampled dataset to generate a *distribution* of the *bootstrapped*[8]parameter(s),
3. Calculate a confidence interval for the bootstrapped parameter values — this will provide an estimate of the confidence you have in your observed parameter. This will provide an *empirical confidence interval.*

2.3.1 An Example: *Chamaedorea* palm demography

Chamaedorea radicalis Mart. (Arecaceae) is an forest understory palm of northern Mexico, and it is one of approximately 100 *Chamaedorea* species, many of which are economically valuable as either small, shade-tolerant potting plants or as harvested leaves in floral arrangements. Its demography is interesting for a number of reasons, including both management and as an example of a population that appears to be maintained through source-sink dynamics [12]. Berry *et al.* modeled *Chamaedorea* demography with five stages (Fig. 2.5). Demography is also influenced by substrate type, by livestock browsing, and harvesting [12]. Here we use a subset of the data to illustrate the generation of demographic parameters and confidence intervals.

This study was conducted in the montane mesophyll forests of Sierra de Guatemala mountain range, near the communities of San José and Alta Cimas within the El Cielo Biosphere Reserve, Tamaulipas, Mexico (22°55'–23°30'N and 99°02'–99°30'W). Villagers within El Cielo (palmilleros) harvest adult *C.*

[8] "Bootstrapped" estimates are thus named because you are picking yourself up by your own bootstraps – a seemingly impossible task.

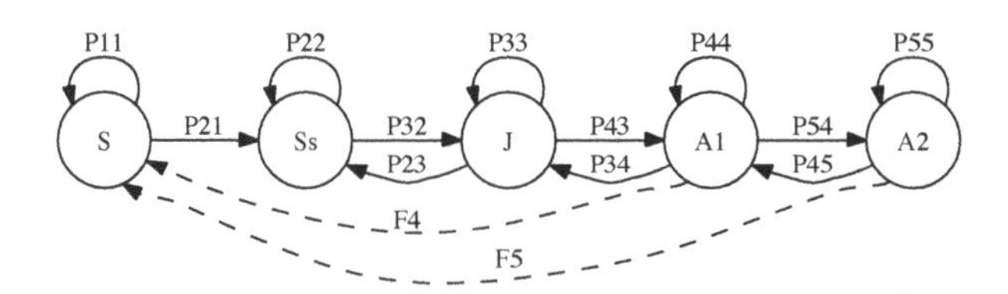

Fig. 2.5: Life cycle graph for *Chamaedorea radicalis*. Classification criteria are based on the number of leaflets on the youngest fully-expanded leaf. Life-history stage transitions are indicated by arrows with solid lines and reproduction is indicated by dashed lines. Abbreviations: S-seed, Ss-seedling (bifid leaves), J-Juvenile (3–9 leaflets), A1-small adult (10–24 leaflets), A2-large adult (> 24 leaflets). Source: [12, 50]

radicalis leaves for sale to international cut-foliage markets. Harvested leaves are usually >= 40 cm in length, and have minimal damage from insects or pathogens [50]. These palm leaves are the only natural resource that these villagers are authorized to harvest, and provide the main source of income for most families. Although *C. radicalis* is dioecious (more complications!), Berry et al. [12] used a one sex model, because its simplifying assumptions were well supported with data. Data collected allowed a postbreeding census model with a birth-pulse dynamic.

2.3.2 Strategy

There are an infinite number of ways to do anything in R, and I am certain that my approach to this bootstrapping is not the very best way, but it is *useful*. It gives valid answers in a reasonable amount of time, and that is what we want from a model.

This is how we proceed in this instance:

1. We import the data and look at it. The appearance of the data, how the data are entered for instance, will influence subsequent decisions about how to proceed.
2. We extract the relevant data and calculate the projection matrix elements (fecundities and transition probabilities). We first do it all piecewise, to figure out what we are doing. Then we can wrap it up inside a function putting `funcname <- function(data1, data2, data3)` at the beginning and collecting and returning all relevant parameters at the end (see sec. B.4.1 for writing functions).
3. We also create a function to generate all the demographic parameters that we will eventually want (λ, elasticities, etc.).
4. Last, we combine these two functions into one that also resamples the original data (with replacement), and then calls the data extraction and calculation functions to generate the new parameters for the bootstrapped data.
5. The bootstrapping is repeated B times.
6. Having generated B bootstrapped estimates of all the parameters, we can then calculate confidence intervals for any parameter that we like.

2.3.3 Preliminary data management

Let's import the data and have a look at it. For these purposes, we will assume that the data are clean and correct. Obviously, if I were doing this for the first time, data-checking and clean-up would be an important first step. Here we simply load them from the `primer` package.

```
> data(stagedat)
> data(fruitdat)
> data(seeddat)
```

Now I look at the structure of the data to make sure it is at least approximately what I think it is.

```
> str(stagedat)

'data.frame':        414 obs. of  4 variables:
 $ PalmNo: int  1 2 3 4 5 6 7 8 9 10 ...
 $ Y2003 : int  4 5 5 4 3 2 4 3 3 4 ...
 $ Y2004 : int  5 4 5 5 4 3 5 3 4 4 ...
 $ Y2005 : int  5 5 5 5 4 3 5 4 4 5 ...
```

The stage data provide the stage of each individual in the study. Each row is an individual, and its ID number is in column 1. Data in columns 2–4 identify its stage in years 2003–2005.

We can count, or tabulate, the number of individuals in each stage in 2004.

```
> table(stagedat[["Y2004"]])

  0   2   3   4   5
 17  58  48 126 165
```

We see, for instance, that in 2004 there were 165 individuals in stage 5. We also see that 17 individuals were dead in 2004 (stage = 0); these were alive in either 2003 or 2005.

The fruit data have a different structure. Each row simply identifies the stage of each individual (col 1) and its fertility (number of seeds) for 2004.

```
> str(fruitdat)

'data.frame':        68 obs. of  2 variables:
 $ Stage: int  4 4 4 4 4 4 4 4 4 4 ...
 $ Y2004: int  6 0 0 0 0 0 0 0 0 0 ...
```

We can tabulate the numbers of seeds (columns) of each stage (rows).

```
> table(fruitdat[["Stage"]], fruitdat[["Y2004"]])

     0  1  2  3  4  5  6  8 15 22 30 37 70 98 107 109
  4 28  0  0  0  0  0  1  0  0  0  0  0  0  0   0   0
  5 23  1  1  1  2  2  0  1  1  1  1  1  1  1   1   1
```

For instance, of the individuals in stage 4 (row 1), 28 individuals had no seeds, and one individual had 6 seeds. Note also that only stage 4 and 5 had plants with *any* seeds.

The seed data are the fates of each seed in a sample of 400 seeds, in a data frame with only one column.

```
> table(seeddat)
```

```
seeddat
  0   1   2
332  11  57
```

Seeds may have germinated (2), remained viable (1), or died (0).

2.3.4 Estimating projection matrix

Now we work through the steps to create the projection matrix from individuals tagged in year 2003 and re-censused in 2004. If we convert the life cycle graph (Fig. 2.5) into a transition matrix.

$$\begin{pmatrix} P_{11} & 0 & 0 & F_4 & F_5 \\ P_{21} & P_{22} & P_{23} & 0 & 0 \\ 0 & P_{32} & P_{33} & P_{34} & 0 \\ 0 & 0 & P_{43} & P_{44} & P_{45} \\ 0 & 0 & 0 & P_{54} & P_{55} \end{pmatrix} \tag{2.15}$$

Along the major diagonal (where $i = j$) the P_{ij} represent the probability that a palm stays in the same stage. In the lower off-diagonal ($i > j$) the P_{ij} represent the probability of growth, that an individual grows from stage j into stage i. In the upper off-diagonal ($i < j$) the P_{ij} represent the probability of regression, that an individual regresses from stage j back into stage i. The F_i represent the fertility of stage i.

As a practical matter, we will use basic data manipulation in R to transform the raw data into transition elements. We had no particular reason for having the data in this form, this is simply how the data were available.

We first create a zero matrix that we will then fill.

```
> mat1 <- matrix(0, nrow = 5, ncol = 5)
```

Fertilities

For each stage, we get mean fertility by applying `mean` to each stage of the 2004 fertility data. Here `Stage` is a factor and `tapply` will caculate a mean for each level of the factor. We will assume that half of the seeds are male. Therefore, we divide fertility by 2 to calculate the fertility associated with just the female seeds.

```
> ferts <- tapply(fruitdat$Y2004, fruitdat$Stage, mean)/2
> ferts
```

```
     4      5
0.1034 6.6667
```

These fertilities, F_4 and F_5, are the transitions from stages 4 and 5 (adults) to stage 1 (seeds). Next we insert the fertilities (`ferts`) into the matrix we established above.

```
> mat1[1, 4] <- ferts[1]
> mat1[1, 5] <- ferts[2]
```

Seed transitions

Now we get the frequencies of each seed fate (die, remain viable but dormant, or germinate), and then divide these by the number of seeds tested (the length of the seed vector); this results in proportions and probabilities.

```
> seed.freqs <- table(seeddat[, 1])
> seedfates <- seed.freqs/length(seeddat[, 1])
> seedfates

     0      1      2
0.8300 0.0275 0.1425
```

The last of these values is P_{21}, the transition from the first stage (seeds) to the stage 2 (seedlings). The second value is the transition of seed dormancy ($P_{1,1}$), that is, the probability that a seed remains a viable seed rather than dying or becoming a seedling.

Next we insert the seed transitions into our projection matrix.

```
> mat1[1, 1] <- seedfates[2]
> mat1[2, 1] <- seedfates[3]
```

Vegetative stage transitions

Here we calculate the transition probabilities for the vegetative stages. The pair of for-loops will calculate these transitions and put them into stages 2–5.The functions inside the for-loops (a) subset the data for each stage in 2003, (b) count the total number of individuals in each stage in 2003 (year j), (c) sum the number of individuals in each stage in 2004, given each stage for 2003, and then (d) calculate the proportion of each stage in 2003 that shows up in each stage in 2004.

```
> for (i in 2:5) {
+     for (j in 2:5) mat1[i, j] <- {
+         x <- subset(stagedat, stagedat$Y2003 == j)
+         jT <- nrow(x)
+         iT <- sum(x$Y2004 == i)
+         iT/jT
+     }
+ }
```

Here we can see the key parts of a real projection matrix.

```
> round(mat1, 2)

     [,1] [,2] [,3] [,4] [,5]
[1,] 0.03 0.00 0.00 0.10 6.67
[2,] 0.14 0.70 0.05 0.01 0.00
[3,] 0.00 0.23 0.42 0.04 0.00
[4,] 0.00 0.00 0.46 0.67 0.07
[5,] 0.00 0.00 0.02 0.26 0.90
```

Compare these probabilities and fertilities to the life cycle graph and its matrix (Fig. 2.5, eq. (2.15)).

The diagonal elements $P_{j,j}$ are stasis probabilities, that an individual remains in that stage. Growth, from one stage to the next, is the lower off-diagonal, $P_{j+1,j}$. Regression, moving back one stage, is the upper off diagonal, $P_{j-1,j}$. The fertilities are in the top row, in columns 4 and 5. Note that there is a transition element in our data that is not in eq. (2.15): P_{53}. This corresponds to very rapid growth — a real event, albeit highly unusual.

A function for all transitions

What a lot of work! The beauty, of course, is that we can put all of those lines of code into a single function, called, for instance, `ProjMat`, and all we have to supply are the three data sets. You could examine this function by typing `ProjMat` on the command line, with no parentheses, to see the code and compare it to our code above. You code also try it with data.

```
> ProjMat(stagedat, fruitdat, seeddat)
```

This provides the observed transition matrix (results not shown).

2.3.5 Eigenanalyses

Next we want to do all those eigenanalyses and manipulations that gave us λ, the stable age distribution,reproductive value, and the sensitivity and elasticity matrices. All of this code is wrapped up in the function `DemoInfo`. Convince yourself it is the same code by typing `DemoInfo` with no parentheses at the prompt. Here we try it out on the projection matrix we created above, and examine the components of the output.

```
> str(DemoInfo(mat1))

List of 6
 $ lambda       : num 1.13
 $ SSD          : num [1:5] 0.5632 0.195 0.0685 0.0811 0.0922
 $ RV           : num [1:5] 1 7.76 14.37 20.18 33.95
 $ Sensitivities: num [1:5, 1:5] 0.072 0.559 1.034 1.452 2.442 ...
 $ Elasticities : num [1:5, 1:5] 0.00174 0.0702 0 0 0 ...
 $ PPM          : num [1:5, 1:5] 0.0275 0.1425 0 0 0 ...
```

We find that `DemoInfo` returns a *list* with six named *components*. The first component is a scalar, the second two are numeric vectors, and the last three are numeric matrices. The last of these is the projection matrix itself; it is often useful to return that to prove to ourselves that we analyzed the matrix we intended to.

2.3.6 Bootstrapping a demographic matrix

All of the above was incredibly useful and provides the best estimates of most or all the parameters we might want. However, it does not provide any idea of the certainty of those parameters. By bootstrapping these estimates by resampling our data, we get an idea of the uncertainty.

Here we work through the steps of resampling our data, as we build a function, step by step, inch by inch. The basic idea of resampling is that we assume that our sample data are the best available approximation of the entire population. Therefore, we draw, with replacement, new data sets from the original one. See the last section in Chapter 1 for ideas regarding simulations and bootstrapping.

We will create new resampled (bootstrapped) data sets, where the rows of the original data sets are selected at random with replacement. We then apply `ProjMat` and `DemoInfo`.

The first step is to get the number of observations in the original data.

```
> nL <- nrow(stagedat)
> nF <- nrow(fruitdat)
> nS <- nrow(seeddat)
```

With these numbers, we will be able to resample our original data sets getting the correct number of resampled observations.

Now we are going to use `lapply` to perform everything multiple times. By "everything," I mean

1. resample the observations to get bootstrapped data sets for vegetative stages, seed fates, and fertilities,
2. calculate the projection matrix based on the three bootstrapped data sets,
3. perform eigenanalysis and calculate λ, stage structure, sensitivities, and elasticities.

All of that is one replicate simulation, $n = 1$.

For now, let's say $n = 5$ times as a trial. Eventually this step is the one we will ask R to do 1000 or more times.

```
> n <- 5
```

Next we use `lapply` to do *everything*, that is, a replicate simulation, n times. It will store the n replicates in a *list*, n *components* long. Each of the n components will be the output of `DemoInfo`, which is itself a list.

```
> n <- 5
> out <- lapply(1:n, function(i) {
+     stageR <- stagedat[sample(1:nL, nL, replace = TRUE), ]
```

```
+     fruitR <- fruitdat[sample(1:nF, nF, replace = TRUE), ]
+     seedR <- as.data.frame(seeddat[sample(1:nS, nS, replace = TRUE), ])
+     matR <- ProjMat(stagedat = stageR, fruitdat = fruitR,
+         seeddat = seedR)
+     DemoInfo(matR)
+ })
```

This code above uses `sample` to draw row numbers at random and with replacement to create random draws of data (`stageR, fruitR, and seedR`). We then use `ProjMat` to generate the projection matrix with the random data, and use `DemoInfo` to perform all the eigenanalysis and demographic calculations.

Let's look at a small subset of this output, just the five λ generated from five different bootstrapped data sets. The object `out` is a list, so using `sapply` on it will do the same thing to each component of the list. In this case, that something is to merely extract the bootstrapped λ.

```
> sapply(out, function(x) x$lambda)
```

```
[1] 1.084 1.137 1.134 1.126 1.158
```

We see that we have five different estimates of λ, each the dominant eigenvalue of a projection matrix calculated from bootstrapped data.

We now have all the functions we need to analyze these demographic data. I have put all these together in a function called `DemoBoot`, whose arguments (inputs) are the raw data, and n, the number of bootstrapped samples.

```
> args(DemoBoot)
```

```
function (stagedat = NULL, fruitdat = NULL, seeddat = NULL, n = 1)
NULL
```

2.3.7 The demographic analysis

Now we are armed with everything we need, including estimates and means to evaluate uncertainty, and we can move on to the ecology. We first interpret point estimates of of demographic information, including λ and elasticities. Then we ask whether λ differs significantly from 1.0 using our bootstrapped confidence interval.

First, point estimates based on original data.

```
> estims <- DemoInfo(ProjMat(stagedat, fruitdat, seeddat))
> estims$lambda
```

```
[1] 1.134
```

Our estimate of λ is greater than one, so the population seems to be growing.

Which transitions seem to be the important ones?

```
> round(estims$Elasticities, 4)
```

```
       [,1]   [,2]   [,3]   [,4]   [,5]
[1,] 0.0017 0.0000 0.0000 0.0009 0.0693
[2,] 0.0702 0.1196 0.0030 0.0005 0.0000
[3,] 0.0000 0.0738 0.0470 0.0049 0.0000
[4,] 0.0000 0.0000 0.0712 0.1234 0.0145
[5,] 0.0000 0.0000 0.0044 0.0793 0.3162
```

It appears that the most important transition is persistence in the largest adult stage ($a_{5,5} = 0.3$). Specifically, proportional changes to the persistence in this stage, neither regressing nor dying, are predicted to have the largest postive effect on the lambda of this population.

We stated above that the population appears to be growing. However, this was based on a sample of the population, and not the entire population. One way to make inferences about the population is to ask whether the confidence interval for λ lies above 1.0. Let's use `DemoBoot` to bootstrap our confidence interval for λ.[9] First, we'll run the bootstrap, and plot the λ's.

```
> system.time(out.boot <- DemoBoot(stagedat, fruitdat, seeddat,
+     n = 1000))

   user  system elapsed
 12.539   0.022  12.561

> lambdas <- sapply(out.boot, function(out) out$lambda)

> hist(lambdas, prob = T)
> lines(density(lambdas))
```

From this it seems clear that the population is probably growing ($\lambda > 1.0$), because the lower limit of the histogram is relatively large (Fig. 2.6). We need to get a real confidence interval, however. Here we decide on a conventional α and then calculate quantiles, which will provide the median (the 50th percentile), and the lower and upper limits to the 95% confidence interval.[10]

```
> alpha <- 0.05
> quantile(lambdas, c(alpha/2, 0.5, 1 - alpha/2))

 2.5%   50% 97.5%
1.062 1.129 1.193
```

From this we see that the 95% confidence interval (i.e. the 0.025 and 0.975 quantiles) does not include 1.0. Therefore, we conclude that under the conditions experienced by this population in this year, this *Chamaedorea* population, from which we drew a sample, could be characterized as having a long-term asymptotic growth rate, λ, that is greaater than 1.0, and therefore would be likely to increase in abundance, if the environment remains the same.

[9] The number of replicates needed for a bootstrap depend in part on how close the interval is to critical points. If, for instance, your empirical P-value seems to be very close to your cutoff of $\alpha = 0.05$, then you should increase the replicates to be sure of your result. These days $n = 1000$ is considered a bare minimum.

[10] Quantiles are ordered points in a cumulative probability distribution function.

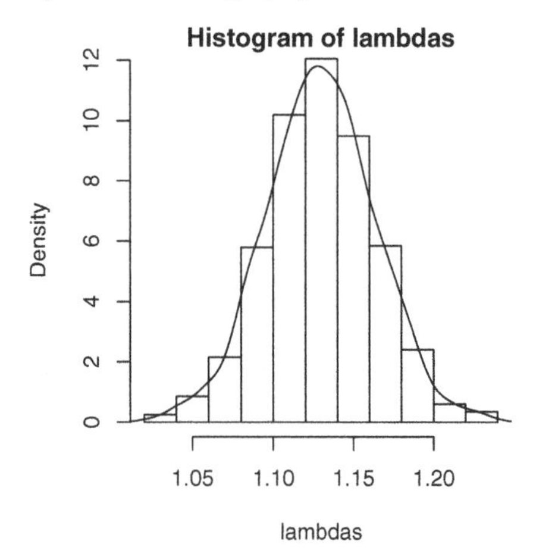

Fig. 2.6: The frequency distribution for our bootstrapped λ. Note that it is fairly symmetrical, and largely greater than 1.0.

A caveat and refinement

Bootstrapping as we have done above, known variously as the basic or percentile bootstrap, is not a cure-all, and it can give inappropriate estimation and inferrence under some circumstances. A number of refinements have been proposed that make bootstrapping a more precise and accurate procedure [46]. The problems are worst when the data are such that the bootstrap replicates are highly skewed, so that the mean and median are quite different. When the data are relatively symmetric, as ours is (Fig. 2.6), the inference is relatively reliable.

Often skewness will cause the mean of the bootstrap samples to differ from our observed estimate, and we refer to this as *bias*. We should adjust the bootstrapped samples for this bias [140]. Here we calculate the bias.

```
> bias <- mean(lambdas) - estims$lambda
> bias
```

```
[1] -0.004208
```

We find that the bias is very small; this gives us confidence the our confidence intervals are pretty good. Nonetheless, we can be thorough and correct our samples for this bias. We subtract the bias from the bootstrapped λ to get our confidence interval.

```
> quantile(lambdas - bias, c(alpha/2, 0.5, 1 - alpha/2))
```

```
 2.5%   50% 97.5%
1.067 1.133 1.197
```

These bias-corrected quantiles also indicate that this population in this year can be characterized by a $\lambda > 1$.

If we want to infer something about the future success of this population, we need to make additional assumptions. First, we must assume that our sample was representative of the population; we have every reason to expect it is. Second, we need to assume that this year was representative of other years. In particular, we need to assume that the weather, the harvest intensity, and the browsing intensity are all representative. Clearly, it would be nice to repeat this for other years, and to try to get other sources of information regarding these factors.

2.4 Summary

Demography is the study of structured populations. Structure may be described by age or stage, and is represented by life cycle graphs and a corresponding projection or transition matrix of transition probabilities and fertilities. The finite rate of increase, λ, and the stable stage/age distribution are key characteristics of a population, and are estimated using eigenalysis; populations will grow geometrically at the per capita rate of λ only when the population has reached its stable stage/age distribution. We measure the importance of transition elements with sensitivities and elasticities, the absolute or relative contributions λ of transition elements. Demographic information is frequently useful for endangered and invasive species.

Problems

2.1. Demographic analysis of a plant population

Goldenseal (*Hydrastis canadensis*) is a wild plant with medicinal properties that is widely harvested in eastern North American. Its rhizome (the thick underground stem) is dug up, and so harvesting can and frequently does have serious negative impacts on populations. A particular population of goldenseal is tracked over several years and investigators find, tag, and monitor several sizes of individuals [57]. After several years of surveys, they identify six relevant stages: dormant seed, seedling, small 1-leaved plant, medium 1-leaved plant, large 1-leaved plant, fertile plant (flowering, with 2 leaves). They determine that the population project matrix is:

$$\mathbf{A} = \begin{pmatrix} 0 & 0 & 0 & 0 & 0 & 1.642 \\ 0.098 & 0 & 0 & 0 & 0 & 0.437 \\ 0 & 0.342 & 0.591 & 0.050 & 0.095 & 0 \\ 0 & 0.026 & 0.295 & 0.774 & 0.177 & 0.194 \\ 0 & 0 & 0 & 0.145 & 0.596 & 0.362 \\ 0 & 0 & 0 & 0.016 & 0.277 & 0.489 \end{pmatrix} \qquad (2.16)$$

(a) Draw a life cycle graph of this population of goldenseal. Include the matrix elements associated with each transition.

(b) Start with $\mathbf{N} = (0\;10\;10\;10\;10\;10)$ and graph population dynamics for all

stages for 10 years.
(c) Determine the stable stage distribution.
(d) Determine λ. Explain what this tells us about the population, including any assumptions regarding the stable stage distribution.
(d) Determine the elasticities. Which transition(s) are most influential in determining growth rate?
(e) Discuss which stages might be most suitable for harvesting; consider this question from both a financial and ecological perspective.

2.2. Demographic analysis of an animal population
Crouse et al. [44] performed a demographic analysis of an endangered sea turtle species, the loggerhead (*Caretta caretta*). Management of loggerhead populations seemed essential for their long term survival, and a popular management strategy had been and still is to protect nesting females, eggs, and hatchlings. The ground breaking work by Crouse[11] and her colleagues compiled data to create a stage-based projection matrix to analyze quantitatively which stages are important and least important in influencing long-term growth rate. This work led to US Federal laws requiring that US shrimp fishermen use nets that include Turtle Excluder Devices (TEDs, http://www.nmfs.noaa.gov/pr/species/turtles/teds.htm). Crouse et al. determined the transition matrix for their loggerhead populations:

$$\mathbf{A} = \begin{pmatrix} 0 & 0 & 0 & 0 & 127 & 4 & 80 \\ 0.6747 & 0.7370 & 0 & 0 & 0 & 0 & 0 \\ 0 & 0.0486 & 0.6610 & 0 & 0 & 0 & 0 \\ 0 & 0 & 0.0147 & 0.6907 & 0 & 0 & 0 \\ 0 & 0 & 0 & 0.0518 & 0 & 0 & 0 \\ 0 & 0 & 0 & 0 & 0.8091 & 0 & 0 \\ 0 & 0 & 0 & 0 & 0 & 0.8091 & 0.8089 \end{pmatrix} \tag{2.17}$$

(a) Draw a life cycle graph of this loggerhead population. Include the matrix elements associated with each transition.
(b) Determine the stable stage distribution.
(c) Determine λ. Explain what this tells us about the population, including any assumptions regarding the stable stage distribution.
(d) Determine the elasticities. Which transition(s) are most influential in determining growth rate?
(e) What is the predicted long-term relative abundance of all stages? What do we call this?
(f) If your interest is to maximize long-term growth rate, in which stage(s) should you invest protection measures? Which stages are least likely to enhance long-term growth rate, regardless of protective measures?
(g) Start with $\mathbf{N} = (0\ 10\ 10\ 10\ 10\ 10\ 10)$ and graph dynamics for all stages for 10 years.

[11] Crouse was a graduate student at the time — graduate students are the life-blood of modern science, doing cutting edge work and pushing their fields forward.

3

Density-dependent Growth

Let's go back to our Song Sparrow (*Melospiza melodia*) data from Chapter 1 on density-independent growth — now we look at *all* the data.

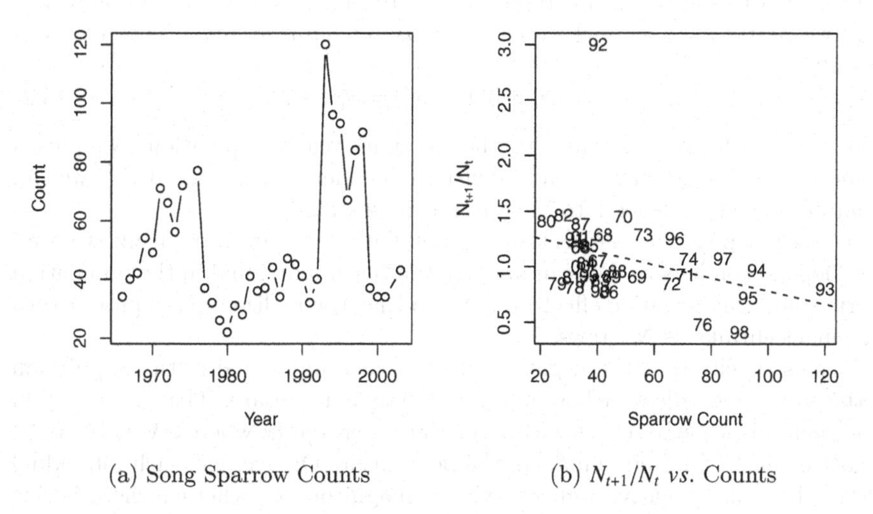

(a) Song Sparrow Counts

(b) N_{t+1}/N_t *vs.* Counts

Fig. 3.1: Song Sparrow *Melospiza melodia* counts from 1966–2003 and the relation between observed counts and annual growth rate determined from those counts, fit with ordinary least squares regression. See Chapter 1 for data source.

In Chapter 1, we modeled Song Sparrow growth without reference to the abundance, or counts, of sparrows, using a small subset of the data. When we look at all the available data (Fig. 3.1a), it seems clear that the annual growth rate ($R_t = N_{t+1}/N_t$) depends on the density of the bird population (Fig. 3.1b). The larger the population, the lower R becomes, until above about 70 counted birds, $R < 1$, indicating population decline. What might limit the population growth of these sparrows? Space available for territories in their successional-scrub type habitat? Availability of seeds, fruits and invertebrates? We don't necessarily know precisely what limits it, but if it is something related to their own abundance, perhaps we can study its *density-dependence*.

Density-dependent population growth is the case where the per capita population growth rate depends statistically on the density of the population. Ecologists consider that negative density-dependence is typically a characteristic of a population undergoing *intraspecific competition*. That is, individuals of the same species are competing for shared resources. So, ... how would we represent this algebraically?[1]

3.1 Discrete Density-dependent Growth

3.1.1 Motivation

To begin, we recall our model of geometric growth for a single time step,

$$N_{t+1} = \lambda N_t. \tag{3.1}$$

We can pull this apart into two sections by recalling that we can decompose λ into two parts, $\lambda = 1 + r_d$, where r_d is the *discrete growth factor*. This allows us to state,

$$N_{t+1} = \lambda N_t = N_t\left(1 + r_d\right) = N_t + r_d N_t. \tag{3.2}$$

Here we see the N_{t+1} is equal to the previous year's population, N_t, plus a proportional change, $r_d N_t$. To add density dependence (e.g., Fig. 3.1b), we can build *density-dependence* into that proportional change.

Density dependence means that the population will grow or shrink at a rate that depends on its size. We can imagine that each individual in the population exerts some tiny negative effect on $r_d N_t$, so that the realized per capita growth increment shrinks as N_t grows.

Let us specify that the negative effect of each individual in the population is the same, regardless of how many individuals there are. That is, we could represent this negative effect with a constant, perhaps α, where αN_t is the total negative effect of all individuals in the population. On average, each individual exerts the same negative impact, with a magnitude α, whether there is one individual or 1000 individuals. A particularly convenient way of implementing this is to keep using r_d but to multiply it by a scaling factor. This scaling factor should equal 1 when the population size is zero so that $r_d \times (\text{scaling factor}) = r_d$. The scaling factor should be zero when the population is so big that per capita growth increment is zero, $r_d \times (\text{scaling factor}) = 0$. One such scaling factor looks like this,

$$\text{Per Capita Increment} = r_d\left(1 - \alpha N_t\right) \tag{3.3}$$

where α is the per capita negative effect of an individual on the per capita growth increment. As N_t shrinks toward zero, this expression grows toward r_d; as N_t grows, this expression shrinks toward zero.

[1] root: Arabic. *Al-jabr*

At precisely what value of N will per capita growth shrink to zero? We can solve this by noting that r_d is a constant and won't change; all that matters is inside the parentheses. We set the per capita increment equal to zero, and solve for N_t.

$$0 = r_d (1 - \alpha N_t) \tag{3.4}$$

$$0 = r_d - r_d \alpha N_t \tag{3.5}$$

$$N_t = \frac{1}{\alpha}. \tag{3.6}$$

When the population reaches the $N = 1/\alpha$, the growth increment will be zero, and the population will stop growing.

In a sense, the algebraic rearrangement (eq. 3.4) is at the core of theoretical ecology. We began with a set of assumptions (constant r_d and α), assumed a relation between them ($r_d [1 - \alpha N_t]$), and examined one of the consequences: the population will stop changing when it reaches a density of $1/\alpha$, *ceretus paribus*.[2] Welcome to theoretical ecology.

Now instead of density-independent per capita growth, r_d, we have density-dependent per capita growth $r_d (1 - \alpha N_t)$, and our population growth equation becomes,

$$N_{t+1} = N_t + r_d (1 - \alpha N_t) N_t \tag{3.7}$$

This describes *discrete logistic growth.* A common alternative representation uses $1/\alpha$ symbolized as K,

$$N_{t+1} = N_t + r_d \left(1 - \frac{N_t}{K}\right) N_t \tag{3.8}$$

where K represents the carrying capacity. The carrying capacity is the population size at which the per capita growth increment has fallen to zero. The context dictates whether we prefer to represent the scaling factor as the per capita effect, α, or the population carrying capacity, K.

Writing a Function For Discrete Logistic Growth

An R *function* will simplify our explorations. It will return a vector of N, given α, r_d, N_0, and t. The function *arguments* can have defaults (e.g., t=10).

```
> dlogistic <- function(alpha = 0.01, rd = 1, NO = 2, t = 15) {
+     N <- c(NO, numeric(t))
+     for (i in 1:t) N[i + 1] <- {
+         N[i] + rd * N[i] * (1 - alpha * N[i])
+     }
+     return(N)
+ }
```

The function first makes a new vector containing N_0, uses a for-loop to implement eq. 3.7 for each time step, and then returns the vector N.

[2] all else being equal

With discrete logistic growth, if we start with a small population ($N << K$), we will see the population rise and gradually approach K or $1/\alpha$ (Fig. 3.2). We refer to K as an *attractor* because N moves in a deterministic fashion toward K. We explore the meanings of *attractor* and related terms throughout the book.

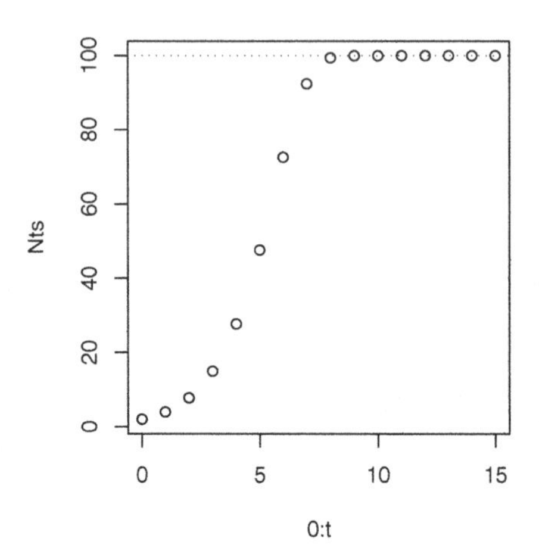

Fig. 3.2: Discrete logistic growth with $r_d = 1$, $\alpha = 0.01$.

Graphing Population Size

We can use the function created above, `dlogistic`, with default settings, to generate a population projection.

```
> Nts <- dlogistic()
```

Now we plot the projection, and put in a dotted line for $1/\alpha$ or K.

```
> t <- 15
> a <- 0.01
> plot(0:t, Nts)
> abline(h = 1/a, lty = 3)
```

3.1.2 Relations between growth rates and density

We already know a lot about one description of density dependent growth, the discrete logistic model. In particular, we know that with a constant per capita negative effect, α, the population size at which growth falls to zero is K. Let us explore further how the *per capita* growth increment, and the *population* growth increment, vary with population size.

Casual examination of Fig. 3.2 suggests that the total population growth increment ($\Delta N_t = N_{t+1} - N_t$) starts out small when both t and N are small, accelerates as N grows, and then over time, slows down and reaches an asymptote of

K. Is this changing rate of population growth a function of time, or a function of density? Let us first consider the growth increment as a function of N.

First consider the relation between the population growth increment and population size (Fig. 3.3a). We see it increase as N grows, and then decrease as N approaches K. The pattern is fairly symmetric. That is, it increases and decreases at about the same rates.

Next consider the per capita growth increment ($\Delta N_t/N_t$; Fig. 3.3b). There is a direct linear relation between the per capita growth increment and the size of the population — this is *linear density dependence.* This linear dependence on N comes from our assumption that the per capita negative effect is a constant, α.

(Per Capita) Population Growth Increment vs. N (Fig. 3.3)

Using the previous projection, we now capture both the total and the per capita growth increment per unit time, from t to $t+1$. We graph these versus N_t, population size at t.

```
> total.incr <- Nts[1:t + 1] - Nts[1:t]
> per.capita.incr <- total.incr/Nts[1:t]

> plot(Nts[1:t], total.incr)

> plot(Nts[1:t], per.capita.incr)
```

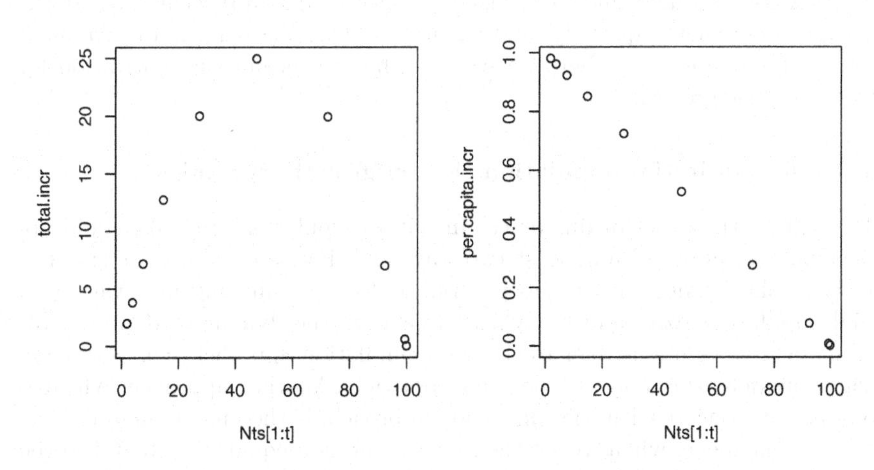

(a) Population Growth Increment (b) Per capita Growth Increment

Fig. 3.3: Relations between the total and per capita discrete growth increments and population size.

Let's use a simple analytical approach to understand Figs. 3.3a and 3.3b a little better. Take a look at eq. 3.7. First let's rearrange the expression so that we can set the *increment* of change equal to zero. How would you do that? Take a minute and try it, before you look at the answer below.

Setting the increment of change equal to zero, we rearrange with the population growth increment on the left, and then set it to zero.

$$N_{t+1} - N_t = r_d N_t (1 - \alpha N_t) \tag{3.9}$$

$$N_{t+1} - N_t = r_d N_t - r_d \alpha N_t^2 \tag{3.10}$$

$$0 = r_d N_t - r_d \alpha N_t^2 \tag{3.11}$$

What do we notice? One observation we could make is that we have a quadratic equation,[3] where the intercept is zero. This tells us that perhaps Fig. 3.3a is symmetric because it is a quadratic expression in terms of N.

What would satisfy this quadratic expression (eq. 3.11), that is, cause the growth increment to equal zero? Well, if r_d or N_t equal zero, those would yield potentially interesting solutions. Assuming neither r_d nor N_t equal zero, we can divide each side by these, and we are left with the solution we found in eq. 3.6, that the growth increment will be zero when $N_t = \frac{1}{\alpha} = K$.

Now let us examine the per capita growth increment (Fig. 3.3b). If we start with the population growth increment eq. 3.9, all we need is to divide through by N_t to get

$$\frac{N_{t+1} - N_t}{N_t} = r_d - r_d \alpha N_t. \tag{3.12}$$

With r_d and α being constants, and N varying, what is this expression? It is the expression for a straight line,[4] just like we observe (Fig. 3.3b). When $N_t = 0$, the per capita increment equals r_d, and when $N_t = 1/\alpha$, the per capita increment is zero. This is precisely where we started when we began the motivation for discrete logistic growth.

3.1.3 Effect of initial population size on growth dynamics

What will be the effect of differences in initial population size? We could approach such a question in at least two ways [21]. For some of us, the simplest way is to play games with numbers, trial and error, plugging in a variety of initial population sizes, and see what happens to the dynamics. If we do this systematically, we might refer to this as a simulation approach. For very complicated models, this may be the *only* approach. Another approach, which is often used in concert with the simulation approach, is the analytical approach. We used this above, when we set the growth equation equal to zero and solved for N. In general, this analytical approach can sometimes give us a definitive qualitative explanation for *why* something happens. This has been used as a justification for using simple models that actually have analytical solutions — they can provide answers [132].

For an analytical approach, first consider the endpoint solutions to the discrete logistc model eq. 3.10. The population will stop changing when $N_t = K$.

---

[3] $ax^2 + bx + c = 0$

[4] $y = mx + b$.

Note that it does not appear to matter what the initial population size was. The only thing that matters is α. Recall also that the population would not grow if for any reason $N_t = 0$ — the population will be stuck at zero. Based on these analyses, it appears that the only thing that matters is whether the initial population size is zero, or something greater than zero. If the latter, then initial population size appears to have no effect on the eventual population size.

It always pays to check our analytical answer with a brute force numerical approach, so we will use a little simple simulation to see if we are right. In this case, we can vary systematically the initial population size, and see what happens (Fig. 3.4a). What our approach shows us is that regardless of the initial conditions (except zero), N converges on K — K is an attractor. We also might notice that sometimes when $N_0 > K$, it crashes below K before converging on K in the end (more on that later). Last, because there is a qualitative shift in the behavior of the population when $N = 0$, we might want to investigate what happens when N gets very very close to zero. However, in this situation, the analytical solution is so straightforward that it seems convincing that as long as $N > 0$, it will grow toward K.

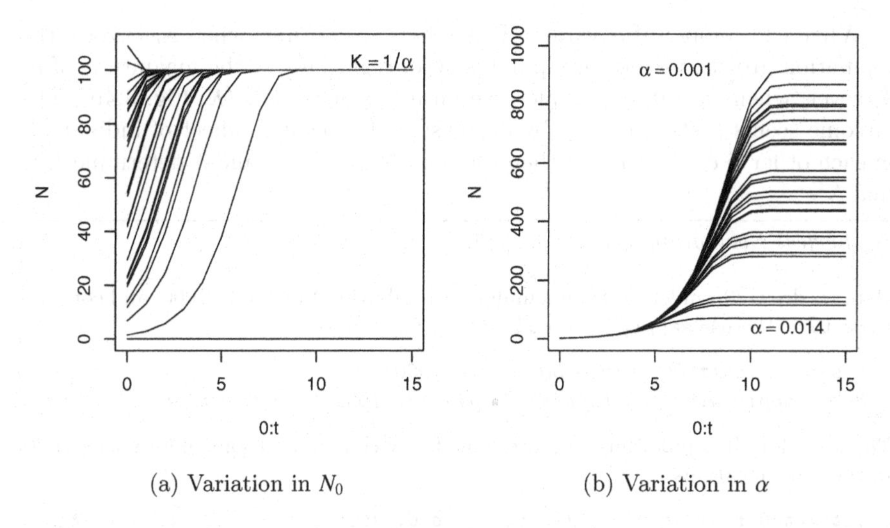

(a) Variation in N_0 (b) Variation in α

Fig. 3.4: (a) Dynamics due to different initial N (zero was also specifically included, and $\alpha = 0.01$). (b) Dynamics due to different α. All N, except $N = 0$, converge on $K = 1/\alpha$, regardless of the particular value of α ($r_d = 1$).

Numerical Evaluation of Initial Conditions (Fig. 3.4a)

Here we draw randomly 30 N_0 from a uniform distribution between zero and $1.2K$. We also include zero specifically. We then use `sapply` to run `dlogistic` for each N_0, using defaults for the other arguments.

```
> N0s <- c(0, runif(30) * 1.1 * 1/a)
> N <- sapply(N0s, function(n) dlogistic(N0 = n))
```

```
> matplot(0:t, N, type = "l", lty = 1, lwd = 0.75, col = 1)
> text(t, 1/a, expression(italic("K") == 1/alpha), adj = c(1,
+     0))
```

A serious simulation might include a much larger number of N_0.

3.1.4 Effects of α

Our conclusions thus far have been based on specific values of α and r_d. Have we been premature? Just to be on the safe side, we should probably vary these also.

What will happen if α varies? This seems easy. First, when N_t is zero, the population growth increment eq. 3.9 is zero, regardless of the magnitude of α. However, when $N_t > 0$, N will increase until it reaches $1/\alpha$ (K; Fig. 3.4b). The outcome seems pretty clear — by decreasing the negative effect of individuals on each other (i.e. decrease α) then the final N increases, and α determines the final N.

Numerical Evaluation of α (Fig. 3.4b)

Here we draw 30 random K from a uniform distribution from 50 to 1000, and convert these to α. We use `sapply` to run `dlogistic` for each α.

```
> a.s <- 1/runif(30, min = 50, max = 1000)
> N <- sapply(a.s, function(a) dlogistic(alpha = a, t = 15))
```

We next plot all populations, and use some fancy code to add some informative text in the right locations.

```
> matplot(0:t, N, type = "l", ylim = c(0, 1000), lty = 1, lwd = 0.75,
+     col = 1)
> text(8, 1/min(a.s), bquote(italic(alpha) == .(round(min(a.s),
+     3))), adj = c(1, 0.5))
> text(10, 1/max(a.s), bquote(italic(alpha) == .(round(max(a.s),
+     3))), adj = c(0, 1.2))
```

Note that we use the minimum and maximum of `a.s` to both position the text, and provide the values of the smallest and largest α.

3.1.5 Effects of r_d

What will variation in r_d do? Probably nothing unexpected, if our exploration of geometric growth is any guide. Our analytical approach indicates that it should have no effect on K (sec. 3.1.2). Nonetheless, let us be thorough and explore the effects of r_d by varying it systematically, and examining the result.

Yikes — what is going on in Fig. 3.5? Perhaps it is a good thing we decided to be thorough. These wild dynamics are real — let's go back and look more carefully at r_d.

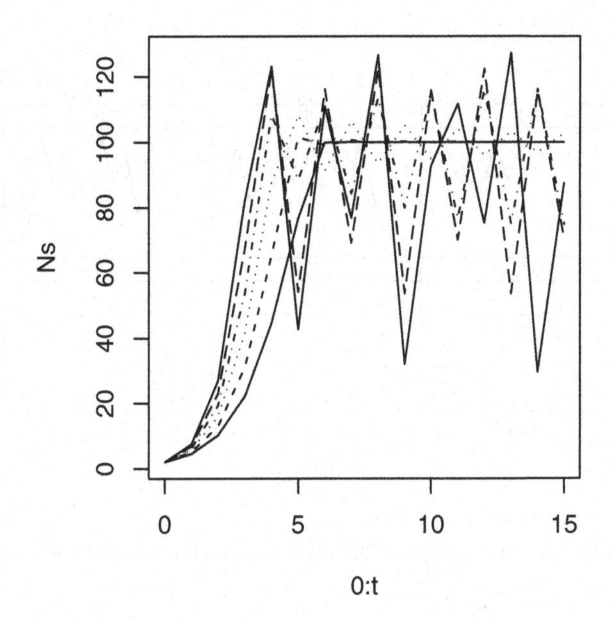

Fig. 3.5: The variety of population dynamics resulting from different values of r_d for the discrete logistic growth model ($r_d = 1, 1.2, \ldots, 3$, $\alpha = 0.01$). See Fig. 3.6 for a more informative view.

Simple Numerical Evaluation of r_d (Fig. 3.5)

Here we vary r_d by creating a short systematic sequence $r_d = 1.3, 1.6, \ldots, 2.8$. We set $t = 50$, and use `dlogistic` to create a trajectory for each of the six r_d.

```
> rd.v <- seq(1.3, 2.8, by = 0.3)
> t <- 15
> Ns <- data.frame(sapply(rd.v, function(r) dlogistic(rd = r,
+      t = t)))

> matplot(0:t, Ns, type = "l", col = 1)
```

Note that many populations do not seem to settle down at K.

If we examine each projection separately, we see a cool pattern is emerging (Fig. 3.6). At the lowest r_d, the population grows gradually toward its carrying capacity, K, and stays there. Once $r_d = 1.6 - 1.9$ it overshoots K just a bit, creating oscillations; these oscillations, however, dampen back down to K. When $r_d = 2.2$, however, the populations seem to bounce back and forth between two values. When $r_d = 2.5$, N bounces around, but now it bounces around between four different values. When $r_d = 2.8$, however, it seems to bounce back and forth around K, but at values that vary every time. This model is just about as simple as a model can be, and includes no random numbers. What is going on?

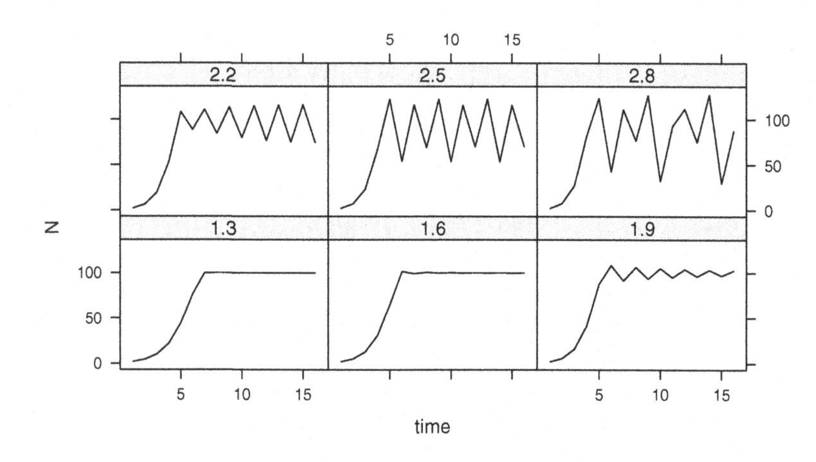

Fig. 3.6: A more informative view of the effects of variation in r_d on population dynamics.

Presentation of Limit Cycles (Fig. 3.6)

First we make a data frame with the six r_d values in the first column, and the respective populations in rows, using `t()` to transpose `Ns`. This puts the data in *wide* format, with a different time step in each column. (This might, for instance, be how you record data in a field experiment with repeated measurements through time).

```
> tmp <- data.frame(rd = as.factor(rd.v), t(Ns))
```

Next, we reshape the data to *long* format, were all N are in the second column, and each is associated with a time step and its r_d value (cols. 3 and 4).

```
> Ns2 <- reshape(tmp, varying = list(2:ncol(tmp)), idvar = "rd",
+     v.names = "N", direction = "long")
> str(Ns2)
```

...(output omitted) We plot each trajectory separately using `xyplot` in a different graphics package, `lattice`. Known as a *conditioning* plot, `xyplot` graphs y *vs.* x conditional on g (`y ~ x | g`).

```
> library(lattice)
> print(xyplot(N ~ time | rd, data = Ns2, type = "l", layout = c(3,
+     2, 1), col = 1))
```

What is going on is the emergence of *stable limit cycles*, and *chaos*.[5] At low r_d, we have simple asymptotic approach to K. As r_d increases, we see the population overshoot the carrying capacity and exhibit *damped oscillations*. When $2 < r_d < 2.449$, the population is attracted to two-point limit cycles. In this case, *these two points are stable attractors*. Regardless where the population starts out, it is attracted to the same two points, for a given r_d. As r_d increases further, the number of points increases to a four-point limit cycle (e.g., at $r_d = 2.5$), then an eight-point cycle, a 16-point limit cycle, and so on. These points are stable attractors. As r_d increases further , however, stable limit cycles shift into *chaos* ($r_d > 2.57$). Chaos is *a non-repeating, deterministic fluctuating trajectory, that is bounded, and sensitive to initial conditions.*

Robert May [128] shocked the ecological community when he first demonstrated stable limit cycles and chaos using this model. His groundbreaking work, done on a hand calculator, showed how very complicated, seemingly random dynamics emerge as a result of very simple deterministic rules. Among other things, it made population biologists wonder whether prediction was possible at all. In general, however, chaos seems to require very special circumstances, including very high population growth.

Is there a biological interpretation of these fluctuations? Consider some simple environment, in which small vegetation-eating animals with high reproductive rates eat almost all the vegetation in one year. The following year, the vegetation will not have recovered, but the animal population will still be very high. Thus the high growth rate causes a disconnect between the actual population size, and the negative effects of those individuals comprising the population.

[5] Not the evil spy agency featured in the 1960's US television show, *Get Smart.*

The negative effects of the actions of individuals (e.g., resource consumption) are felt by the offspring of those individuals, rather than the individuals themselves. We won't belabor the point here, but it is certainly possible to extend this delayed density dependence to a wide variety of populations. The discrete logistic model has a built in delay, or *time lag*, of one time step, because the growth increment makes a single leap of one time step. This delay is missing from the analogous continuous time model because the growth increment covers an infinity small time step, thanks to the miracles of calculus.[6]

Bifurcations

Up until now, we have examined N as a function of time. We have graphed it for different α and N_0, but time was always on the X-axis. Now we are going to examine N as a function of r_d, so r_d is on the X-axis. Specifically, we will plot the stable limits or attractors *vs.* r_d (Fig. 3.7). What does it mean? For $r_d < 2$, there is only a single N. This is what we mean by a stable point equilibrium, or point attractor — as long as r_d is small, N always converges to a particular point.[7] When $2 < r_d < 2.45$, then all of a sudden there are two different N; that is, there is a two-point stable limit cycle. Note that when $r_d \approx 2$ these oscilliations between the two point attractors around K are small, but as we increase r_d, those two points are farther apart. The point at which the limit cycle emerges, at $r_d = 2$, is called a *bifurcation*; it is a splitting of the single attractor into two attractors. At $r_d \approx 2.45$, there is another bifurcation, and each the two stable attractors split into two, resulting in a total of four unique N. At $r_d \approx 2.53$, there are eight N. All of these points are *periodic* attractors because N is drawn to these particular points at regular intervals. As r_d increases the number of attractors will continue to double, growing geometrically. Eventually, we reach a point when there becomes an infinite number of unique points, *that are determined by* r_d.[8] This completely deterministic, non-repeating pattern in N is a property of *chaos*. Chaos is not a random phenomenon; rather it is the result of deterministic mechanisms generating non-repeating patterns.

[6] A time lag can be explicitly built in to a continuous time model, with only a small effort.

[7] It need not be the same N for each r_d, although in this case it is.

[8] They are also determined by the initial N, but we will get to that later.

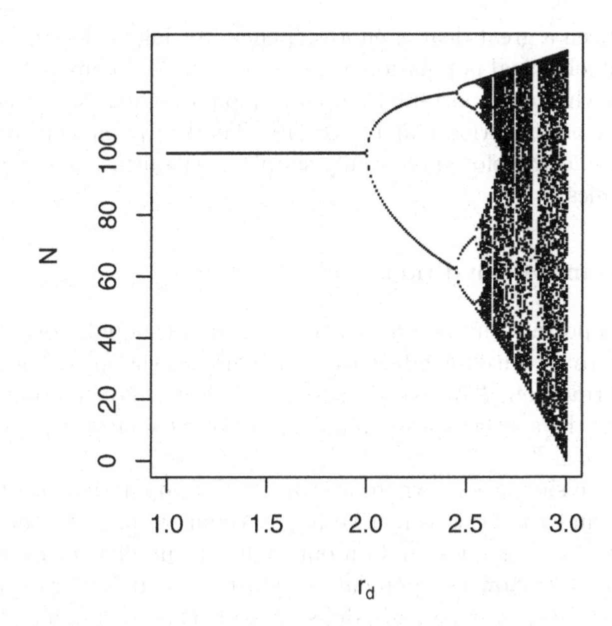

Fig. 3.7: Illustration of the long term dynamics of discrete logistic population growth. When a small change in a continuous parameter results in a change in the number of attractors (e.g. a single point equilibrium to a stable 2-point limit cycle), we call this a bifurcation.

Bifurcation Plot: Attractors as a Function of r_d (Fig. 3.7)

Here we perform more comprehensive simulations, and plot the point and periodic attractors *vs.* r_d. First we pick some constraints for the simulation: the number of different r_d, the sequence of r_d values, and the number of time steps.

```
> num.rd <- 201; t <- 400
> rd.s <- seq(1, 3, length = num.rd)
```

Next we use `sapply` for the simulations.

```
> tmp <- sapply(rd.s, function(r) dlogistic(rd = r, N0 = 99,
+     t = t))
```

Next we convert the output to a data frame and stack up the N in one column. We also rename each of the stacked columns, and add new columns for the respective r_d and time steps.

```
> tmp.s <- stack(as.data.frame(tmp))
> names(tmp.s) <- c("N", "Old.Column.ID")
> tmp.s$rd <- rep(rd.s, each = t + 1)
> tmp.s$time <- rep(0:t, num.rd)
```

We save just the later dynamics in order to focus on the N after they have converged on the periodic attractors. Here we select the last 50% of the time steps. (Your figure will look a little different than Fig. 3.7 because I used more r_d and time steps.)

```
> N.bif <- subset(tmp.s, time > 0.5 * t)
> plot(N ~ rd, data = N.bif, pch = ".", xlab = quote("r"["d"]))
```

There has been a great deal of effort expended trying to determine whether a particular model or real population exhibits true chaos. In any practical sense, it may be somewhat unimportant whether a population exhibits true chaos, or merely a higher order periodic attractor [49]. The key point here is that very simple models, and therefore potentially simple mechanisms, can generate very complex dynamics.

Sensitivity to initial conditions

Another very important characteristic feature of chaotic populations is that they are very sensitive to initial conditions. Thus emerges the idea that whether a butterfly in Sierra Leone flaps its wings twice or thrice may determine whether a hurricane hits the southeastern United States in New Orleans, Louisiana, or in Galveston, Texas.[9]

If we generate simulations where we vary initial population size by a single individual, we find that this can have an enormous impact on the similarity of two populations' dynamics, and on our ability to predict future population sizes (Fig. 3.8). Note how the populations start with similar trajectories, but soon diverge so that they experience different sequences of minima and maxima (Fig. 3.8). This is part of what was so upsetting to ecologists about May's 1974 paper — perhaps even the simplest deterministic model could create dynamics so complex that we could not distinguish them from random [128]. Over time, however, we came to learn that (i) we could distinguish random dynamics from some chaos-like dynamics, and (ii) the hunt for chaos could be very exciting, if most frequently disappointing [8, 39, 90].

Sensitivity to Intitial Conditions

We start with three populations, all very close in initial abundance. We then propogate with a r_d to generate chaos for 100 time steps.

```
> N.init <- c(97, 98, 99); t <- 30
> Ns <- sapply(N.init, function(n0) dlogistic(rd = 2.7, N0 = n0,
+     t = t))
```

Now we would like to graph them over the first 12 times, and look at the correlations between N_1 and the other two populations.

```
> matplot(0:t, Ns, type = "l", col = 1)
```

Boundedness, and other descriptors

One last issue that we should note is the extent to which our populations are *bounded.* A population may have complex dynamics, but we may be able to characterize a given population by its upper and lower bounds. In spite of the

[9] Clearly, this suggests that the solution to increased storm severity due to global warming is to kill all butterflies.

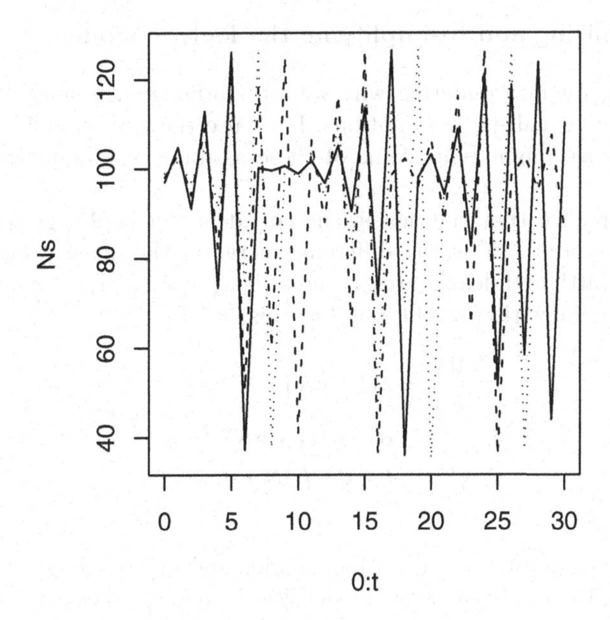

Fig. 3.8: Effects of differences in initial population size on the short term and long term dynamics, and their correspondence, of three populations.

differences created by the initial conditions, the upper and lower bounds of our chaotic populations were very similar (Fig. 3.8). Note also as r_d increases (Fig. 3.7) the oscillations increase very systematically.

In general, we can describe many characteristics of populations, even if we cannot predict exactly what the population size will be five years hence. For instance, we can describe the shape of density dependence (linear, nonlinear), and characteristics of N, such as the average, the variance, the periodicity, the upper and lower bounds, and the color (i.e. degree of temporal auto-correlation). Indeed, these characteristics may vary among types of organisms, body sizes, or environments.

3.2 Continuous Density Dependent Growth

The classic version of continuous density dependent growth [93] is the continuous logistic growth equation, the continuous version of eq. 3.8 above,

$$\frac{\mathrm{d}N}{\mathrm{d}t} = rN\left(\frac{K-N}{K}\right) \tag{3.13}$$

which, as we saw above, is no different than

$$\frac{\mathrm{d}N}{\mathrm{d}t} = rN\left(1-\alpha N\right). \tag{3.14}$$

Take a moment to see the parallels between these and eq. 3.7.

3.2.1 Generalizing and resimplifying the logistic model

By now you might be wondering why we are studying such simplistic models that make such unrealistic assumptions. Let's use the next couple of pages to figure out how we might relax some of these assumptions, and generalize eq. 3.14.

First let's realize that the density independent per capita growth rate, r, is really a net rate — it is the difference between the density independent, instantaneous birth and death rates, b and d. If we start with per capita logistic growth, the generalization would look like this.

$$\frac{dN}{Ndt} = r(1 - \alpha N) \tag{3.15}$$

$$= (b - d)(1 - \alpha N) \tag{3.16}$$

$$= b - d - b\alpha N + d\alpha N \tag{3.17}$$

where $b - d = r$.

Note that the positive, density-independent effect on growth rate of b is counterbalanced a bit by a negative density-dependent effect of b. As N increases, increasing births tend to rein in growth rate a bit, because more births mean a larger negative density-dependent effect. Similarly, mortality, d includes a small *positive* density-dependent effect that helps enhance growth rate because it reduces the negative effect of αN — death frees up resources.

Density dependence is the effect of density on growth rate, and thus far we have let that be $1 - \alpha N$. Let's represent this as the function $F(N) = 1 - \alpha N$. We can now note that α is also a net effect — N will have separate effects on the birth and death rates, and α could be just the sum of more arbitrary constants (e.g., $\alpha N = (x + y)N$); $1 - \alpha N$ is merely the simplest form of $F(N)$. We can generalize further and let density affect the birth and death rates separately,

$$\frac{dN}{Ndt} = rN\,F(N) \tag{3.18}$$

$$F(N) = B(N) + D(N) \tag{3.19}$$

We might anticipate that density has no effect on death rates when N is low, because there is no particular limiting factor (Fig. 3.9a). Paradoxically, however, when N becomes large, the ensuing mortality benefits growth rate a bit because death frees up limiting resources (Fig. 3.9a). Such an idea might have the simple form

$$D(N) = gN^2 \tag{3.20}$$

where g is a constant. This function starts out at zero (no effect) and increases rapidly (Fig. 3.9a). This means that at low density, mortality does not influence growth rate, whereas at high density, mortality enhances growth rate.

The density dependence for birth rates could also be more complicated. The *Allee effect* [192] arises when a population gets very small and mating success declines because mates can't find each other, leading to a large negative per capita impact of N. We can describe this with a quadratic function,

$$B(N) = -aN^2 + eN - f \tag{3.21}$$

where a, e, and f are constants. Here, the negative effect of N on birth rates, $B(N)$, will be important at low N because when $N = 0$, $B(N) = -f$. The negative effect will also be large at very high N because $-aN^2$ becomes a large negative number at large N (Fig. 3.9a). The point here is that we can alter density dependence to incorporate specific biology.

If we sum the density dependences, we have a function $F(N) = B(N) + D(N)$ that is analogous to $1 - \alpha N$, but which is a different shape, because it describes a particular underlying biology (Fig. 3.9b). Note the differences between the simple linear density dependence of the logistic model, and the nonlinear density dependence of our generalization.

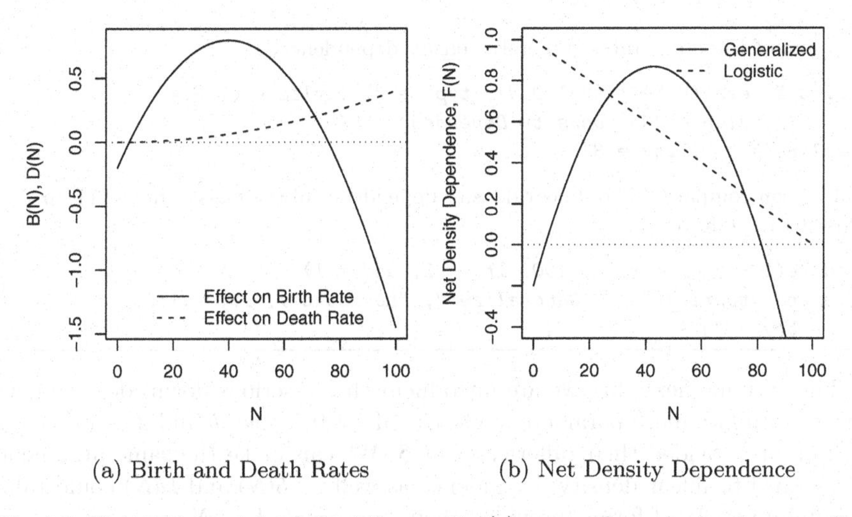

(a) Birth and Death Rates (b) Net Density Dependence

Fig. 3.9: Elaborating on simple logistic growth. (a), a specific example of a generalized density dependence via effects on birth rate and death rate. (b) Net density dependence for our generalized model and logisitc linear density dependence.

Density Dependence on Birth and Death Rates (Figs. 3.9a, 3.9b)

To make expressions for eqs. 3.21, 3.20, use `expression`.

```
> B.N <- expression(-a * N^2 + e * N - f)
> D.N <- expression(g * N^2)
```

We then provides constants and evaluate the expressions to plot the density dependence on birth and death rates.

```
> a <- 1/1600; e <- 80 * a; f <- 0.2; g <- 4e-05; N <- 0:100

> plot(N, eval(B.N), type = "l", ylab = "B(N), D(N)")
> lines(N, eval(D.N), lty = 2)
> abline(h = 0, lty = 3)
> legend("bottom", c("Effect on Birth Rate", "Effect on Death Rate"),
+     lty = 1:2, bty = "n")
```

The sum of these two rates is merely density dependence,

```
> plot(N, eval(B.N) + eval(D.N), type = "l", ylim = c(-0.4,
+     1), ylab = "Net Density Dependence, F(N)")
> abline(h = 0, lty = 3)
```

and we can compare this to linear density dependence of the logistic model $(1-\alpha N)$. If $\alpha = 0.01$, we have

```
> curve(1 - 0.01 * x, 0, 100, lty = 2, add = T)
> legend("topright", c("Generalized", "Logistic"), lty = 1:2,
+     bty = "n")
```

The extreme flexibility we are introducing has a serious downside—we now have to estimate more parameters. Clearly, if $r = 0.5$, then b and d can take on any values, provided their difference is 0.5. We can make the same argument with regard to linear density dependence as well — $B(N)$ and $D(N)$ could take an infinite variety of forms, provided their sum equals $1 - \alpha N$.

In an attempt to simplify our lives a bit, and in the absence of any additional information, let us impose the simplest function for B and D, a linear constant

$$\frac{\mathrm{d}N}{N\mathrm{d}t} = (b - m_b N) - (d - m_d N) \tag{3.22}$$

where m_b and m_d are constants that describe the effect of an additional individual on the base density-independent per capita birth and death rates, respectively. If we really wanted to simplify our lives, we could let $m_b + m_d = m$, $b - d = r$, and simplify to

$$\frac{\mathrm{d}N}{N\mathrm{d}t} = r - mN \tag{3.23}$$

Thus, the *net* effect of N is linear, even though the effects of N on birth and death rates may not be. For reasons that probably have more to do with history than with mathematics, we create another constant, $\alpha = m/r$, and we find ourselves back to

$$\frac{\mathrm{d}N}{N\mathrm{d}t} = r(1 - \alpha N). \tag{3.24}$$

Let's review the circular path of elaboration and simplification upon which we have travelled. We started with a historical representation of logistic growth, using K to represent an equilibrium population size toward which N is attracted. We represented this in a tad more mechanistic fashion (I would argue) by rephrasing the limitation factor in terms of a per capita effect of an individual on the growth rate. We then expanded and generalized this expression to see that the *net* effects, r and α, could each be thought of as two components related to birth rates and death rates. We further expanded on density dependence, allowing it to be curvilinear. We then collapsed this back to a simple difference ($r - mN$, eq. 3.23) and finally recast it in the original expression.

The integral of logistic growth

We might as well add that there is an expression for N_t, analogous to $N_t = N_0 \exp(rt)$ for exponential growth. If we integrate eq. 3.24, we have

$$N_t = \frac{N_0 e^{rt}}{1 + \alpha N_0 \left(e^{rt} - 1\right)}. \tag{3.25}$$

Note the resemblance to the exponential equation. We have the exponential equation in the numerator, divided by a correction factor which starts at 1 when $t = 0$. The correction factor then grows with time, at a rate which is virtually the same as exponential growth but adjusted by α and scaled by N_0. Thus when $t = 0$, and if N_0 is small, we have near-exponential growth, decreased merely by a factor of αN_0. We will take advantage of this integral below, when we describe some data in terms of the logistic growth model.

3.2.2 Equilibria of the continuous logistic growth model

As we did with the discrete model, we can determine the *equilibria* of the continuous logistic growth model, that is the values of N for which $\mathrm{d}N/\mathrm{d}t = 0$. Consider again eq. 3.14

$$\frac{\mathrm{d}N}{\mathrm{d}t} = rN\left(1 - \alpha N\right). \tag{3.26}$$

What values of N will cause the growth rate to go to zero, thus causing the population to stop changing? As with the discrete model, we would find that if $N = 0,\ 1/\alpha$, the growth rate would be zero, and the population stops changing. Therefore the equilibria for continuous logistic growth are $N^* = 0,\ 1/\alpha$.

3.2.3 Dynamics around the equilibria — stability

The *stability* of a system[10] is a measure of how much it tends to stay the same, in spite of external disturbances or changes in the state of the system.

[10] A system might be nearly any set of interacting parts, such as a population, or an economy, or the internet.

The term *stability* has been given many more specific meanings, including resilience, resistant, reactivity, and permanence. We won't go into these here, but one could consider them different facets of stability [25].[11] We will focus on resilience, the tendency for a population to return an equilibrium, or be *attracted toward* an equilibrium, if it is perturbed from it.

Consider the stability of a marble inside a wok. If the wok doesn't move, then the marble just sits in the lowest spot on the bottom. If the wok is bumped, the marble jiggles and rolls around a bit, but settles back down to the bottom. This is a stable system because there is a point at the bottom of the bowl (the attractor) toward which the marble rolls — all points inside the wok allow the marble to find the lowest point (the attractor). For this reason, we call the collection of points on the inside surface of the wok the *basin of attraction.* The steeper the sides, the more quickly the marble moves toward the bottom. This rate is referred to as its *resilience.*

To translate the notion of a basin into a population, let the position of marble inside the wok be the value of N at any particular point. Bumping the wok is any disturbance that causes a change in N. The slope of the sides of the wok reflects the biology and its mathematics that cause N to move quickly or slowly back toward the bottom. For the logistic model, this rate is determined by r and α, and also by N itself. The attractor at the very bottom of the bowl is the population's carrying capacity, $K = 1/\alpha$.[12]

When we imagine a marble in a wok, it becomes easy to envision K as an attractor at the bottom of the bowl. That is why we consider the carrying capacity a stable equilibrium point, or attractor, even if other factors, such as a disturbance, may cause N to deviate from it.

The equilibrium we have focused on has been K, but recall that $N = 0$ is also an equilibrium. This equilibrium is actually the edge of the wok — the slightest jiggle, and the marble falls in and moves toward the center of the wok, K. The biological analog of this "jiggle" is the additional of one or a very small numbers of individuals to an otherwise extinct population. For example, consider that a sterile petri dish with nutrient agar has an *E. coli* population size of zero. If that *E. coli* $N = 0$ gets "perturbed" with a single added cell, the population will move quickly toward its carrying capacity K. In this sense, $N = 0$ is an equilibrium, but it is an *unstable* equilibrium. We also call such an unstable equilibrium a *repeller.*

Analytical linear stability analysis

We are interested in the dynamics or stability of N at each of the equilibria, N^*. Here we use *analytical linear stability analysis* to show mathematically what we

[11] Resistance refers to the tendency to resist change in the first place; resilience refers to the long term rate of return to an equilibrium; reactivity refers to the rate of change immediately following such a disturbance [148], and permanence is the degree to which a system does not simplify by entering one of its boundary equilibria.

[12] The marble's inertia, which contributes to its momentum, is analogous to a time lag in density dependence that we saw in the discretet model; with a time lag, a population can overshoot the equilibrium.

described above with the marble and the wok. While somewhat limited in its scope, linear stability is nonetheless a powerful technique for dynamic model analysis.

In a nutshell, what we do is determine whether the growth rate, which is zero at each equilibrium, becomes positive or negative in response to a small change in N. That is, if N takes a tiny step away from the equilbrium, does that tiny step shrink, or grow? If a tiny step results in a shrinking of that step, back toward the equilibrium, that demonstrates stability. If a tiny step results in growth and a bigger step, that demonstrates instability. That is what analytical stability analysis tells us.

Consider a plot of the growth rate, $\mathrm{d}N/\mathrm{d}t$ *vs.* N (Fig. 3.10). The equilibria, N^*, are the N (x-axis) at which $\mathrm{d}N/\mathrm{d}t = 0$ (points a, d). Note where population growth rate is positive and where it is negative. What will happen to this population if it finds itself at $N = 50$? It will grow, moving along the x-axis, until population growth rate slows so much that it comes to rest where $N = 1/\alpha = K = 100$. Thus, N changes until it converges on the equilibrium, N^*, where $\mathrm{d}N/\mathrm{d}t = 0$. Alternatively at $N = 110$, $\mathrm{d}N/\mathrm{d}t$ is negative, and so N shrinks back down to K (point d). This return toward K means that K is an attractor and a stable equilibrium.

Next, consider $N = 0$, at point a; it cannot change on its own. However, if "perturbed" at all, with the addition of one or more individuals, then this "perturbation" will grow, sending N across the x-axis, away from $N = 0$, toward $N = K$. This stasis at $N = 0$, and movement away from $N = 0$ with the slightest perturbation means that $N = 0$ is a repeller and an unstable equilibrium.

Linear stability analysis will calculate the degree of attraction or repulsion.

Growth rate vs. N

We first define an expression, and constants.

```
> pop.growth.rate <- expression(r * N * (1 - alpha * N))
> r <- 1; alpha <- 0.01; N <- 0:120
```

A basic plot.

```
> plot(N, eval(pop.growth.rate), type = "l",
+      ylab = "Population Growth Rate (dN/dt)", xlab = "N")
> abline(h = 0); legend("topright", "r=1", lty = 1)
```

Add a few points with labels,

```
> N <- c(0, 10, 50, 100, 115)
> points(N, eval(pop.growth.rate), cex = 1.5)
> text(N, eval(pop.growth.rate), letters[1:5], adj = c(0.5, 2))
```

and arrows for the direction of change in N.

```
> arrows(20, 2, 80, 2, length = 0.1, lwd = 3)
> arrows(122, -2, 109, -2, length = 0.1, lwd = 3)
```

How fast will N return to $N^* = K$, if perturbed? Let's start by imagining that we have two populations with different intrinsic rates of increase ($r = 1, 0.5$),

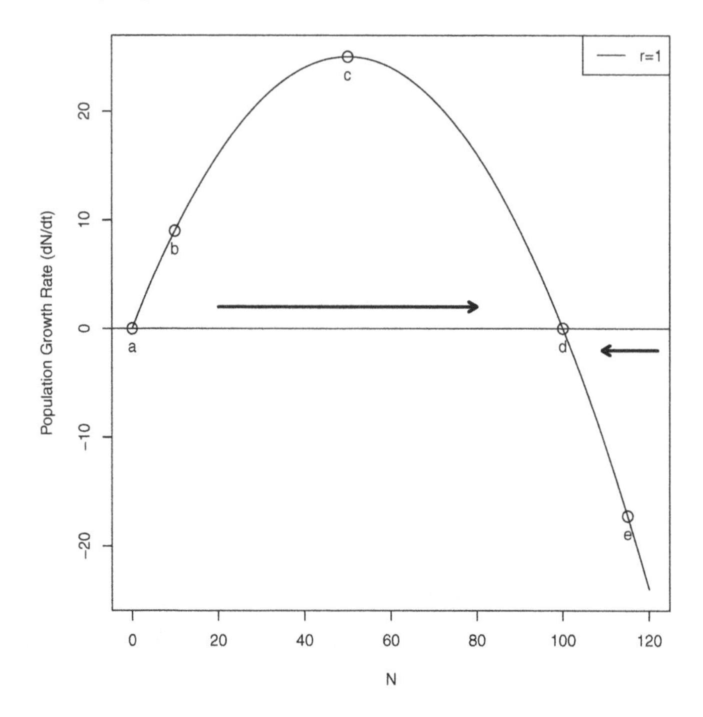

Fig. 3.10: Population growth rate, $\mathrm{d}N/\mathrm{d}t$, as a function of N. Points a–e are labelled for convenience. At values of N associated with points a and d, population growth rate equals zero. At values of N associated with b and c growth rate is positive, and for e growth rate is negative. *Note this is growth rate as a function of N* (time is not explicitly part of this graph).

and focus in on the growth rate at $N^* = K$ (Fig. 3.11a). For which one of these populations will $\mathrm{d}N/\mathrm{d}t$ go to zero more quickly? If each population loses a few individuals and declines by x to $N^* - x$, which population will return to N^* more quickly? Not surprisingly, it is the population with the higher r — when $r = 1$, its population growth rate (y-axis) is greater than when $r = 0.5$, and therefore will return at a faster rate.

Note also that this same population ($r = 1$) has a negative growth rate of greater magnitude at $N^* + x$. Regardless of whether it gains or loses individuals, it will return to N^* more quickly than the population with the lower r.

How do we quantify the rate of return at N^* so that we can compare two or more populations? The slope of the curve at N^* quantifies the rate of return, because the slope is the exponential return rate. To calculate the slope of a curve we use calculus, because the slope of a curve is a derivative. In this case, we need the derivative of the growth rate ($\mathrm{d}N/\mathrm{d}t$) with respect to the variable on the x-axis, N. Such a derivative is called a *partial derivative*. Therefore we need the partial derivative of growth rate, with respect to N. This will be the slope of the growth rate with respect to N.

To find the partial derivative, let us denote the growth rate as $\dot{N}$ ("N-dot"),

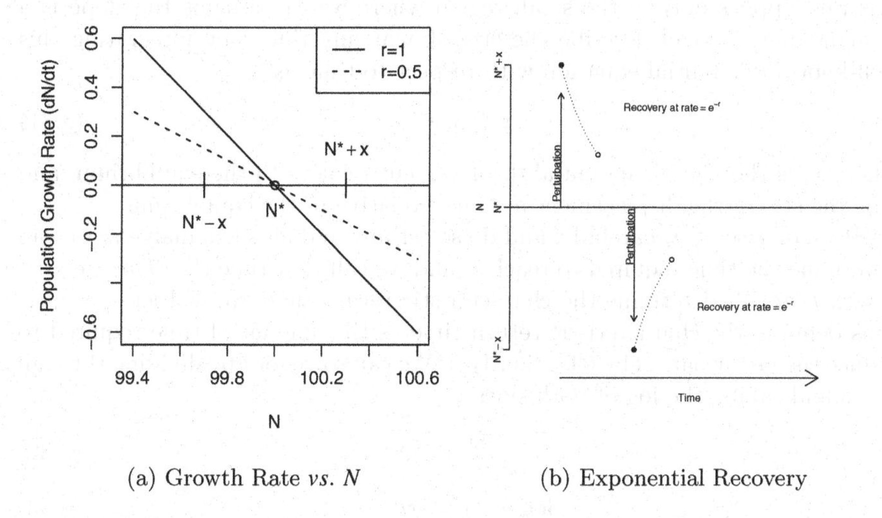

(a) Growth Rate *vs.* N (b) Exponential Recovery

Fig. 3.11: Very close to an equilibrium, a population recovers from a perturbation, moving toward the equilibrium attractor, at rate e^{-r}. (a), Slopes of population growth *vs.* N near the equilibrium; around the equilibrium attractor, small decreases in N lead to positive growth rate, whereas small increases lead to negative growth rate; the population with the steeper slope changes faster. (b), Regardless of the particular r of a population, a population recovers from a perturbation, moving toward the equilibrium attractor, at rate e^{-r}.

$$\dot{N} = rN(1 - \alpha N). \tag{3.27}$$

$\dot{N}$ is a common symbol for a time derivative such as a growth rate.

Given $\dot{N}$, its partial derivative with respect to N is

$$\frac{\partial \dot{N}}{\partial N} = r - 2r\alpha N. \tag{3.28}$$

Further, we are interested in the equilibrium, where $N = 1/\alpha$. If we substitute this into eq. 3.28 and simplify, this reduces to

$$\frac{\partial \dot{N}}{\partial N} = -r. \tag{3.29}$$

At $N^* = 1/\alpha$, the slope is $-r$. *This negative slope at the equilibrium demonstrates the stability of this equilibrium.*

Near N^*, population growth rate is $-rx$.[13] That is, the rate of change of x (Figs. 3.11a, 3.11b) is

$$\frac{\mathrm{d}x}{\mathrm{d}t} = -rx \tag{3.30}$$

This allows us to pretend that the perturbation will diminish exponentially, because the growth rate is constant. We say "pretend," because we realize that

[13] Because $\Delta y/\Delta x = -r$, i.e. the change in y equals the change in x times the slope.

this rate applies only in the small region where we can assume the slope is a straight line. We can describe the size of x at any time t by integrating this well-known differential equation with respect to time as

$$x_t = x_0 e^{-rt} \tag{3.31}$$

where x_0 is the size of the initial displacement relative to the equilibrium, and x_t is the size of the displacement at time t relative to the equilibrium.

If we choose an x_t carefully, and do so for any and all such analyses, we can determine the time required to reach x_t and we call that time the *characteristic return time*. To determine the characteristic return time, we will let $x_t = x_0/e$, thus defining the characteristic return time as the amount of time required to reduce the perturbation by 63% (i.e. $1/e$). We can solve for t by dividing through by x_0 and taking the log of both sides,

$$\frac{x_0}{e} = x_0 e^{-rt} \tag{3.32}$$

$$\log\left(e^{-1}\right) = -rt \tag{3.33}$$

$$t = -\frac{1}{(-r)}. \tag{3.34}$$

Thus we see return time, t, here is a positive number, with the same units as r, and depends on the slope of the the growth rate with respect to N. It also depends on the assumption of linearity very close to the equilibrium.

To review, we took the partial derivative of $\mathrm{d}N/\mathrm{d}t$ with respect to N, and then evaluated the partial derivative at N^* to get the rate of change of the perturbation, x. A negative value indicates that the perturbation will decline through time, indicating a stable equilibrium or attractor. A positive value would indicate that the perturbation will grow over time, thus indicating an unstable equilibrium, or a repellor.

Symbolic differentiation

We can use R's minimal symbolic capabilities to get derivatives. Here we get the partial derivative and evaluate for the two equilibria (Fig. 3.10).

```
> dF.dN <- deriv(pop.growth.rate, "N")
> N <- c(0, 1/alpha)
> eval(dF.dN)
```

```
[1] 0 0
attr(,"gradient")
      N
[1,]  1
[2,] -1
```

The first value, 1, corresponds to the first value of N, which is 0. Because it is positive, this indicates that the perturbation will increase with time, meaning that $N = 0$ is a repellor. The second value, -1, is negative, and so indicates that the perturbation will decrease with time, meaning that $N = 1/\alpha$ is an attractor.

3.2.4 Dynamics

How does the continuous logistic growth equation behave? It is very well behaved, compared to the discrete version. It has a very characteristic shape (Fig. 3.12a), and regardless of r or N_0, it moves converges boringly toward $K = 1/\alpha$ (Fig. 3.12b).

Function for an ODE

Making a function to use with R's ODE solver is pretty easy, provided we follow the rules (see Appendix, secs. , B.10). To make the code transparent, I translate the vector parameters and the vector of populations (in this single species model, we have only one population).

```
> clogistic <- function(times, y, parms) {
+     n <- y[1]
+     r <- parms[1]
+     alpha <- parms[2]
+     dN.dt <- r * n * (1 - alpha * n)
+     return(list(c(dN.dt)))
+ }
```

We create vectors for the parameters and the initial densities for all of the populations in the model. We also specify the time.

```
> prms <- c(r = 1, alpha = 0.01)
> init.N <- c(1)
> t.s <- seq(0.1, 10, by = 0.1)
```

We load the `deSolve` library, and run `ode`. The output is a matrix that includes the time steps and the N (Fig. 3.12a).

```
> library(deSolve)
> out <- ode(y = init.N, times = t.s, clogistic, parms = prms)

> plot(out[, 1], out[, 2], type = "l", xlab = "Time", ylab = "N")
```

So why is the continuous version so boring, while the discrete version is so complex? Remind yourself what is going on with the discrete version. The model could only take steps from one generation to the next. The step into generation $t + 1$ was a function of N at t, and not a function of N_{t+1}. Therefore the rate of change in N was not influenced by a contemporaneous value of N. There was a delay between N and the effect of N on the population growth rate. For many organisms, this makes sense because they undergo discrete reproductive events, reproducing seasonally, for instance. In contrast, the continuous logistic population growth is always influenced by a value of N that is updated continuously, instantaneously. That is the nature of simple differential equations. They are instantaneous functions of continuous variables. We can, if we so choose, build in a delay in the density dependence of continuous logistic growth. This is referred to as "delayed density dependence" or "time-lagged logistic growth"

$$\frac{dN}{dt} = rN\left(1 - \alpha N_{t-\tau}\right) \tag{3.35}$$

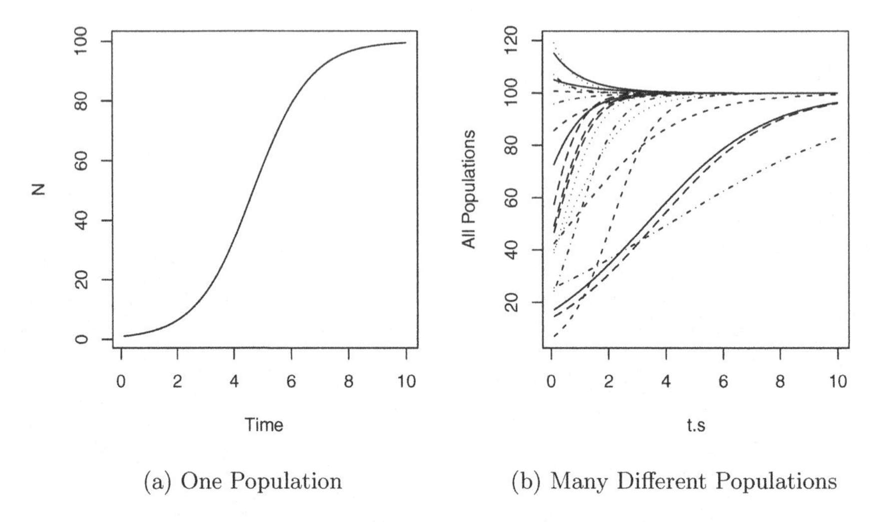

(a) One Population (b) Many Different Populations

Fig. 3.12: Dynamics of continuous logistic population growth. (a) The characteristic shape of logistic growth. (b) Regardless of N_0 or r, populations converge slowly (small r) or quickly (high r) on $K = 1/\alpha$.

where τ is the degree of the delay, in time units associated with the particular model. Just as with the discrete model, the dynamics of this continuous model can get very complicated, with sufficient lag and sufficient r.

Plotting Random Populations ((Fig. 3.12b))

We use the above function to create 20 populations with different traits. We start with an empty matrix, and then for each of the populations, we draw random N_0 and r, run the ODE solver, keeping just the column for N. Last we plot the output.

```
> outmat <- matrix(NA, nrow = length(t.s), ncol = 20)
> for (j in 1:20) outmat[, j] <- {
+     y <- runif(n = 1, min = 0, max = 120)
+     prms <- c(r = runif(1, 0.01, 2), alpha = 0.01)
+     ode(y, times = t.s, clogistic, prms)[, 2]
+ }

> matplot(t.s, outmat, type = "l", col = 1, ylab = "All Populations")
```

3.3 Other Forms of Density-dependence

There are commonly used types of density-dependence other than logistic. Some, like the Richards model, are more flexible and have more parameters. Others, like the Gompertz model, are used more widely in other fields, such as cancer research for tumor growth, von Bertlanaffy for body size growth, and the Ricker model for fisheries; the Richards model provides even more flexibility, at the cost of more paramteters. Here we explore a simple extension of the logistic model, the *theta-logistic* model, which adds a parameter to increase flexibility and generality.

$$\frac{\mathrm{d}N}{\mathrm{d}t} = rN\left(1 - (\alpha N)^{\theta}\right) \tag{3.36}$$

Here θ is strictly positive ($\theta > 0$); $\theta = 0$ means zero growth, and $\theta < 0$ would mean negative growth below K, and unbounded positive growth above K. Approximations of eq. 3.36 that allow $\theta \leq 0$ are possible [188], but the interpretation becomes strained.

Theta-logistic function

Here we make a function that we can use with `ode`, the numerical integration function.

```
> thetalogistic <- function(times, y, parms) {
+     n <- y[1]
+     with(as.list(parms), {
+         dN.dt <- r * n * (1 - (alpha * n)^theta)
+         return(list(c(dN.dt)))
+     })
+ }
```

Using `with()` and `as.list()` creates an environment in which R will look inside `parms` for named elements. This will work as long as we name the parameters in `parms`.

By varying θ, we can change the linear density dependence of the simple logistic model to curvilinear density dependence (Fig. 3.13a). This curvilinearity arises because when $\alpha N < 1.0$ (i.e. $0 < N < K$),

$$(\alpha N)^{\theta} < \alpha N, \quad \theta > 1 \tag{3.37}$$

$$(\alpha N)^{\theta} > \alpha N, \quad \theta < 1. \tag{3.38}$$

In contrast, when $\alpha N = 1.0$, then $(\alpha N)^{\theta} < \alpha N$. This means that θ does not affect K.

When $\theta > 1$, this weakens density dependence at low N, so the population grows faster than logistic, all else being equal. When $\theta < 1$, this strengthens density dependence at low N, causing the population to grow more slowly than logistic, all else being equal.

The effects of θ on density dependence controls the shape of relation between growth rate *vs.* N (a.k.a. the production function, Fig. 3.13b). First, note that for a given r, growth rate for $N < K$ increases with θ. Second, note that the position of peak of the production function shifts to the right as θ increases. That is, as θ increases, the N at which the maximum growth rate occurs also increases. If we wanted to shift the peak growth rate to a higher N without also increasing the height of the peak, we could decrease r simultaneously.

We could speculate on biological meaning of θ and the shape of the denisty dependence. For instance, very high θ suggests a threshold, wherein the population exhibits little density dependence until very close to K. Perhaps this is

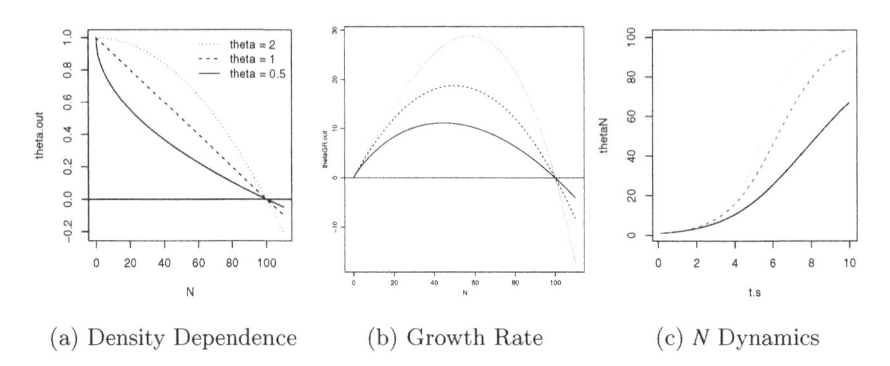

(a) Density Dependence (b) Growth Rate (c) N Dynamics

Fig. 3.13: Theta-logistic growth. In a review of population dynamics, Sibly et al. [188] use theta-logistic density dependence, $1-(N/K)^{\theta}$, to show that populations most frequently have a concave-up $\theta < 1$ pattern.

related to territoriality, or a spatial refuge from predators. Alternatively, low θ might suggest resource preemption, wherein a small number of individuals can sequester large amounts of resources, but increasing N results in reduced preemption. For organisms with very plastic body sizes (plants, fish), this could mean that at low N, average body size is large, but as $N \rightarrow K$, average body size decreases. While θ, *per se*, is devoid of mechanism, discovering the magnitude of θ for different types of species [188] could lead to the generation of new testable hypotheses about what controls populations.

Theta-logistic density dependence

We first graph theta-logistic, for $\theta < 1$, $\theta = 1$, and $\theta > 1$ (Fig. 3.13a).

```
> r <- 0.75
> alpha <- 0.01
> theta <- c(0.5, 1, 2)
> N <- 0:110
> theta.out <- sapply(theta, function(th) {
+     1 - (alpha * N)^th
+ })
> matplot(N, theta.out, type = "l", col = 1)
> abline(h = 0)
> legend("topright", legend = paste("theta =", c(2, 1, 0.5)),
+     lty = 3:1, bty = "n")
```

Theta-logistic growth rate

We plot the growth rate (a.k.a. the production function) for the theta-logistic model with $\theta < 1$, $\theta = 1$, and $\theta > 1$ (Fig. 3.13b).

```
> thetaGR.out <- sapply(theta, function(th) {
+     r * N * (1 - (alpha * N)^th)
+ })
> matplot(N, thetaGR.out, type = "l", col = 1)
> abline(h = 0)
```

We also add an example of growth with low θ, but higher r.

Theta-logistic dynamics

We solve numerically for N, and plot the dynamics for N with $\theta < 1$, $\theta = 1$, and $\theta > 1$ (Fig. 3.13c).

```
> prms <- c(r = 0.75, alpha = 0.01, theta = 1)
> thetaN <- sapply(theta, function(th) {
+     prms["theta"] <- th
+     ode(y = 1, t.s, thetalogistic, prms)[, 2]
+ })
> matplot(t.s, thetaN, type = "l")
```

3.4 Maximum Sustained Yield

The classical model of harvesting fisheries and wildlife populations is based on a particular conceptualization of *maximum sustained yield* (MSY). Maximum sustained yield is historically defined as the population size where both population growth rate and harvest rate are at a maximum. For the logistic growth model, this is half the carrying capacity. We can solve this by finding the maximum of the production function. To do this we use calculus (differentiation) to find where the slope of the production function equals zero. Starting with the partial derivative with respect to N (eq. 3.28), we solve for zero.

$$0 = r - 2r\alpha N \tag{3.39}$$

$$N = \frac{r}{2r\alpha} = \frac{K}{2} \tag{3.40}$$

Similarly, we could determine this peak for the θ-logistic model as well.

$$\frac{\partial \dot{N}}{\partial N} = r - (\theta + 1)\, r\, (\alpha N)^{\theta} \tag{3.41}$$

$$0 = \frac{K}{(\theta + 1)^{1/\theta}} \tag{3.42}$$

When $\theta = 1$, this reduces to $K/2$, as for the logistic model. As we saw above, $\theta < 1$ causes this value to be less than $K/2$.

Even if we assume the logistic model is appropriate for a given population, harvests in practice seldom reduce populations to $K/2$, because often N and K vary with time and are difficult to estimate. Moreover, there are economic forces that often lead to over-harvesting. More contemporary models of population harvesting incorporate these concepts [182]. We should note that when more detailed information is available on age- or stage-specific survivorship and reproduction, age- or stage-structured models of harvesting are the preferred method (e.g., Chapter 2). For many populations, such as marine fisheries, however, data are often available only on stock size (N) and total catch (H). In some fisheries models, the total catch is fixed and does not vary with N. This leads to highly unstable population dynamics and fortunately is no longer practiced in most fisheries. Usually the total catch H is modeled as a function of N.

Let us assume that total catch or harvest, H, is a simple linear function of N; the harvest rate FN is usually described this way.[14] In this case, a maximum sustained yield, MSY, can be determined from the logistic growth-harvest model. This basic model is then

$$\frac{\mathrm{d}N}{\mathrm{d}t} = rN(1 - \frac{N}{K}) - FN \tag{3.43}$$

where $FN = H$ harvest rate, and F is the per capita fishing mortality rate. Here we assume a fixed per capita catch mortality F so that total harvest is $H = F \times N$. MSY occurs at $K/2$ (Fig. 3.14b). Therefore would like to know the value of F for which a new $N^* = K/2$ rather than K. We can determine the value of F for which this is true by finding when $\mathrm{d}N/\mathrm{d}t = 0$ and $N = K/2$.

$$0 = r\frac{K}{2}\left(1 - \frac{\frac{K}{2}}{K}\right) - F\frac{K}{2} \tag{3.44}$$

$$F = \frac{r}{2} \tag{3.45}$$

MSY *and harvesting (Fig. 3.14a)*

Here we illustrate the interaction harvesting at a rate associated with MSY for logistic growth. We set logistic model parameters, and first plot logistic growth without harvest.

```
> r <- 0.5; alpha <- 0.01; N <- 0:105
> plot(N, eval(pop.growth.rate), type = "l", ylim = c(-3, 15),
+     ylab = "dN/dt and FN")
> abline(h = 0)
```

We then calculate F based on our solution eq. 9.6, and plot the linear harvest function with an intercept of zero, and slope of F.

```
> F <- r/2
> abline(c(0, F), lty = 2)
```

[14] We could, but will not, further describe this by $F = qE$, where q is the catchability coefficient for the fishery, and E is fishing effort in hours.

Equilibrium solution for logistic growth with harvesting (Fig. 3.14b)

When we add harvesting at rate $F = r/2$, we get a new equilibrium. Here we illustrate this using the same parameters as above, but now using the entire function with both growth and harvest.

```
> pgr.H <- expression(r * N * (1 - alpha * N) - F * N)
> N <- 0:55
> plot(N, eval(pgr.H), type = "l", ylab = "dN/dt (growth - harvesting)")
> abline(h = 0)
```

This merely represents the new equilibrium (where dN/dt crosses the x-axis) with a harvest of rate $F = r/2$.

If harvesting occurs at a rate of $F < r/2$, the slope of the harvest line is shallower (cf. Fig. 3.14a), and we refer to this as *under-harvesting.* If the rate is $F > r/2$, then the slope is steeper (cf. Fig. 3.14a), and we refer to this as *over-harvesting.* Roughgarden and Smith [182] show that under-harvesting is probably the most ecologically stable, whereas over-harvesting is frequently thought to be the economically optimal approach.

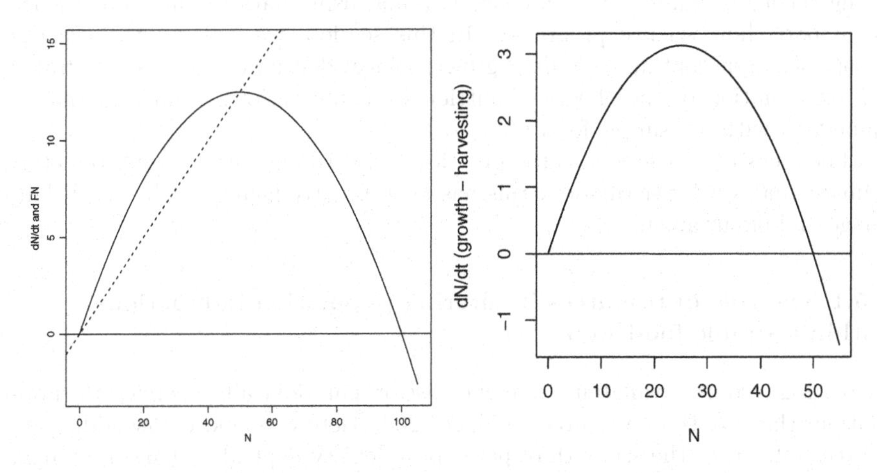

(a) Logistic Growth and Harvest Functions

(b) Logistic Growth *with* Harvesting

Fig. 3.14: Illustration of maximum sustained yield under harvesting. (a) both the logistic growth model eq. 3.13 and the linear harvest model eq. 9.6, and (b) the combined growth model that includes harvesting as a source of mortality. Equilibria occur when the growth rate, y, equals zero, crossing the x axis.

Typically the economically optimal approach leads to over-harvesting because the marginal income on a fishery is related to *the interest rate on invested income*, $F = (r + \rho)/2$, where ρ is the interest rate (usually ρ is set to 0.05). We can substitute this into eq. 3.43 and solve for N (ignoring $N = 0$).

$$0 = rN(1 - \frac{N}{K}) - \frac{r+\rho}{2}N \tag{3.46}$$

$$N\frac{r}{K} = r - \frac{r}{2} - \frac{\rho}{2} \tag{3.47}$$

$$N = \frac{K}{2}\left(1 - \frac{\rho}{r}\right). \tag{3.48}$$

Thus, unless $\rho = 0$, the typical economic optimum N_e^* will be smaller than $N* = K/2$. This makes intuitive economic sense if we consider income from invested profits. The margin of error, however, is much smaller, putting the fisheries population, and the investment of fishing fleets and associated econmoies at greater risk. Roughgarden and Smith [182] showed that ecological stability of a fishery is likely to beget economic and cultural stability. They show that the supposed "economic optimal" harvesting rate was so low that it increased the extinction risk of the harvested population to unacceptably high levels. The collapse of a fishery is, of course, both economically and culturally devastating.

3.5 Fitting Models to Data

Using theory to guide our data collection, and using data to guide our theory is, in part, how science progresses. In this section, we will use the use the theory of simple continuous logistic growth (linear density dependence) to frame our examination of the effects of nutrients on interactions of an alga that is embedded within a simple food web.

The focus of this section is the practice of describing data in terms of deterministic and stochastic models; this has been greatly facilitated by the R language and environment [13].

3.5.1 The role of resources in altering population interactions within a simple food web

Increasing resource availability may increase or, paradoxically, decrease the populations that use those resources [176, 179, 200]. Here we explore how adding resources influences the strength of per capita denisty dependence, α, of an alga, *Closterium acerosum*, embedded in a simple food web [195].

The variety of observed responses to resource availability is sometimes mediated by the configuration of the food web in which a population is embedded. To explore the interaction between food web configuration and edibility, Stevens and Steiner [195] performed a simple experimental resource enrichment using a food web with one predator (*Colpidium striatum*) and two competing consumers which differed in their edibility; bacteria, which were edible and *Closterium acerosum*, which was inedible. Resources were added at two concentrations, a standard level, and a ten-fold dilution. We were interested in whether the inedible competitor *Closterium* could increase in abundance in the face of competition from bacteria. Although *Closterium* is inedible, it may be unable to increase in response to resource additions because it is competing with many bacterial taxa which appear to also be predator resistant. In addition, many bacteria are able to grow quickly in enriched environments. However, keystone predation theory [104] suggested that *Closterium* might prosper because the bacteria which coexist with the predator may be poor competitors, and the bacteria with the greatest capacity for utilizing higher resource levels may be kept in check by a predator which also might increase in abundance. Stevens and Steiner [195] presented time-averaged responses, perhaps related to K, because the predictions of keystone predation theory were focused on equilibrium predictions.

Here, we will explore the effects of the resource supply rate on two parameters of linear density dependence (logistic growth) r, and α of *Closterium acerosum*.

In the rest of this section, we will follow a generally cautious and pragmatic step-wise scheme which we adapt for our purposes here. We will

- import the data and examine its structure.
- look at the population dynamics of each replicate microcosm.
- consider whether we have outliers due to sampling.
- estimate parameters for one of the populations.
- fit logistic models to each population.
- fit a model to all populations and test the effects of nutrients on logistic growth parameters.

3.5.2 Initial data exploration

First we load two more libraries, the data, and check a summary.

```
> library(nlme)
> library(lattice)
> data(ClostExp)
> summary(ClostExp)
```

```
 Nutrients   No.per.ml             Day           rep           ID
 high:72   Min.   :    1.0   Min.    : 1.0   a:36   a.low   :18
 low :72   1st Qu.:   16.2   1st Qu.:11.0   b:36   d.low   :18
           Median :   42.0   Median :26.0   c:36   c.low   :18
           Mean   :  131.5   Mean   :25.9   d:36   b.low   :18
           3rd Qu.:  141.5   3rd Qu.:39.0          c.high :18
           Max.   :1799.1   Max.    :60.0          a.high :18
                                                   (Other):36
```

Next we plot each replicate of the high and low nutrient treatments.

```
> xyplot(No.per.ml ~ Day | Nutrients, ClostExp, groups = rep,
+     type = "b", scales = list(relation = "free"),
+     auto.key = list(columns = 4,    lines = T))
```

Right away, we might notice several things (Fig. 3.15). First, the high nutrient populations achieve higher abundances (note values on the y-axes). This seems consistent with the ten-fold difference in resource concentration [195]. Perhaps

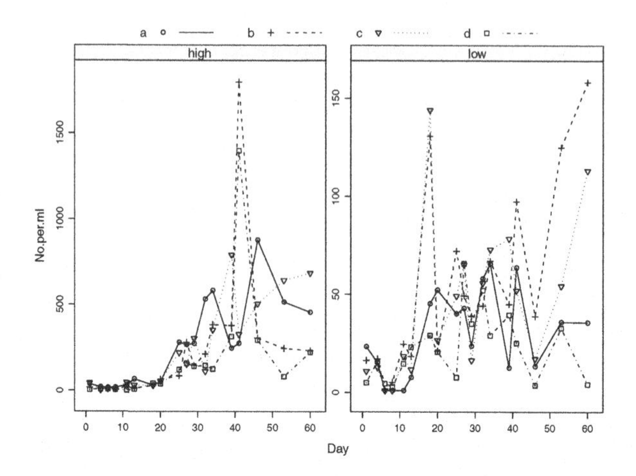

Fig. 3.15: Raw data for *Closterium* time series. Each line connects estimates of the *Closterium* popoulation in a single replicate microcosm. Each estimate is based on a ~ 0.3 mL sample drawn from the microcosm after it has been mixed. Low nutrient concentration is 1/10× the standard (high) concentration.

the second thing we notice is that the low nutrient populations seem more variable, relative to their own population sizes. Third, several observations seem remarkably high — which days?

```
> subset(ClostExp, Nutrients == "high" & No.per.ml > 1000)

Grouped Data: No.per.ml ~ Day | ID
    Nutrients No.per.ml Day rep      ID
114      high      1799  41   b b.high
116      high      1398  41   d d.high

> subset(ClostExp, Nutrients == "low" & No.per.ml > 100)

Grouped Data: No.per.ml ~ Day | ID
    Nutrients No.per.ml Day rep     ID
54        low     130.6  18   b b.low
55        low     144.0  18   c c.low
134       low     124.8  53   b b.low
142       low     158.2  60   b b.low
143       low     113.0  60   c c.low
```

What might we make of these observations? Are they mistakes, or important population dynamics, or some combination thereof? There may be several reasonable explanations for these data which we revisit at the end of this section. For now, we will assume that these points are no less accurate than the other values, and so will retain them.

Finally, we might note that the populations seem to decline before they increase (days 0–5, Fig. 3.15). These declines might well represent a bit of evolutionary change, under the new environment into which these populations have been introduced [56] from their stock cultures; we might consider removing the first two sample points to avoid confound mechanisms of evolution versus simple growth. In general, however, we should be cautious about throwing out data that do not conform to our expectations — let us leave these data in, for the time being.

Now we estimate the parameters of deterministic logistic growth, first without, and then with, an explicit consideration of time.

3.5.3 A time-implicit approach

How might we estimate the parameters without explicitly considering the temporal dynamic? Recall the relation between per capita population growth *vs.* population size,

$$\frac{\mathrm{d}N}{N\mathrm{d}t} = r - r\alpha N. \tag{3.49}$$

Here we see that per capita growth rate is a linear function (i.e., $y = mx + b$) of population density, N. Therefore, all we have to do is calculate per capita growth rate from our data and fit a simple linear regression of those growth rates versus N. The y-intercept (when $N = 0$) will be our estimate of r, and the slope will be an estimate of $r\alpha$.

We will estimate per capita growth rate, pgr, with

$$pgr = \frac{\log\left(N_{t+i}/N_t\right)}{i} \tag{3.50}$$

where i is the time interval between the two population densities [188]. In this and other contexts, logarithm of two populations is sometimes referred to simply as the *log ratio*.

Let's estimate this for one of the microcosms, high nutrient, replicate c.

```
> Hi.c <- subset(ClostExp, Nutrients == "high" & rep == "c")
```

We calculate the differences in density, the time intervals over which these changes occured, and the per capita growth rate.

```
> n <- nrow(Hi.c)
> N.change <- Hi.c$No.per.ml[-1]/Hi.c$No.per.ml[-n]
> interval <- diff(Hi.c$Day)
> pgr <- log(N.change)/interval
```

We can then (i) plot these as functions of N_t (as opposed to N_{t+1}), (ii) fit a linear regression model through those points, and (iii) plot that line (Fig. 3.18a).

```
> Nt <- Hi.c$No.per.ml[-n]
> plot(pgr ~ Nt)
> mod1 <- lm(pgr ~ Nt)
> abline(mod1)
```

The linear regression model defines a best fit straight line through these points,

```
> summary(mod1)
```

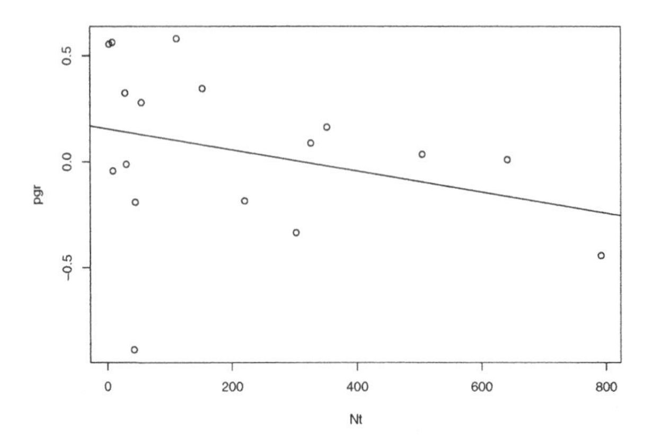

Fig. 3.16: Per capita growth rate *vs*. population size at the beginning of the interval (data from one high nutrient replciate).

```
Call:
lm(formula = pgr ~ Nt)

Residuals:
   Min     1Q Median     3Q    Max
-1.021 -0.206  0.129  0.182  0.480

Coefficients:
             Estimate Std. Error t value Pr(>|t|)
(Intercept)  0.154556   0.125327    1.23     0.24
Nt          -0.000495   0.000395   -1.25     0.23

Residual standard error: 0.382 on 15 degrees of freedom
Multiple R-squared: 0.0944,      Adjusted R-squared: 0.0341
F-statistic: 1.56 on 1 and 15 DF,  p-value: 0.23
```

where y-intercept is the first coefficient of the linear regression; this is an estimate of r (eq. 3.49). The slope of the line is the second coefficient and is $r\alpha$ (eq. 3.49).

Now we can do this with all the replicates. Rearranging the data is a bit complicated, but this will suffice. Essentially, we split up the `ClostExp` data frame into subsets defined by nutrient and replicate ID, and then apply a function that performs all of the steps we used above for a single microcosm. The output will be a *list* where each component is a new small data frame for each microcosm.

```
> EachPop <- lapply(split(ClostExp, list(ClostExp$Nutrients,
+     ClostExp$rep)), function(X) {
+     n <- nrow(X)
+     N.change <- (X$No.per.ml[-1]/X$No.per.ml[-n])
```

```
+     interval <- diff(X$Day)
+     data.frame(Nutrients = as.factor(X$Nutrients[-n]),
+     rep = as.factor(X$rep[-n]),
+         pgr = log(N.change)/interval, Nt = X$No.per.ml[-n])
+ })
```

Next we just stack all those individual data frames up into one.

```
> AllPops <- NULL
> for (i in 1:length(EachPop)) AllPops <- rbind(AllPops, EachPop[[i]])
```

Finally, we plot all of the microcosms separately.

```
> xyplot(pgr ~ Nt | rep * Nutrients, AllPops, layout = c(4,
+     2, 1), scales = list(x = list(rot = 90)), panel = function(x,
+     y) {
+     panel.grid(h = -1, v = -4)
+     panel.xyplot(x, y, type = c("p", "r"))
+ })
```

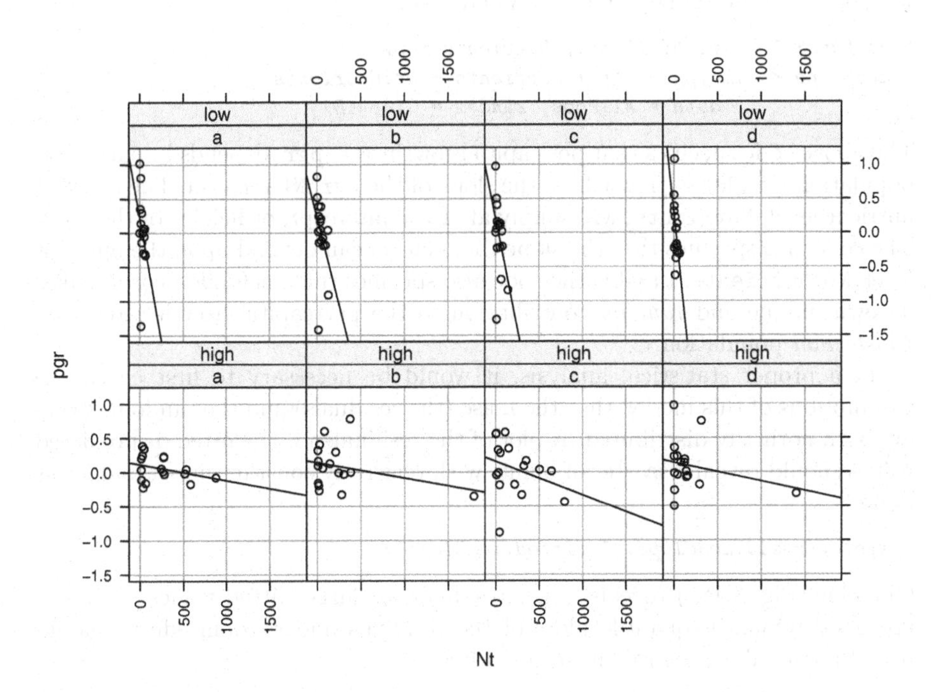

Fig. 3.17: Per capita growth rates *vs.* population size, N, at the beginning of the interval for which the growth rate is calculated (includes all microcosms). Note that y-intercepts are all fairly similar, while the maximum N, slopes and x-intercepts differ markedly.

The biggest difference about these populations appears to be that the carrying capacities differ between the high and low nutrients (Fig. 3.17). First note that the y-intercepts are all remarkably similar, indicating that their r are all very similar. The big differences between the high and low nutrients are that the slopes of the relations and the x-intercepts are both very different. The slope is $r\alpha$, while the x-intercept is K. Next we fit a statistical model to these data.

A statistical model of per capita growth

A *mixed model*[15] is a statistical model that includes both random effects (e.g., blocks, or subjects) and fixed effects (e.g., a treatment) [161]. Let's not worry about the details right now, but we'll just use it to test for differences in slope, where we have measure the same microcosms repeatedly through time.

Previously, we showed that it makes at least a modicum of sense to fit eq. 3.49 to each microcosm (Fig. 3.17). Having done that, we next fit all of these to a single mixed model, in which the microcosm is our *block* or *subject* which may exert some unknown random effect on the populations. First we make sure we have a unique ID label for each microcosm.

```
> AllPops$ID <- with(AllPops, Nutrients:rep)
> modSlope <- lme(pgr ~ Nt + Nutrients + Nt:Nutrients,
+                 data = AllPops, random = ~1 | ID)
```

The above code specifies that per capita growth rate, `pgr`, depends in part upon population size, `Nt` — this will be the slope of the `pgr Nt` relation. The effect of nutrient level, `Nutrients`, will simply alter the intercept, or height, of the lines. The code also specifies that the slope the relation can depend upon the nutrient level, `Nt:Nutrients`. Finally, the code also specifies that each microcosm makes its own unique and *random* contribution to the per capita growth rate of its *Closterium* population.

In a proper statistical analysis, it would be necessary to first check the assumptions of this model, that the noise (the residuals) and the random effects are both normally distributed. A plot of the residuals *vs.* the fitted or predicted values should reveal a scatter of points with relatively constant upper and lower bounds.

```
> xyplot(resid(modSlope) ~ fitted(modSlope))
```

Our plot (Fig. 3.18a) reveals larger residuals for larger fitted values — this is not good. Quantile-quantile plots of the residuals and random effects should reveal points along a straight diagonal line.

```
> qqmath(~resid(modSlope) | ID, data = AllPops, layout = c(8,
+     1, 1))

> qqmath(~ranef(modSlope))
```

[15] This section is advanced material. You may want to skim over the parts you don't understand, and focus on conclusions, or perhaps the general process.

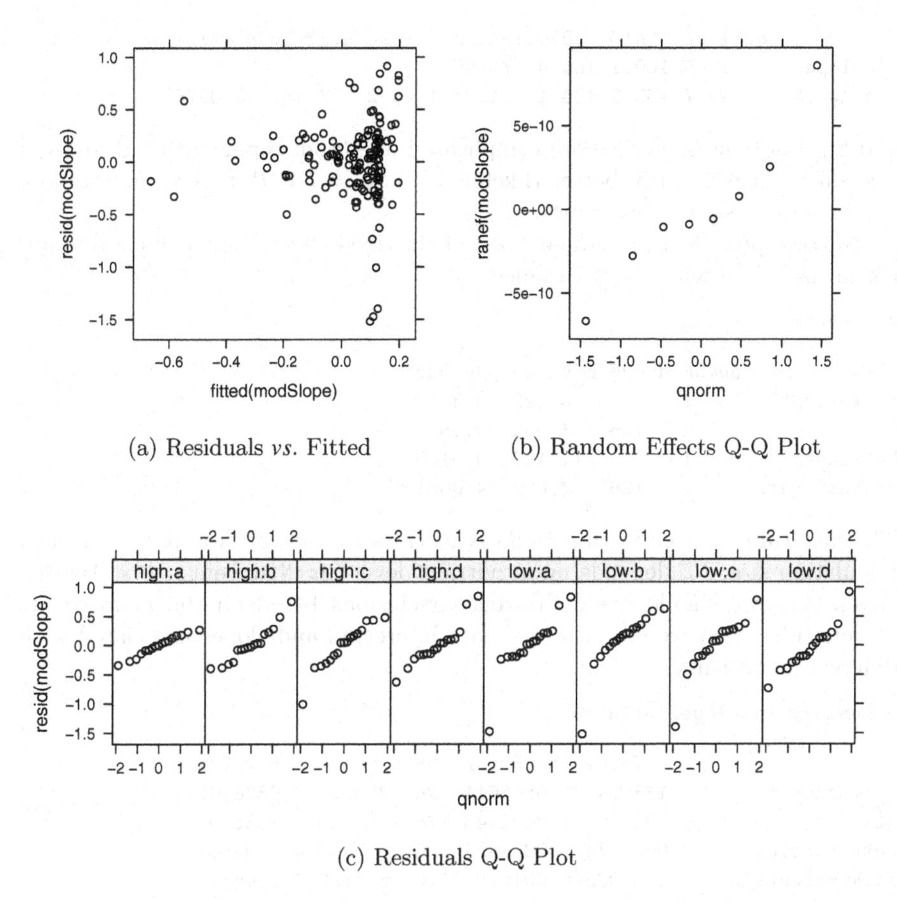

(a) Residuals *vs.* Fitted

(b) Random Effects Q-Q Plot

(c) Residuals Q-Q Plot

Fig. 3.18: Diagnostic plots for fitting the per capita growth rate to population size.

Our random effects (Fig. 3.18b) are very small, indicating only small systematic differences among microcosms that are not accounted for by the treatment. Nonetheless, their scatter of points along the diagonal are fairly consistent with a normal distribution. The residuals (Fig. 3.18c) for each microcosm seem fairly normally distributed except that several populations have outliers (replicates low:a-c, high:c). Taken as a group (Fig. 3.18a), we see that the residual variation increases the mean (fitted). This is at least in part due to having more observations at larger values, but is also a typical and common problem — variables with small absolute values simple cannot vary as much as variables with large absolute value. Therefore, we specify that that model takes that into account by weighting residuals observations differently, depending upon the expected (i.e. fitted or mean) value.

```
> modSlope2 <- update(modSlope, weights = varExp())
```

We then compare these two models,

```
> anova(modSlope, modSlope2)

          Model df   AIC   BIC logLik   Test L.Ratio p-value
modSlope      1  6 169.1 186.4 -78.55
modSlope2     2  7 163.9 184.1 -74.97 1 vs 2   7.153  0.0075
```

and find that the slightly more complicated one (`modSlope2` with 7 degrees of freedom) is significantly better (likelihood ratio = 7.1, $P < 0.05$) and is more parsimonious (smaller AIC) [18].

Now we provide a preliminary test of the fixed effects (nutrients and population size) with analysis of variance.

```
> anova(modSlope2)

             numDF denDF F-value p-value
(Intercept)      1   126   0.823  0.3659
Nt               1   126   3.665  0.0578
Nutrients        1     6  11.198  0.0155
Nt:Nutrients     1   126  27.659  <.0001
```

The first thing we see in this *analysis of variance* output is that the effect of population size, N_t, depends upon nutrient level (`Nt:Nturients`, $P < 0.0001$). Given this, we should avoid drawing conclusions based on the main effects of nutrients. Next we go on to estimate intercepts and slopes associated with different treatments.

```
> summary(modSlope2)$tTable

                      Value Std.Error  DF t-value   p-value
(Intercept)       0.1336241 0.0568611 126  2.3500 2.033e-02
Nt               -0.0002861 0.0001146 126 -2.4965 1.383e-02
Nutrientslow      0.0641174 0.0878452   6  0.7299 4.930e-01
Nt:Nutrientslow  -0.0056023 0.0010652 126 -5.2592 6.014e-07
```

Given R's default contrasts (linear model parameters), this output indicates that the intercept (our estimate of r) for the high nutrient group is ~ 0.13, and the slope of the high nutrient microcosms is −0.000286. The intercept for the low nutrient level is calculated as 0.13 – 0.07, and we find it is not significantly smaller ($P > 0.05$). On the other hand, the slope of the low nutrient microcosms is −0.000278 – 0.00573, which is significantly smaller than the slope of the high nutrient microcosms ($P < 0.05$). This information can guide another take on these data.

```
> cfs <- fixef(modSlope2)
> cfs

    (Intercept)              Nt    Nutrientslow Nt:Nutrientslow
      0.1336241      -0.0002861       0.0641174      -0.0056023

> -cfs[2]/cfs[1]

      Nt
0.002141
```

```
> -(cfs[2] + cfs[4])/(cfs[1] + cfs[3])

     Nt
0.02978
```

indicating that the per capita effects are much smaller in the high nutrient treatment. (The label Nt is an unimportant artifact, retained from using the coefficients to calculate the values.) Next, we use these estimates to help us fit a times series model.

3.5.4 A time-explicit approach

Here we use the above information to help us fit the time-explicit version of the logistic growth model. In this section, we

1. create and examine a nonlinear logistic model,
2. fit the model to each microcosm using nlsList to confirm it makes sense, and
3. fit a single nonlinear mixed model that tests which logistic growth parameters varies between treatments.

At each step, we examine the output and diagnose the fit; it should give you a very brief introduction to this process. For more thorough treatments, see [13, 161].

First, we code in R a model of eq. 3.25, that is, logistic growth with parameters α, r, and N_0.

```
> ilogistic <- function(t, alpha, N0, r) {
+     N0 * exp(r * t)/(1 + alpha * N0 * (exp(r * t) - 1))
+ }
```

Let's make sure this makes sense by plotting this curve on top of the raw data, focusing on the raw data and model parameters for the high nutrient microcosms.

```
> plot(No.per.ml ~ Day, ClostExp, subset = Nutrients == "high")
> curve(ilogistic(x, alpha = -cfs[2]/cfs[1], N0 = 6, r = cfs[1]),
+     1, 60, add = T, lty = 2)
```

Our model and the coefficients we generated above seem to be a reasonable approximation of the data (Fig. 3.19). Now let's use nlsList to fit the logistic model to each microcosm separately.

```
> Cmod.list <- nlsList(No.per.ml ~ ilogistic(Day, alpha, N0,
+     r) | ID, data = ClostExp, start = c(alpha = 1/100, N0 = 5,
+     r = 0.15), control = list(maxiter = 1000))
```

We note that models cannot be fit for three microcosms. For now, we plot the coefficients for each microcosm for which a model was fit.

```
> plot(coef(Cmod.list), layout = c(3, 1, 1))
```

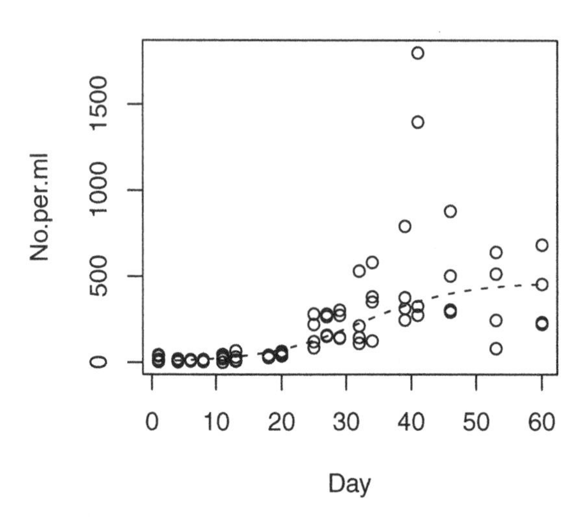

Fig. 3.19: High nutrient microcosms and the logistic growth model using coefficients from `modSlope2`.

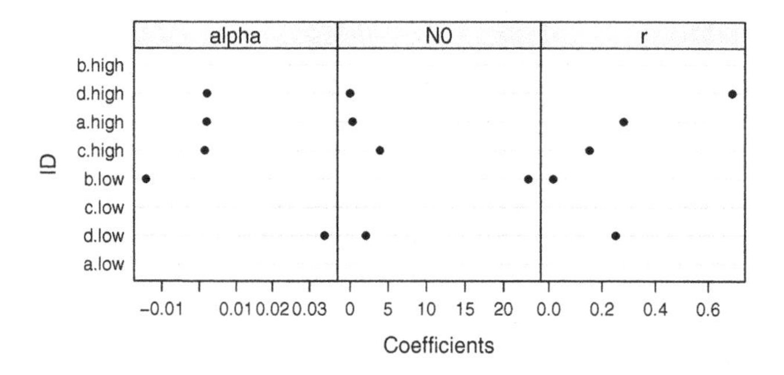

Fig. 3.20: Coefficients estimated for each microcosm separately. Missing coefficients occurred when model fits could not be easily acheived.

We can see that the coefficients (Fig. 3.20) are somewhat similar to those we estimated above with `modSlope2`. We note, however, that for one replicate $\alpha < 0$, which does not make sense. If we check the residuals,

```
> plot(Cmod.list)
```

we see that the residuals increase dramatically with the fitted values (graph not shown). This is consistent with our previous observations.

In this next step, we create a single nonlinear mixed model using `nlme`. We use `pdDiag` to specify that each microcosm has its own unique effect on each of the parameters, aside from nutrient concentrations. We use `weights` to specify that the variance of the residuals is a power function (a^{2t}) of the mean, and which fits an extra parameter, t.

```
> Cmod.all <- nlme(Cmod.list, random = pdDiag(form = alpha +
+     NO + r ~ 1), weights = varPower())
```

We saw above that the nutrients caused a big difference in the slopes of the *pgr vs. N* relations. Now we would like to estimate the *alpha* for each level of nutrients. In estimating these, we can also test whether they are different.

First we capture the fixed effect coefficients of the current model.

```
> cfs2 <- fixef(Cmod.all)
> cfs2

   alpha       NO        r
0.009266 7.143118 0.130551
```

Next we update the model, specifying that `alpha` should depend on nutrient level, that is, estimate one α for each level of nutrient. We specify that the other paramters are the same in both treatments. This will mean that we have one additional coefficent, and we need to provide the right number of starting values, and in the correct order (`alpha1, alpha2, NO, r`).

```
> Cmod.all2 <- update(Cmod.all, fixed = list(alpha ~ Nutrients,
+     NO + r ~ 1), start = c(cfs2[1], 0, cfs2[2], cfs2[3]))
```

It would be good practice to plot diagnostics again, as we did above. Here, however, we push on and check the confidence intervals for our parameters summary.

```
> summary(Cmod.all2)

Nonlinear mixed-effects model fit by maximum likelihood
  Model: No.per.ml ~ ilogistic(Day, alpha, NO, r)
 Data: ClostExp
   AIC  BIC logLik
  1549 1575 -765.3

Random effects:
 Formula: list(alpha ~ 1, NO ~ 1, r ~ 1)
 Level: ID
 Structure: Diagonal
        alpha.(Intercept)    NO        r Residual
StdDev:         6.533e-20 1.392 2.56e-09    1.488

Variance function:
 Structure: Power of variance covariate
 Formula: ~fitted(.)
 Parameter estimates:
 power
```

```
0.8623
Fixed effects: list(alpha ~ Nutrients, N0 + r ~ 1)
                     Value Std.Error  DF t-value p-value
alpha.(Intercept)    0.002    0.0003 133   5.021       0
alpha.Nutrientslow   0.016    0.0026 133   6.185       0
N0                   6.060    1.4245 133   4.254       0
r                    0.139    0.0144 133   9.630       0
 Correlation:
                   al.(I) alph.N N0
alpha.Nutrientslow -0.019
N0                 -0.289  0.048
r                   0.564  0.058 -0.744

Standardized Within-Group Residuals:
    Min      Q1     Med      Q3     Max
-1.3852 -0.6010 -0.2038  0.3340  4.5236

Number of Observations: 144
Number of Groups: 8
```

If we look first at the random effects on the parameters, we see that the variance components for alpha or r are vanishingly small, we should therefore remove them from the analysis, and then compare the two models.

```
> Cmod.all3 <- update(Cmod.all2, random = N0 ~ 1)
> anova(Cmod.all2, Cmod.all3)

          Model df  AIC  BIC logLik   Test  L.Ratio p-value
Cmod.all2     1  9 1548 1575 -765.3
Cmod.all3     2  7 1544 1565 -765.3 1 vs 2 0.0008463  0.9996
```

We see that the less complex model is not significantly different ($P > 0.05$) and is more parsimonious (smaller AIC, Akaike's information criterion, [18]).

At last, we can examine the confidence intervals to get the estimates on the intraspecific competition coefficents, α_{ii}.

```
> intervals(Cmod.all3)$fixed

                      lower     est.   upper
alpha.(Intercept)  0.000978 0.001599 0.00222
alpha.Nutrientslow 0.011095 0.016206 0.02132
N0                 3.279937 6.057370 8.83480
r                  0.110739 0.138860 0.16698
attr(,"label")
[1] "Fixed effects:"
```

see that the analysis, such as it is, indicates that the confidence intervals for the estimated α do not overlap, indicating that the expected values are very, very different from each other. We could calculate the bounds on K_{high} and K_{low} as well, based on our lower and upper confidence limits on α_{high}, α_{low}.

```
> 1/intervals(Cmod.all3)$fixed[1:2, c(1, 3)]
```

```
                    lower  upper
alpha.(Intercept) 1022.41 450.43
alpha.Nutrientslow  90.13  46.91
```

This aproximately ten-fold difference in K seems surprisingly consistent with the ten-fold difference in nutrient concentration initially established in the experiment. This ten-fold increase in *Closterium* may be merely an odd coincidence because it implies that all species in the food web should have increased ten-fold, but that was not the case [195]. On the other hand, it may guide future experiments in keystone predation.

We can plot the fixed effects, and also the added variation due random variation in N_0 and r due to the ID of the microcosms.

```
> plot(augPred(Cmod.all3, level = 0:1), layout = c(4, 2, 1),
+     scales = list(y = list(relation = "free")))
```

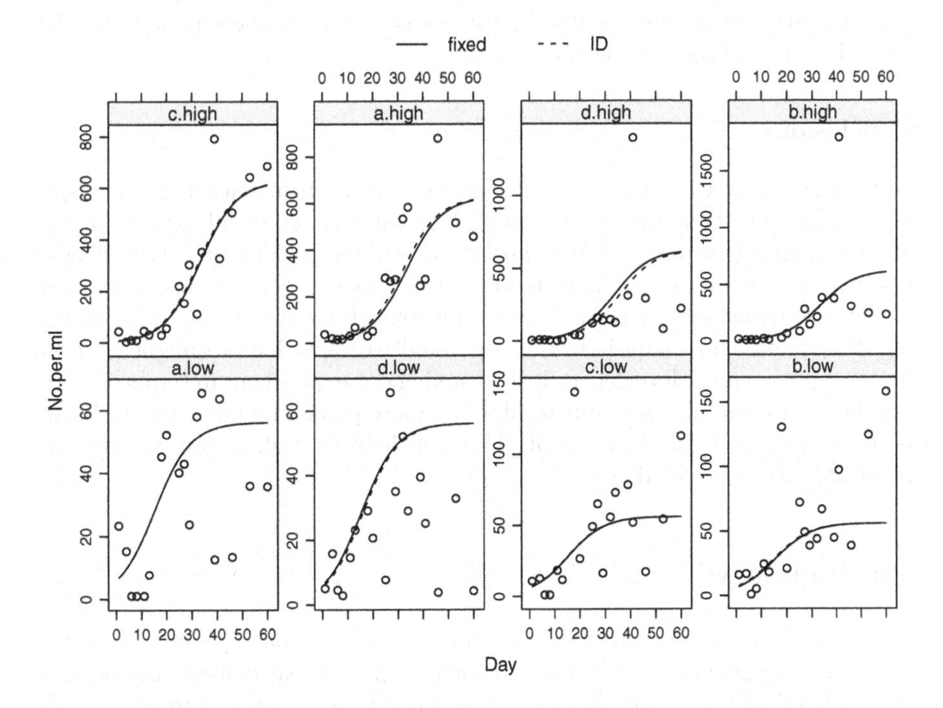

Fig. 3.21: Times series plots augmented with the overall predictions (solid lines for the fixed effects) and predictions for individual microcosms (dashed lines).

When we plot our data along the with the model (Fig. 3.21), we can see the disparity between the predictions and the data — the source of those big residuals we've been ignoring.

There may be at least a couple issues to consider when we think about the disparities.

Sampling error We do not have much idea about how accurate or precise our sampling is; this might be remedied in the next experiment by using a mean with multiple samples per time interval.

Experimental Design In this system, 10% of the medium was replaced each week (to remove waste products and replenish nutrients), and this is likely to introduce variability.

Dynamics Species interactions may drive dynamics, derived from feedback between (i) bacteria recycling nutrients, (ii) bacteria consuming nutrients, (iii) the predator *Colpidium* consuming bacteria, and (iv) *Closterium* consuming nutrients (Chapters 5, 6).

Perhaps the first steps for an improved experiment might include sampling at least twice for each sample period, to at least find out what our sampling error is. It may, or may not, be a problem, but a few trials would let us know. We also might increase the frequency and decrease the magnitude of the medium replacement, and to make sure the timing of the sampling corresponds appropriately with the medium replace. After we take care of these issues, then we can take a harder look at the dynamics.

Conclusions

Our results show that lowering nutrient concentration increased the intraspecific interaction strengths, α_{ii}, ten-fold. Alternatively, we could express this as a ten-fold change in the carrying capacity K of these populations. The effect of resources on interaction strength is sometimes a heated topic of discussion, and our precise definitions, derived from simple growth models, provides an unambiguous result. Our analysis has also pointed to ways that we might perform better experiments. More generally, by framing our experiment in terms of existing theory, we can interpret our results in a more precise manner, and facilitate better advances in the future [13]. This will help us make more constructive contributions to the field.

3.6 Summary

In this chapter, we motivated density dependence, focusing specifically on discrete and continuous logistic growth which assumes linear density dependence. We explored how algebraic rearrangements can yield useful and interesting insights, and how we might think about generalizing or simplifying logistic growth. We showed how the time lag built into the discrete model results in chaos. We explored ideas of maximum sustained yield, and how we might fit the continuous logistic growth model to real data.

Problems

3.1. Dynamics of an annual plant
(a) Calculate r_d of an annual plant that has a maximum growth rate of $N_{t+1}/N_t =$ 2 at very, very small population sizes.
(b) Calculate the appropriate per capita density dependent effect of an annual plant with a carrying capacity K of 100 inds·m^{-2}.
(c) Write the appropriate logistic growth equation that incorporates the intrinsic growth of (a) and the density dependence of (b).
(d) Graph the 10 y dynamics ($t = 0, \ldots, 10$) of the annual plant in (a) and (b), starting with $N_0 = 1$.

3.2. Dynamics of *E. coli*
(a) Calculate r of *E. coli* that has a doubling time of 30 min. Express this rate in hours.
(b) Calculate the per capita density dependent effect of an *E. coli* culture that grows logistically over a 24 h period, and which levels off at a density of 10^7 CFU·mL^{-1} (CFU is colony forming units — for *E. coli* its is equivalent to individuals).
(c) Graph the 50 h dynamics ($t = 0, \ldots, 50$) of the *E. coli* population in (a) and (b), starting with $N_0 = 1000$.

3.3. Nonlinear Density Dependence
Sibly et al. [188] found that most species have nonlinear and concave-up density dependence. They use the θ-logistic growth model. (a) Create a theta-logistic continuous growth model for use with the `ode()` function in the `deSolve` package.
(b) Show that with $\theta = 1$, it is identical our function `clogistic` above.
(c) Graph N for $t = 0, \ldots, 100$ using a few different values of θ and explain how θ changes logistic growth.

3.4. Harvested Populations
The logistic growth equation and other similar equations have been used and abused in efforts to achieve a *maximum sustained yield* of a harvested population. The immediate goal of maximum sustained yield management practices is to kill only the number of individuals that reduces the population to half of its carrying capacity, assuming that eq. 3.13 describes the population growth. Answer the questions below to help you explain why this would be a goal.
(a) Find expressions for population growth rates when $N = K/4, K/2, 3K/4$ (substitute these values for N in eq. 3.13, and show your work). Which of these results in the highest population growth rate? How do these relate to the management of a harvested population?
(b) Show the derivation of the partial derivative of the continuous logistic growth model, with respect to N (i.e., $\partial \dot{N}/\partial N$). Solve for zero to determine when total population growth rate reaches a maximum. Show your work.
(c) What would be the ecological and economic rationale for *not* killing more individuals, and keeping $N > K/2$?

(d) What would the consequences be for the population if you *assume* linear density dependence $(1 - \alpha/N)$, but in fact the population is governed by non-linear density dependence where $\theta < 1$ and $\theta > 1$ (Figs. 3.13a-3.13c)?
(e) What economic benefit would you gain if you harvested the entire population all at once (and eliminated it from the face of the planet)? What could you do with all that money?
(f) How would you incorporate both harvesting and economic considerations into your logistic growth model?

3.5. Environmental Variability
Most environments change continually. Temperature, resource availability, changes in predator or pathogen abundance all influence the carrying capacity of the environment.
(a) Use the discrete version of the logistic growth equation to model a population in a variable environment. Do this by creating a discrete logistic growth function that adds (or subtracts) a random amount to K in each time step. Use one of the many functions that can draw random numbers from particular distributions (e.g., `rpois()`, `rlnorm()`, `runif()`). You might start by playing with one of the random number generators:

```
Kstart <- 100; time <- 1:20; K <- numeric(20);
for(i in 1:20) K[i] <- Kstart + rnorm(1, m=0, sd=5);
plot(time, K).
```

(b) All distributions are characterized by their *moments*. For instance, the Normal distribution is typically defined by its mean, μ, and variance, σ^2. Focus on just one moment of your selected distribution in (a), and use your simulations to determine quantitatively the relation between K and the resulting N derived from the discrete growth model. For instance, you might vary the standard deviation of the random normal distribution that contributes to K, and examine how the standard deviation of K, σ_K relates to mean N, μ_N.
(c) Create a reddened time series for K.[16] (Hint: What are the values of x and y when you do `x <- sample(c(-1,0,1), 20, replace=TRUE); y <- cumsum(x)` ?). Use this time series to create a reddened population dynamic. (Hint: First create the vector of reddened K's equal in length to your time series. Then create a growth function that can access the vector, e.g. `DLG.RK <- func-`

[16] Environmental factors such as temperature frequently vary in a gradual fashion, such that if the weather is hot today, it is likely to be hot tomorrow. Such variation is described in terms of auto-correlation or a spectral distribution or color [67,155]. Spectral decomposition of a times series involves a description of the series in terms of waves of different wavelengths or frequencies. White noise is variation that is described by equal contributions of all wavelengths (hence the term "white"), and it is a series that is completely random from one time step to the next. Reddened environmental variation is variation that is described by a predominance of longer wavelengths (hence the term "red"), and it is a series in which values tend to change gradually, where values that are close together in time tend to be more similar, or auto-correlated. Spectral variation is also referred to as $1/f$ noise ("one over f noise").

tion(alpha=0.001, rd=1, N0=10, gens=20, K=K.red). Then, inside the for loop, use an indexed K, such as 1-N[t]/K[t].

4

Populations in Space

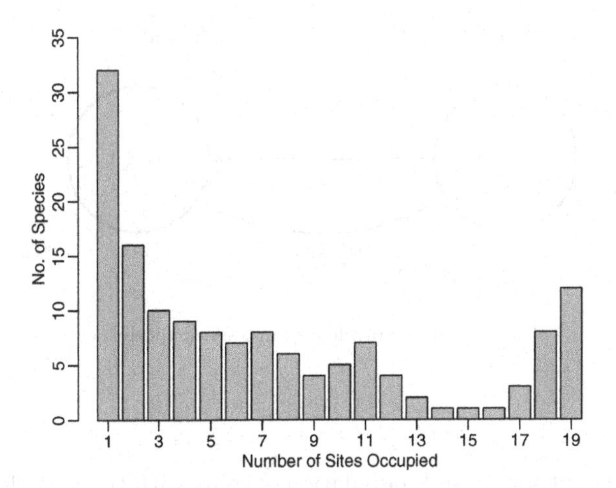

Fig. 4.1: A frequency distribution of the number of plant species (y-axis) that occupy different numbers of grassland remnants (x-axis). Note the U-shaped (bimodal) distribution of the number of sites occupied. Other years were similar [35]

Over relatively large spatial scales, it is not unusual to have many species that seem to occur everywhere, and even more species that seem to be found in only one or a few locations. For example, Scott Collins and Susan Glenn [35] showed that in grasslands, each separated by up to 4 km, there were more species occupying only one site (Fig. 4.1, left-most bar) than two or more sites, and also that there are more species occupying all the sites than most intermediate numbers of sites (Fig. 4.1, right-most bar), resulting in a U-shaped frequency distribution. Illke Hanski [70] coined the rare and common species "satellite" and "core" species, respectively, and proposed an explanation. Part of the answer seems to come from *the effects of immigration and emigration in a spatial context.* In this chapter we explore mathematical representations of individuals and populations that exist in space, and we investigate the consequences for populations and collections of populations.

4.1 Source-sink Dynamics

In Chapters 1-3, we considered *closed* populations. In contrast, one could imagine a population governed by *b*irths plus *i*mmigration, and *d*eaths plus *e*migration (a *BIDE* model). Ron Pulliam [172] proposed a simple model that includes all four components of *BIDE* which provides a foundation for thinking about connected subpopulations. We refer to the dynamics of these as *source-sink dynamics*. Examples of such linked populations might include many different types of species. For instance, a source-sink model could describe linked populations of a single species might occupy habitat patches of different quality, where organisms might disperse from patch to patch.

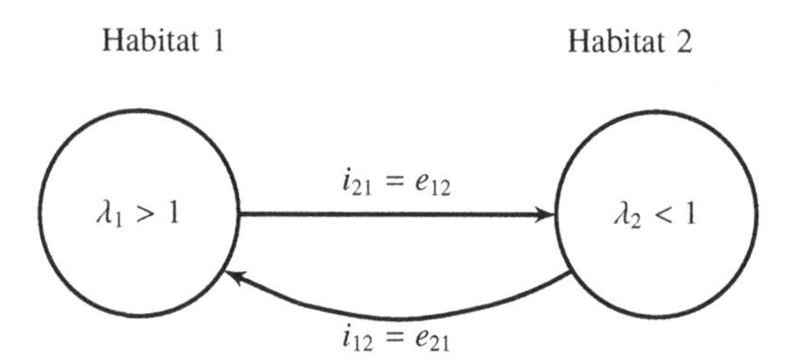

Fig. 4.2: The simplest source-sink model.

The concept

The general idea of source-sink populations begins with the idea that spatially separated subpopulations occupy distinct patches, and each exhibit their own intrinisic dynamics due to births and deaths; that is, we could characterize a λ for each subpopulation. In addition, individuals move from one patch to another; that is, they immigrate and emigrate from one patch (or subpopulation) to another. Therefore, the number of individuals we observe in a particular patch is due, not only to the λ in that population, but also to the amount of immigration, i, and emigration, e.

Subpopulations with more births than deaths, $\lambda > 1$, and with more emigration than immigration, $e > i$, are referred to as *source populations*. Subpopulations with fewer births than deaths, $\lambda < 1$, and with more immigration than emigration, $i > e$, are referred to as *sink populations*.

When we think about what might *cause* variation in λ, we typically refer to the *quality* of patches or habitats. Quality might be inferred from λ, or it might actually be the subject of investigation and independent of λ — typically we think of high quality habitat as having $\lambda > 1$ and poor quality habitat as having $\lambda < 1$.

The equations

Pulliam envisioned two linked bird populations where one could track adult reproduction, and adult and juvenile survival and estimate λ, per capita growth rate separately for each population. For the first population, the number of birds in patch 1 at time $t+1$, $n_{1,t+1}$, is the result of adult survival P_A, reproduction β_1, and survival of the juveniles P_J. Thus,

$$n_{1,t+1} = P_A n_t + P_J \beta_1 n_{1,t} = \lambda_1 n_1. \tag{4.1}$$

Here $\beta_1 n_{1,t}$ is production of juveniles, and P_J is the survival of those juveniles to time $t+1$. Pulliam described the second population in the same fashion as

$$n_{2,t+1} = P_A n_t + P_J \beta_2 n_{1,t} = \lambda_2 n_2. \tag{4.2}$$

Pulliam then assumed, for simplicity's sake, that the two populations vary only in fecundity (β), which created differences in λ_1 and λ_2. He called population 1 the *source population* ($\lambda_1 > 1$) and population 2 the *sink population* ($\lambda_2 < 1$). He also assumed that birds in excess of the number of territories in the source population emigrated from the source habitat to the sink habitat. Therefore, the source population held a constant density (all territories filled), but the size of the population in the sink depended on both its own growth rate $\lambda_2 < 1$ and also the number of immigrants.

A result

One of his main theoretical findings was that *population density can be a misleading indicator of habitat quality* (Fig. 4.3). If we assume that excess individuals in the source migrate to the sink, then as habitat quality and reproduction increase in the source population, the source population comprises *an ever decreasing proportion* of the total population! That is, as λ_1 gets larger, $n_1/(n_1+n_2)$ gets smaller. Thus, density can be a very misleading predictor of long-term population viability, if the source population is both productive and exhibits a high degree of emigration.

A model

We can use a matrix model to investigate source-sink populations [12]. Let us mix up typical demographic notation (e.g., Chapter 2) with that of Pulliam [172], so that we can recognize Pulliam's quantities in a demographic matrix model setting. Further, let us assume a pre-breeding census, in which we count adults. The population dynamics would thus be governed by **A**

$$\mathbf{A} = \begin{pmatrix} P_{A1} + P_{J1}\beta_1 & M_{12} \\ M_{21} & P_{A2} + P_{J2}\beta_2 \end{pmatrix} \tag{4.3}$$

where the upper left element (row 1, column 1) reflects the within-patch growth characteristics for patch 1. The lower right quadrant (row 2, and column 2) reflects the within-patch growth characteristics of patch 2.

We then assume, for simplicity, that migration, M, is exclusively from the source to the sink ($M_{21} > 0$, $M_{12} = 0$). We further assume that $\lambda_1 > 1$ but all excess individuals migrate to patch 2, so $M_{21} = \lambda_1 - 1 > 0$. Then $\mathbf{A}$ simplifies to

$$\mathbf{A} = \begin{pmatrix} 1 & 0 \\ \lambda_1 - 1 & \lambda_2 \end{pmatrix} \tag{4.4}$$

The spatial demographic Pulliam-like model

We first assign λ for the source and sink populations, and create a matrix.

```
> L1 <- 2
> L2 <- 0.4
> A <- matrix(c(1, 0, L1 - 1, L2), nrow = 2, byrow = TRUE)
```

We can then use eigenanalysis, as we did in Chapter 2 for stage structured populations. The dominant eigenvalue will provide the long term asymptotic total population growth. We can calculate the stable "stage" distribution, which in this case is the distribution of individuals between the two habitats.

```
> eigen(A)

$values
[1] 1.0 0.4

$vectors
       [,1] [,2]
[1,] 0.5145    0
[2,] 0.8575    1
```

From the dominant eigenvalue, we see Pulliam's working assumption that the total population growth is set at $\lambda = 1$. We also see from the dominant eigenvector that the sink population actually contains more individuals than the source population ($0.51/(0.51+0.86) < 0.5$).
We could graph these results as well, for a range of λ_1 (Fig. 4.3). Here we let `p1` be the proportion of the population in the source.

```
> L1s <- seq(1, 3, by = 0.01)
> p1 <- sapply(L1s, function(l1) {
+     A[2, 1] <- l1 - 1
+     eigen(A)$vectors[1, 1]/sum(eigen(A)$vectors[, 1])
+ })
> plot(L1s, p1, type = "l", ylab = "Source Population",
+      xlab = expression(lambda[1]))
```

4.2 Two Types of Metapopulations

Our logistic model (Chapter 3) is all well and good, but it has no concept of *space* built into it. In many, and perhaps most circumstances in ecology, space has the potential to influence the dynamics of populations and ecosystem

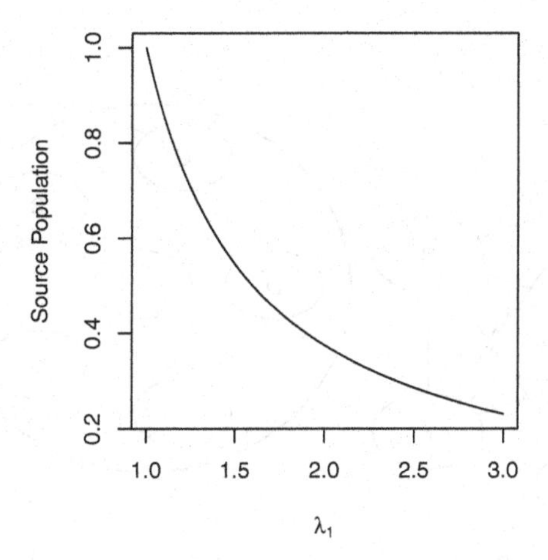

Fig. 4.3: The declining *relative* abundance in the high quality habitat in a source-sink model. The proportion of the total population $(n_1/(n_1 + n_2))$ in the *source* population may decline with increasing habitat quality and growth rate λ_1 habitat.

fluxes [101,102,116]. The logistic equation represents a *closed* population, with no clear accounting for emigration or immigration. In particular cases, however, consideration of space may be essential. What will we learn if we start considering space, such that sites are open to receive immigrants and lose emigrants?

First we consider ideas associated with different types of "collections;" we then consider a mathematical framework for these ideas.

A single spatially structured population

One conceptual framework that we will consider below is that of a single closed population, where individuals occupy sites in an implicitly spatial context (Fig. 4.4). Consider a population in space, where a *site* is the space occupied by one individual. One example might be grasses and weeds in a field. In such a population, for an individual within our population to successfully reproduce and add progeny to the population, the individual must first actually occupy a site. For progeny to establish, however, a propagule must arrive at a site that is unoccupied. Thus the more sites that are already occupied, the less chance there is that a propagule lands on an unoccupied site. Sites only open up at some constant per capita rate as individuals die at a per capita death rate.

A metapopulation

The other conceptual framework that we consider here is that of *metapopulations.* A metapopulation is a population of populations, or a collection of populations (Fig. 4.4). Modeling metapopulations emerged from work in pest management when Levins [110] wanted to represent the dynamics of the proportion of fields infested by a pest. He assumed that a field was either occupied by the pest, or not. The same models used to represent a population of individuals

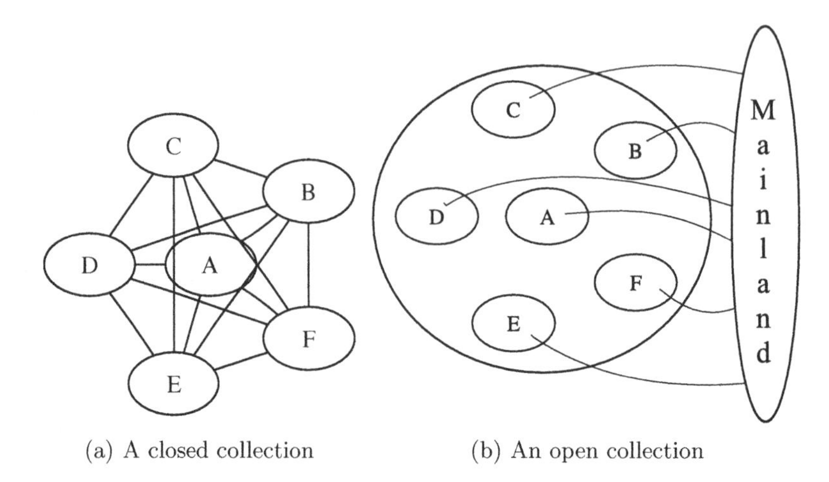

(a) A closed collection (b) An open collection

Fig. 4.4: Collections of sites. (a) Sites may be recolonized via internal propagule production and dispersal only, or (b) sites may receive immigrants from an outside source that is not influenced by the collection. Each site (A-F) may be a spot of ground potentially occupied by a single plant, or it may be an oceanic island potentially occupied by a butterfly population. Sites may also be colonized via both internal and external sources.

that occupy sites (above) can also be used to represent populations that occupy sites, with conceptually similar ecological interpretation. In this case, each *site* is a location that either contains a population or not. In this framework, we keep track of the proportion of all populations that remain extant, that is, the proportion of sites that are occupied. As with a single population (above), the metapopulation is closed, in the sense that there exists a finite number of sites which may exchange migrants.

Whether we consider a single spatial population, or single metapopulation, we can envision a *collection of sites* connected by dispersal. Each site may be a small spot of ground that is occupied by a plant, or it may be an oceanic island that is occupied by a population. All we know about a single site is that it is occupied or unoccupied. If the site is occupied by an individual, we know nothing of how big that individual is; if the site is occupied by a population, we know nothing about how many indiviuals are present. The models we derive below keep track of the proportion of sites that are occupied. These are known loosely as *metapopulation models*. Although some details can differ, whether we are modeling a collection of spatially discrete individuals in single population or a collection of spatially discrete populations, these two cases share the idea that there are a collection of sites connected by migration, and each is subject to extinction.

The most relevant underlying biology concerns colonization and extinction in our collection of sites (Fig. 4.4). In this chapter, we will assume that all sites experience equal rates; when we make this assumption, we greatly simplify everything, and we can generalize across all sites. All of the models we consider are simple elaborations of what determines colonization and extinction. Another useful concept to consider is whether the collection of sites receives propagules from the outside, from some external source that is not influenced by the collection of sites (Fig. 4.4).

4.3 Related Models

Here we derive a single mathematical framework to describe our two types of models. In all cases, we will consider how total rates of colonization, C, and extinction, E, influence the the rate of change of p, the proportion of sites that are occupied,

$$\frac{\mathrm{d}p}{\mathrm{d}t} = C - E. \tag{4.5}$$

We will consider below, in a somewhat orderly fashion, several permutations of how we represent colonization and extinction of sites (e.g., [62, 63]).

4.3.1 The classic Levins model

Levins [110] proposed what has come to be known as the classic metapopulation model,

$$\frac{\mathrm{d}p}{\mathrm{d}t} = c_i p\,(1-p) - ep. \tag{4.6}$$

This equation describes the dynamics of the proportion, p, of a set of fields invaded by a pest (Fig. 4.5a). The pest colonizes different fields at a total rate governed by the rate of propagule production, c_i, and also on the proportion of patches that contain the pest, p. Thus, propagules are being scattered around the landscape at rate $c_i p$. The rate at which p changes, however, is also related to the proportion of fields that are unoccupied, $(1-p)$, and therefore available to become occupied and increase p. Therefore the total rate of colonization is $c_i p(1-p)$. The pest has a constant local extinction rate e, so the total extinction rate in the landscape is ep.

The parameters c_i and e are very similar to r of continuous logistic growth, insofar as they are dimensionless instantaneous rates. However, they are sometimes thought of as probabilities. The parameter c_i is approximately the proportion of open sites colonized per unit time. For instance, if we created or found 100 open sites, we could come back in a year and see how many became occupied over that time interval of one year, and that proportion would be a function of c_i. The parameter e is often thought of as the probability that a site becomes unoccupied per unit time. If we found 100 occupied sites in one year, we could revisit them a year later and see how many became *un*occupied over that time interval of one year.

We use the subscript i to remind us that the colonization is coming from within the sites that we are studying (i.e. internal colonization). With internal colonization, we are modeling a closed spatial population of sites, whether "site" refers to an entire field (as above), or a small patch of ground occupied by an individual plant [202].

The Levins metapopulation model (Fig. 4.5a)

A function for a differential equation requires arguments for time, a vector of the state variables (here we have one state variable, p), and a vector of parameters.

```
> levins <- function(t, y, parms) {
+     p <- y[1]
+     with(as.list(parms), {
+         dp <- ci * p * (1 - p) - e * p
+         return(list(dp))
+     })
+ }
```

By using `with`, we can specify the parameters by their names, as long as `parms` includes names. The function returns a list that contains a value for the derivative, evaluated at each time point, for each state variable (here merely dp/dt). We then use `levins` in the numerical integration function `ode` in the `deSolve` package.

```
> library(deSolve)
> prms <- c(ci = 0.15, e = 0.05)
> Initial.p <- 0.01
> out.L <- data.frame(ode(y = Initial.p, times = 1:100, func = levins,
+     parms = prms))
```

We then plot the result (Fig. 4.5a).

```
> plot(out.L[, 2] ~ out.L[, 1], type = "l", ylim = c(0, 1),
+     ylab = "p", xlab = "time")
```

Can we use this model to predict the eventual equilibrium? Sure — we just set eq. 4.6 to zero and solve for p. This model achieves and equilibrium at,

$$0 = c_i p - c_i p^2 - ep$$
$$p^* = \frac{c_i - e}{c_i} = 1 - \frac{e}{c_i}.$$

When we do this, we see that $p^* > 0$ as long as $c_i > e$ (e.g., Fig. 4.5a). When is $p^* = 1$, so that all the sites are filled? In principle, all sites cannot be occupied simultaneously unless $e = 0$!

4.3.2 Propagule rain

From where else might propagules come? If a site is not closed off from the rest of the world, propagules could come from outside the collection of sites that we are actually monitoring.

For now, let us assume that our collection of sites is continually showered by propagules from an external source. If only those propagules are important, then we could represent the dynamics as,

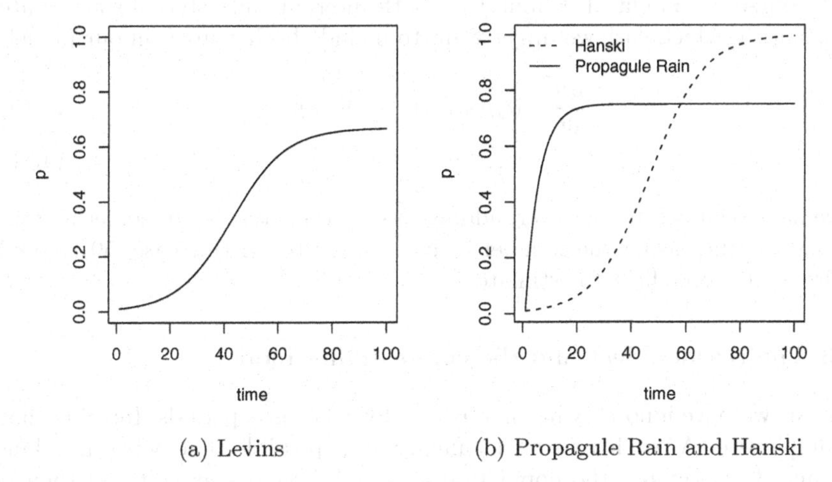

(a) Levins (b) Propagule Rain and Hanski

Fig. 4.5: Three metapopulation models, using similar parameters ($c_i = 0.15$, $c_e = 0.15$, $e = 0.05$).

$$\frac{\mathrm{d}p}{\mathrm{d}t} = c_e(1-p) - ep \tag{4.7}$$

where c_e specifies rate of colonization coming from the external source. Gotelli [63] refers to this model as a metapopulation model with "propagule rain" or the "island–mainland" model. He calls it this because it describes a constant influx of propagules which does not depend on the proportion, p, of sites occupied for propagule production. Extinction here is mediated only by the proportion of sites occupied, and has a constant per site rate.

The propagule rain metapopulation model (Fig. 4.5b)

A function for a differential equation requires arguments for time, a vector of the state variables (here we have one state variable, p), and a vector of parameters.

```
> gotelli <- function(t, y, parms) {
+     p <- y[1]
+     with(as.list(parms), {
+         dp <- ce * (1 - p) - e * p
+         return(list(dp))
+     })
+ }
```

The function returns a list that contains a value for the derivative, evaluated at each time point, for each state variable (here merely $\mathrm{d}p/\mathrm{d}t$).

We can solve for this model's equilibrium by setting eq. 4.7 equal to zero.

$$0 = c_e - c_e p - ep \tag{4.8}$$

$$p^* = \frac{c_e}{c_e + e}. \tag{4.9}$$

Of course, we might also think that both internal and external sources are important, in which case we might want to include both sources in our model,

$$\frac{\mathrm{d}p}{\mathrm{d}t} = (c_i p + c_e)(1 - p) - ep \tag{4.10}$$

$$\tag{4.11}$$

As we have seen before, however, adding more parameters is not something we take lightly. Increasing the number of parameters by, in this case, 50% could require a lot more effort to estimate.

4.3.3 The rescue effect and the core-satellite model

Thus far, we have ignored what happens between census periods. Imagine that we sample site "A" each year on 1 January. It is possible that between 2 January and 31 December the population at site A becomes extinct and then is subsequently recolonized, or "rescued" from extinction. When we sample on 1 January in the next year, we have no way of knowing what has happened in the intervening time period. We would not realize that the population had become extinct and recolonization had occurred.

We can, however, model total extinction rate E with this *rescue effect*,

$$E = -ep(1 - p)\,. \tag{4.12}$$

Note that as $p \to 1$, the total extinction rate approaches zero. Total extinction rate declines because as the proportion of sites occupied increases, it becomes increasingly likely that dispersing propagules will land on all sites. When propagules happen to land on sites that are on the verge of extinction, they can "rescue" that site from extinction.

Brown and Kodric-Brown [17] found that enhanced opportunity for immigration seemed to reduce extinction rates in arthropod communities on thistles. They coined this effect of immigration on extinction as the "rescue effect." MacArthur and Wilson [121] also discussed this idea in the context of island biogeography. We can even vary the strength of this effect by adding yet another parameter q, such that the total extinction rate is $-ep(1 - qp)$ (see [62]).

Assuming only internal propagule supply and the simple rescue effect results in what is referred to as the the core-satellite model,

$$\frac{\mathrm{d}p}{\mathrm{d}t} = c_i p(1 - p) - ep(1 - p) \tag{4.13}$$

This model was made famous by Illka Hanski [70]. It is referred to as the *core-satellite* model, for reasons we explore later.

The core-satellite metapopulation model

A function for a differential equation requires arguments for time, a vector of the state variables (here we have one state variable, p), and a vector of parameters.

```
> hanski <- function(t, y, parms) {
+     p <- y[1]
+     with(as.list(parms), {
+         dp <- ci * p * (1 - p) - e * p * (1 - p)
+         return(list(dp))
+     })
+ }
```

The function returns a list that contains a value for the derivative, evaluated at each time point, for each state variable (here merely dp/dt).

Graphing propagule rain and core-satellite models (Fig. 4.5b)

First, we integrate the models using the same parameters as for the Levins model, and collect the results.

```
> prms <- c(ci <- 0.15, ce <- 0.15, e = 0.05)
> out.IMH <- data.frame(ode(y = Initial.p, times = 1:100,
+     func = gotelli, parms = prms))
> out.IMH[["pH"]] <- ode(y = Initial.p, times = 1:100, func = hanski,
+     parms = prms)[, 2]
```

We then plot the result (Fig. 4.5a).

```
> matplot(out.IMH[, 1], out.IMH[, 2:3], type = "l", col = 1,
+     ylab = "p", xlab = "time")
> legend("topleft", c("Hanski", "Propagule Rain"), lty = 2:1,
+     bty = "n")
```

Core-satellite equilibria

What is the equilibrium for the Hanski model (eq. 4.13)? We can rearrange this to further simplify solving for p^*.

$$\frac{dp}{dt} = (c_i - e)\, p\, (1 - p) \tag{4.14}$$

This shows us that for any value of p between zero and one, the sign of the growth rate (positive or negative) is determined by c_i and e. If $c_i > e$, the rate of increase will always be positive, and because occupancy cannot exceed 1.0, the metapopulation will go to full occupancy ($p^* = 1$), and stay there. This equilibrium will be a *stable attractor* or stable equilibrium. What happens if for some reason the metapopulation becomes globally extinct, such that $p = 0$, even though $c_i > e$? If $p = 0$, then like logistic growth, the metapopulation stops changing and cannot increase. However, the slightest perturbation away from $p = 0$ will lead to a positive growth rate, and increase toward the stable attractor, $p^* = 1$. In this case, we refer to $p^* = 0$ as an *unstable equilibrium* and a repellor.

If $c_i < e$, the rate of increase will always be negative, and because occupancy cannot be less than 0, the metapopulation will become extinct ($p^* = 0$), and stay there. Thus $p^* = 0$ would be a stable equilibrium or attractor. What is predicted to happen if, for some odd reason this population achieved full occupancy, $p = 1$, even though $c_i < e$? In that case, $(1 - p) = 0$, and the rate of change goes to zero, and the population is predicted to stay there, even though extinction is greater than colonization. How weird is that? Is this fatal flaw in the model, or an interesting prediction resulting from a thorough examination of the model? How relevant is it? How could we evaluate how relevant it is? We will discuss this a little more below, when we discuss the effects of habitat destruction.

What happens when $c_i = e$? In that case, $c_i - e = 0$, and the population stops changing. What is the value of p when it stops changing? It seems as though it could be any value of p, because if $c_i - e = 0$, the rate change goes to zero. What will happen if the population gets perturbed — will it return to its previous value? Let's return to question in a bit.

To analyze stability in logistic growth, we examined the slope of the partial derivative at the equilibrium, and we can do that here. We find that the partial derivative of eq. 4.13 with respect to p is

$$\frac{\partial \dot{p}}{\partial p} = c - 2cp - e + 2ep \qquad (4.15)$$

where $\dot{p}$ is the time derivative (eq. 4.13). A little puzzling and rearranging will show

$$\frac{\partial \dot{p}}{\partial p} = (c_i - e)(1 - 2p) \qquad (4.16)$$

and make things simpler. Recall our rules with regard to stability (Chapter 3). If the partial derivative (the slope of the time derivative) is negative at an equilibrium, it means the the growth rate approaches zero following a perturbation, meaning that it is stable. If the partial derivative is positive, it means that the change accelerates away from zero following the perturbation, meaning that the equilibrium is unstable. So, we find the following guidelines:

- $c_i > e$
  - $p = 1$, $\partial \dot{p}/\partial p < 0$, stable equilibrium.
  - $p = 0$, $\partial \dot{p}/\partial p > 0$, unstable equilibrium.
- $c_i < e$
  - $p = 1$, $\partial \dot{p}/\partial p > 0$, unstable equilibrium.
  - $p = 0$, $\partial \dot{p}/\partial p < 0$, stable equilibrium.

What if $c_i = e$? In that case, both the time derivative ($\mathrm{d}p/\mathrm{d}t$) and the partial derivative ($\partial \dot{p}/\partial p$) are zero *for all values of* p. Therefore, if the population gets displaced from any arbitrary point, it will remain unchanged, not recover, and will stay displaced. We call this odd state of affairs a *neutral equilibrium*. We revisit neutral equilibrium when we discuss interspecific competition and predation.

We can also explore the stability of one of these equilibria by plotting the metapopulation growth rate as a function of p (Fig. 4.6). When we set $c_i > e$, and examine the slope of that line at $p^* = 1$, we see the slope is negative, indicating a stable equilibrium.

An equilibrium for the core-satellite metapopulation model (Fig. 4.6)

We first create an expression for the growth itself, $\mathrm{d}p/\mathrm{d}t$. We then plot it, while we evaluate it, on the fly.

```
> dpdtCS <- expression((ci - e) * p * (1 - p))
> ci <- 0.15
> e <- 0.05
> p <- seq(0, 1, length = 50)

> plot(p, eval(dpdtCS), type = "l", ylab = "dp/dt")
```

Levins *vs.* Hanski

Why would we use Levins' model instead of Hanski's core-satellite model? To explore this possibility, let's see how the Hanski model might change gradually into the Levins model. First we define the Hanski model with an extra parameter, a,

$$\frac{\mathrm{d}p}{\mathrm{d}t} = c_i p(1-p) - ep(1-ap). \tag{4.17}$$

Under Hanski's model, $a = 1$ and under Levins' model $a = 0$. If we solve for the equilibrium, we see that

$$p^* = \frac{c-e}{c-ae} \tag{4.18}$$

so that we can derive either result for the two models. In the context of logistic growth, where $K = Hp^*$, this result, eq. 4.18, implies that for the Hanski model, K fills all available habitat, whereas the Levins model implies that K fills only a fraction of the total available habitat. That fraction results from the dynamic balance between c_i and e.

4.4 Parallels with Logistic Growth

It may have already occurred to you that the closed spatial population described here sounds a lot like simple logistic growth. A closed contiguous population, spatial or not, reproduces in proportion to its density, and is limited by its own density. Here we will make the connection a little more clear. It turns out that a simple rearrangement of eq. 4.6 will provide the explicit connection between logistic growth and the spatial population model with internal colonization [181].

Imagine for a moment that you are an avid birder following a population of Song Sparrows in Darrtown, OH, USA (Fig. 3.1a). If Song Sparrows are limited by the number of territories, and males are competing for territories,

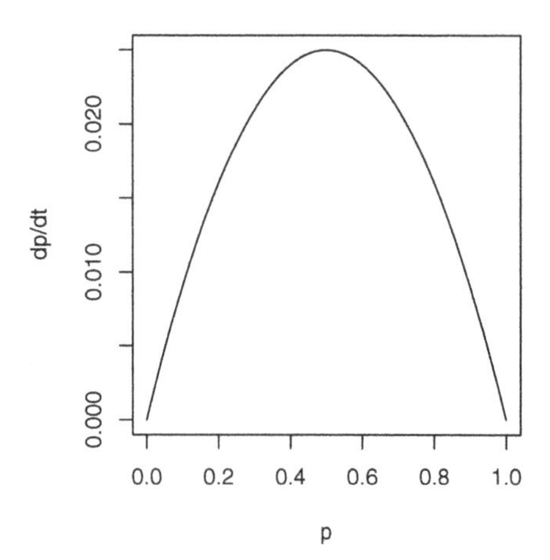

Fig. 4.6: Metapopulation growth rate as a function of p, in the core-satellite model ($c_i = 0.15$, $e = 0.05$). When we plot population growth rate for the core-satellite model, for arbitrary parameter values where $c_i > e$, we see that growth rate falls to zero at full occupancy (i.e., at $p^* = 1$). We also see that the slope is negative, indicating that this equilibrium is stable.

then you could think about male Song Sparrows as "filling up" some proportion, p, of the available habitat. You have already described this population with the logistic growth model ($\mathrm{d}N/\mathrm{d}t = rN(1 - \alpha N)$). Lately, however, you have been thinking about how territories, spatially arranged in the landscape, may limit this population. You therefore decide that you would like to use Levins' spatially-implicit metapopulation model instead (eq. 4.6). How will you do it? You do it by *rescaling* logistic growth.

Let us start by defining our logistic model variables in other terms. First we define N as

$$N = pH$$

where N is the number of males defending territories, H is the total number of possible territories, and p is the proportion of possible territories occupied at any one time. At equilibrium, $N^* = K = p^*H$, so $\alpha = 1/(p^*H)$. Recall that for the Levins model, $p^* = (c_i - e)/c_i$, so therefore,

$$\alpha = \frac{c_i}{(c_i - e)\,H}.$$

We now have N, α, and K in terms of p, H, c_i and e, so what about r? Recall that for logistic growth, the per capita growth rate goes to r as $N \rightarrow 0$ (Chapter 3). For the Levins metapopulation model, the per patch growth rate is

$$\frac{1}{p}\frac{\mathrm{d}p}{\mathrm{d}t} = c_i(1 - p) - e. \tag{4.19}$$

As $p \rightarrow 0$ this expression simplifies to $c_i - e$, which is equivalent to r. Summarizing, then, we have,

$$r = c_i - e \tag{4.20}$$

$$N = pH \tag{4.21}$$

$$\alpha = \frac{1}{K} = \frac{1}{p^*H} = \frac{c_i}{H(c_i - e)} \tag{4.22}$$

$$\tag{4.23}$$

Substituting into logistic growth ($\dot{N} = rN(1 - \alpha N)$), we now have

$$\frac{\mathrm{d}(pH)}{\mathrm{d}t} = (c_i - e)\,pH\left(1 - \frac{c_i}{H(c_i - e)}Hp\right) \tag{4.24}$$

$$= (c_i - e)\,pH - \frac{c_i - e}{c_i - e}c_i p^2 H \tag{4.25}$$

$$= H(c_i p(1 - p) - ep) \tag{4.26}$$

which is almost the Levins model. If we note that H is a constant, we realize that we can divide both sides by H, ending up with the Levins model eq. 4.6.

4.5 Habitat Destruction

Other researchers have investigated effects of habitat loss on metapopulation dynamics [88, 146, 202]. Taking inspiration from the work of Lande [95, 96], Karieva and Wennergren [88] modeled the effect of habitat destruction, D, on overall immigration probability. They incorporated this into Levins' model as

$$\frac{\mathrm{d}p}{\mathrm{d}t} = c_i p(1 - D - p) - ep \tag{4.27}$$

where D is the amount of habitat destroyed, expressed as a fraction of the original total available habitat.

Habitat destruction model

To turn eq. 4.27 into a function we can use with `ode`, we have,

```
> lande <- function(t, y, parms) {
+     p <- y[1]
+     with(as.list(parms), {
+         dp <- ci * p * (1 - D - p) - e * p
+         return(list(dp))
+     })
+ }
```

Habitat destruction, D, may vary between 0 (= Levins model) to complete habitat loss 1.0; obviously the most interesting results will come for intermediate values of D (Fig. 4.7).

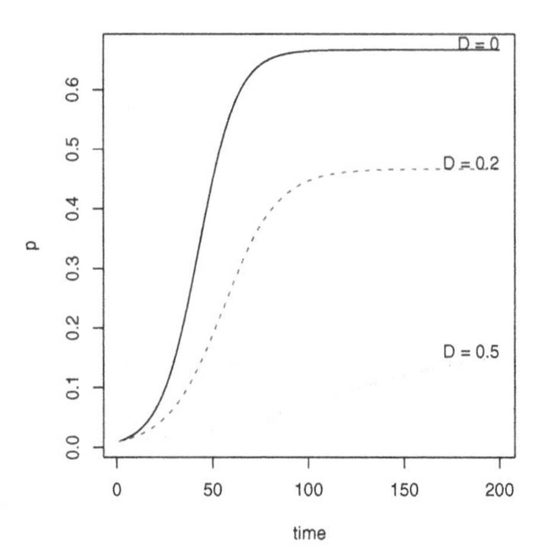

Fig. 4.7: Metapopulation dynamics, combining the Levins model and habitat destruction ($c_i = 0.15$, $e = 0.05$).

Illustrating the effects of habitat destruction (Fig. 4.7)

We can plot the dynamics for three levels of destruction, including none. We first set all the parameters, and time.

```
> library(deSolve)
> prmsD <- c(ci = 0.15, e = 0.05, D = 0)
> Ds <- c(0, 0.2, 0.5)
> Initial.p <- 0.01
> t <- 1:200
```

We then create an empty matrix of the right size to hold our results, and then integate the ODE.

```
> ps <- sapply(Ds, function(d) {
+     prmsD["D"] <- d
+     ode(y = Initial.p, times = t, func = lande, parms = prmsD)[,
+         2]
+ })
```

Last, we plot it and add some useful labels.

```
> matplot(t, ps, type = "l", ylab = "p", xlab = "time")
> text(c(200, 200, 200), ps[200, ], paste("D = ", Ds, sep = ""),
+     adj = c(1, 0))
```

What is the equilibrium under this model? Setting eq. 4.27 to zero, we can then solve for p.

$$0 = c_i - c_iD - c_ip - e \tag{4.28}$$

$$p^* = \frac{c_i - c_iD - e}{c_i} = 1 - \frac{e}{c_i} - D \tag{4.29}$$

Thus we see that habitat destruction has a simple direct effect on the metapopulation.

A core-satellite habitat loss scenario

Let us return now to that odd, but logical, possibility in the core-satellite model where $c_i < e$ and $p = 1$. Recall that in this case, $p = 1$ is an *unstable* equilibrium ($p = 0$ is the stable equilibrium for $c_i < e$). We discuss this in part for greater ecological understanding, but also to illustrate why theory is sometimes useful — because it helps us explore the logical consequences of our assumptions, even when, at first blush, it seems to make little sense.

Imagine that at one time, a metapopulation is regulated by the mechanisms in the core-satellite model, including the rescue effect, and $c_i > e$. We therefore pretend that, the metapopulation occupies virtually every habitable site (let $p = 0.999$). Now imagine that the environment changes, causing $c_i < e$. Perhaps human urbanization reduces colonization rates, or climate change enhances extinction rates. All of a sudden, our metapopulation is poised on an unstable equilibrium. What will happen and how might it differ with and without the rescue effect?

When $c_i > e$, we see that $p^* = 1$ is the stable attractor (Fig. 4.8). However, when $c_i < e$, we see the inevitable march toward extinction predicted by the Hanski model (core-satellite) (Fig. 4.8). Last, when we compare it to the Levins model, we realize something somewhat more interesting. While the Levins model predicts very rapid decline, the Hanski model predicts a much more gradual decline *toward extinction.* Both models predict extinction, but the rescue effect delays the appearance of that extinction. It appears that the rescue effect (which is the difference between the two models) may act a bit like the "extinction debt" [202] wherein deterministic extinction is merely delayed, but not postponed indefinitely. Perhaps populations influenced by the rescue effect might be prone to unexpected collapse, if the only stable equilibria are 1 and 0. Thus simple theory can provide interesting insight, resulting in very different predictions for superficial similar processes.

The unexpected collapse of core populations

Here we plot the dynamics of metapopulations starting at or near equilbrium. The first two use the Hanski model, while the third uses Levins. The second and third use $c_i < e$.

```
> C1 <- ode(y = 0.999, times = t, func = hanski, parms = c(ci = 0.2,
+     e = 0.01))
> C2 <- ode(y = 0.999, times = t, func = hanski, parms = c(ci = 0.2,
+     e = 0.25))
> L2 <- ode(y = 0.95, times = t, func = levins, parms = c(ci = 0.2,
+     e = 0.25))
```

Next, we plot these and add a legend.

```
> matplot(t, cbind(C1[, 2], C2[, 2], L2[, 2]), type = "l",
+     ylab = "p", xlab = "Time", col = 1)
> legend("right", c("c > e", "c < e", "c < e (Levins)"), lty = 1:3,
+     bty = "n")
```

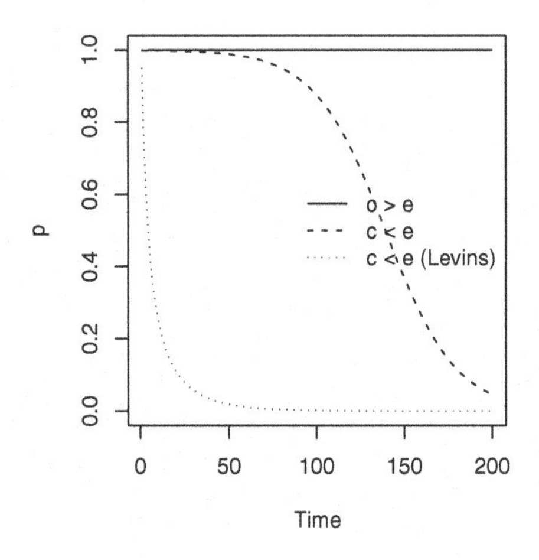

Fig. 4.8: Metapopulation dynamics, starting from near equilibrium for $c_i = 0.20$ and $e = 0.01$. If the environment changes, causing extinction rate to increase until it is greater than colonization rate, we may observe greatly delayed, but inevitable, extinction (e.g., $c_i = 0.20, e = 0.25$).

4.6 Core-Satellite Simulations

Here[1] we explore a simple question that Hanski posed long ago: what would communities look like if all of the populations in the community could be de-

[1] This section relies extensively on code

scribed by their independent own core-satellite model? To answer this question, he created communities as collections of independent (non-interacting) populations that behave according to his metapopulation model with internal colonization and the rescue effect [70]. He found that such simulated communities predict that many species will be in almost all sites ("core species"), and even more species will exist at very few sites ("satellite species"). This seems to be a relatively common phenomenon [35], and an observation we described at the beginning of the chapter (Fig. 4.1).

Hanksi's goal was to simulate simultaneously a substantive number of species, to create a community. Each species is assumed to be governed by internal propagule production only, and the rescue effect. Further, he assumed that the long term average density independent growth rate ($r = c_i - e$) was zero. That is, the populations were not *systematically* increasing or decreasing. However, he allowed for stochastic year-to-year variation in probabilities c_i and e.

In these simulations here, we will select the mean for each parameter, c_i and e, and the proportion, ϕ ("phi") by which they are allowed to vary. The realized values of $c_{i,t}$ and e_t at any one point in time are random draws from a uniform distribution within the ranges $i \pm \phi i$ and $e \pm \phi e$. (This requires that we do numerical integration at each integer time step since there is no obvious analytical solution to an equation in which the parameters vary through time. This will keep these parameters constant for an entire year, and yet also allow years to vary.)

We start by using the `args()` function to find out what arguments (i.e. options) are available in the simulation function, `MetaSim`.

```
> args(MetaSim)
```

```
function (Time = 50, NSims = 1, method = "hanski", ci = 0.25,
    e = 0.25, phi = 0.75, p0 = 0.5, D = 0.5)
NULL
```

What options (or arguments) can you vary in `MetSim`? The 'method' may equal `CoreSatellite`, `Levins`, `IslandMainland`, or `HabitatDestruction`. The default is `CoreSatellite`; if an argument has a value to begin with (e.g. `method='CoreSatellite'`), then it will use that value unless you replace it.

Let's start with an initial run of 10 simulations (produces dynamics for 10 populations) to reproduce Hanski's core-satellite pattern by using the rescue effect with equal i and e.

```
> out.CS.10 <- MetaSim(method = "hanski", NSims = 10)
> matplot(out.CS.10$t, out.CS.10$Ns, type = "l", xlab = "Time",
+     ylab = "Occupancy", sub = out.CS.10$method)
```

These dynamics (Fig. 4.9) appear to be completely random. A *random walk* is a dynamic that is a random increase or decrease at each time step. Such a process is not entirely random because the abundance at time t is related to the abundance at time $t-1$, so observations in random walks are correlated through time; they are temporally autocorrelated.

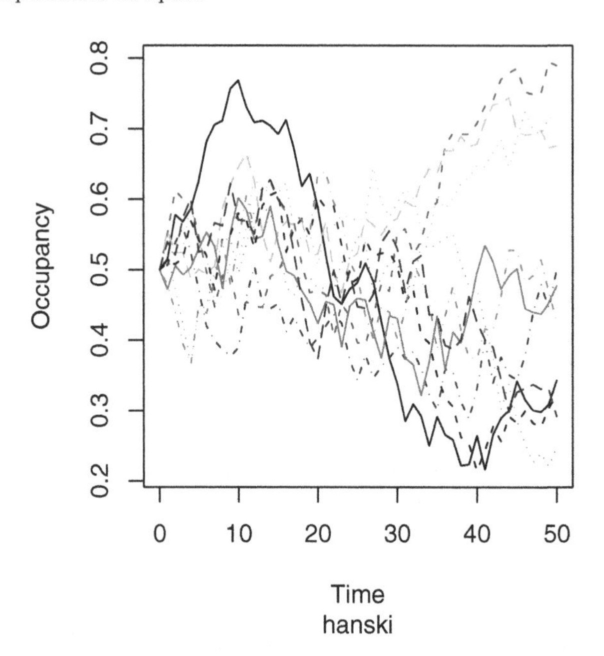

Fig. 4.9: Core-satellite species dynamics with stochasticity ($\bar{i} = \bar{e} = 0.2$).

Does a single metapopulation growth rate appear related to p, the metapopulation size? What would a deterministic dynamic look like if $c_i > e$? It would increase rapidly at first, and then slow down as it approached 1.0. Can you detect that slow-down here? Similarly, as a metapopulation declines toward extinction, its progression toward $p = 0$ slows down. As a result, we tend to accumulate a lot of common and rare species, for which p is close to one or zero.

Now we will do more simulations (50 species), and run them for longer (500 time intervals *vs.* 50). Doing many more simulations will take a little longer, so be patient[2].

```
> system.time(out.CS.Lots <- MetaSim(method = "hanski", NSims = 50,
+     Time = 1000))

   user  system elapsed
 49.628   0.112  49.737
```

time series, although this may not tell you much. Alternatively, we can plot a histogram of the 50 species' final abundances, at $t = 500$.

```
> hist(out.CS.Lots$Ns[501, ], breaks = 10, main = NULL,
+     xlab = expression("Occupancy (" * italic("p") * ")"),
+     ylab = "Number of Species",
+     sub = paste(out.CS.Lots$method, " Model", sep = ""))
```

[2] `system.time` merely times the process, in secs.

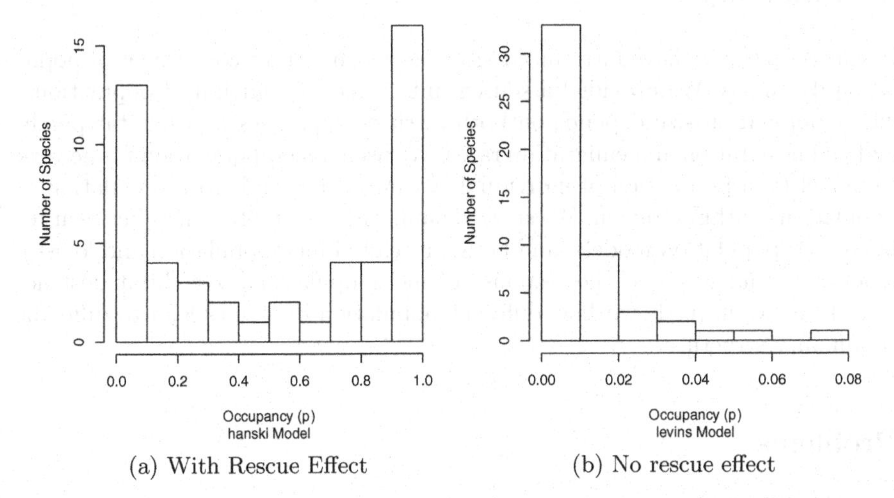

(a) With Rescue Effect (b) No rescue effect

Fig. 4.10: The species-abundance distribution resulting from dynamics for 50 independent metapopulations with internal colonization. (a) includes the rescue effect (Hanski's model), and note that most species are either common ($p > 0.8$) or rare ($p < 0.2$). Levins model (b) does not include the rescue effect, and there are very few core species ($p > 0.8$).

Our simulations (Fig. 4.10) should be consistent with the core-satellite hypothesis — are they? In Hanski's model, we see that most metapopulations are either core species ($p > 0.8$) or satellite species ($p < 0.2$) (Fig. 4.10a). This is not to imply that there should be hard rules about what constitutes a core and satellite species, but rather merely shows we have a plethora of both common and uncommon species.

What does the Levins model predict? Let's run the simulations and find out.

```
> system.time(out.L.Lots <- MetaSim(NSims = 50, Time = 500,
+     method = "levins"))

  user  system elapsed
23.921   0.036  23.958
```

Now we plot a histogram of the 50 species' final abundances, at $t = 500$.

```
> hist(out.L.Lots$Ns[501, ], breaks = 10,
+     xlab = expression("Occupancy (" * italic("p") * ")"),
+     ylab = "Number of Species", main = NULL,
+     sub = paste(out.L.Lots$method, " Model", sep = ""))
```

In contrast to the core-satellite model, the Levins model predicts that many fewer species are common (Fig. 4.10b). Thus these two population models make contrasting predictions regarding the structure of communities (i.e. relative species abundances), and provide testable alternatives [35].

4.7 Summary

In this chapter, we have introduced *space* as an important component of population dynamics. We provided a source-sink framework for linked populations, where population size depends on both intrinsic capacities of a habitat patch, and on immigration and emigration rates. We used a metapopulation framework to model (i) a population of individuals within a site, and (ii) a population of populations within a region. We showed similarities and differences between related metapopulation models, and between related metapopulation and logistic models. We investigated the response of metapopulations to habitat destruction. Last, we have shown how different population dynamics lead to different community structure.

Problems

4.1. Equilibria

Derive expressions and calculate equilibria for the following metapopulation models, with $c_i = 0.05$, $e = 0.01$. Show your work — start with the differential equations, set to zero, and solve p^*; then substitute in values for c_i, e.
(a) Levins model.
(b) Propagule rain model (`gotelli`).
(c) Propagule rain model that also includes both external and internal propagule production and dispersal.
(d) Hanski model.
(e) Lande (habitat destruction) model (with `D=0.1`).

4.2. Habitat destruction

Compare different levels of habitat destruction.
(a) Use the habitat destruction model (`lande`) to compare 9 levels of destruction (`ds <- seq(0,.8, by=.1)`), using $c_i = 0.1$, $e = 0.01$. Plot of graph of the dynamics through time, and calculate the equilibria directly.
(b) Write an ODE function for a habitat destruction model with rescue effect. Let the "rescue" have an additional parameter, a, such that extinction rate is $ep(1 - ap)$.
(c) Let $D = 0.5$, $c_i = 0.1$, $e = 0.02$, and vary a over five levels (including $a = 0,\ 1$) to investigate the effects of "relative rescue effect" on the equilibria and dynamics of a metapopulation.

Part II

Two-species Interactions

5

Lotka–Volterra Interspecific Competition

Different species frequently compete for limiting resources, and as a result have negative impacts on each other. For example, change in species composition during secondary succession (Fig. 5.1) appears mediated by, among other things, a species' ability to intercept light, grow, and cast shade over other species. This chapter addresses very simple ways to represent such interactions between species.

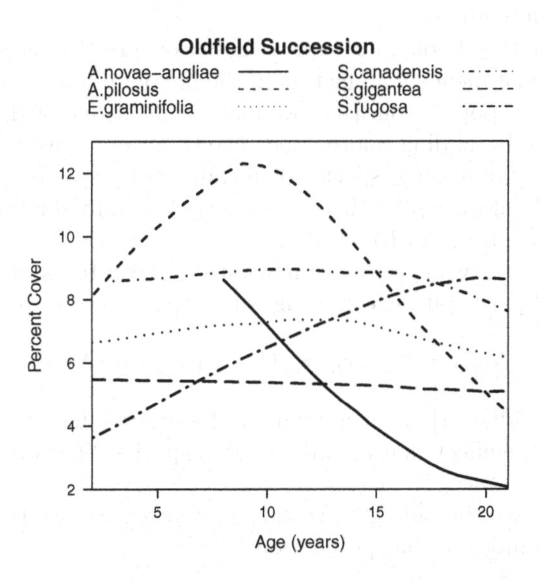

Fig. 5.1: Changes in abundances of six species of *Aster*, *Euthamia*, and *Solidago* during early secondary succession. This turnover of perennial weeds appears mediated by competition for light, among other factors [7] (data from the long-term Buell-Small Succession study [http://www.ecostudies.org/bss/]).

5.1 Discrete and Continuous Time Models

Chapter 3 (Density-dependent Growth) was built upon the assumption that individuals *within a single population* had negative effects on each other. Here we assume that individuals of different species also have negative effects on each other. This negative effect may arise as a direct effect via their behavior, or indirect effects via their uptake of limiting resources.

In this chapter, we learn how to keep track of two species that compete, that is, that have mutually negative effects upon each other. We begin with a model of discrete growth, and then switch to a continuous growth model.

5.1.1 Discrete time model

We pick up from Chapter 3 with the discrete logistic growth model

$$N_{t+1} = N_t + r_d N_t (1 - \alpha N_t) \tag{5.1}$$

where the population size in one year, N_{t+1}, is equal to the previous year's population size, N_t, plus a growth increment. That growth increment includes a proportional change, the discrete growth factor, r_d. Last, we have the density dependence term, $(1-\alpha N_t)$, in which α is the per capita effect of each individual upon all other individuals.

In Part 1 of this book, per capita effects on growth rates attempted to encapsulate simulataneously all factors in the life of an organism that influence the growth of that population. Here we make explicit one of those many other negative impacts by adding another per capita effect — we add the negative effect of another, competing, species. If members of a species' own population can have a per capita negative effect, then certainly individuals of other species might have per capita negative effects.

Because we have two species, we now have to keep track of their particular populations and per capita effects using subscripts. We now have

$$N_{1,t+1} = N_{1,t} + r_{1,d} N_{1,t} (1 - \alpha_{11} N_{1,t} - \alpha_{12} N_{2,t}), \tag{5.2}$$

where α_{11} is the effect that an individual of species 1 has on its own growth rate, and α_{12} is the effect that an individual of species 2 has on the growth rate of species 1 (Fig. 5.2).

Now that we are including a second population, we need an equation describing the dynamics of that population

$$N_{2,t+1} = N_{2,t} + r_{2,d} N_{2,t} (1 - \alpha_{21} N_{1,t} - \alpha_{22} N_{2,t}), \tag{5.3}$$

where α_{21} is the per capita effect of species 1 on species 2, and α_{22} is the per capita effect that species 2 has on itself (Fig. 5.2).

Code for a model of discrete logistic competition

This will calculate N_{t+1}, given N_t, r_d and a matrix of competition coefficients $\boldsymbol{\alpha}$.

```
> dlvcomp2 <- function(N, alpha, rd = c(1, 1)) {
+     N1.t1 <- N[1] + rd[1] * N[1] * (1 - alpha[1, 1] * N[1] -
+         alpha[1, 2] * N[2])
+     N2.t1 <- N[2] + rd[2] * N[2] * (1 - alpha[2, 1] * N[1] -
+         alpha[2, 2] * N[2])
+     c(N1.t1, N2.t1)
+ }
```

Note the indices for `alpha` match the subscripts in eqs. 5.2, 5.3.

5.1.2 Effects of α

Before proceeding it might be good to understand where these subscripts come from. Why are they read from right to left — why not left to right? As we saw in Chapter 2, it comes from the underlying linear algebra used to work with all these equations. We define all of the α's together as a matrix,

$$\boldsymbol{\alpha} = \begin{pmatrix} \alpha_{11} & \alpha_{12} \\ \alpha_{21} & \alpha_{22} \end{pmatrix} = \begin{pmatrix} 0.010 & 0.005 \\ 0.008 & 0.010 \end{pmatrix} \tag{5.4}$$

The subscripts on the αs represent the row and column of the coefficient; α_{12} is in the first row, second column. This merely reflects how mathematicians describe matrix elements and dimensions — row × column. When we use matrix multiplication (Chapter 2), α_{12} becomes the effect of species 2 (column) on species 1 (row). In this case, $\alpha_{11} = \alpha_{22} = 0.01$, $\alpha_{21} = 0.008$, and $\alpha_{12} = 0.005$. Thus, both species have greater effects on themselves than on each other. Remember, the larger the α, the larger the effect.

You can think of a row of coefficients as part of a growth rate equation that includes those coefficients, so row 1 represents the equation governing the growth rate of species 1, and using α_{11} and α_{12}. In general, we can refer to the intraspecific coefficients as α_{ii} and the interspecific coefficients as α_{ij}.

If we model two species using $\boldsymbol{\alpha}$ above (eq. (5.4)), we can see the negative effects of interspecific competition (Fig. 5.2). Species 1 (solid line, Fig. 5.2) has a larger negative affect on species 2 than species 2 has on species 1. As a result, species 1 is able to maintain a positive growth rate at higher population sizes, and thereby grow to a larger N than can species 2. Put another way, species 1 suppresses species 2 to a larger degree than vice versa. Nonetheless, each has a smaller effect on the other than either does on itself ($\alpha_{ij} < \alpha_{ii}$).

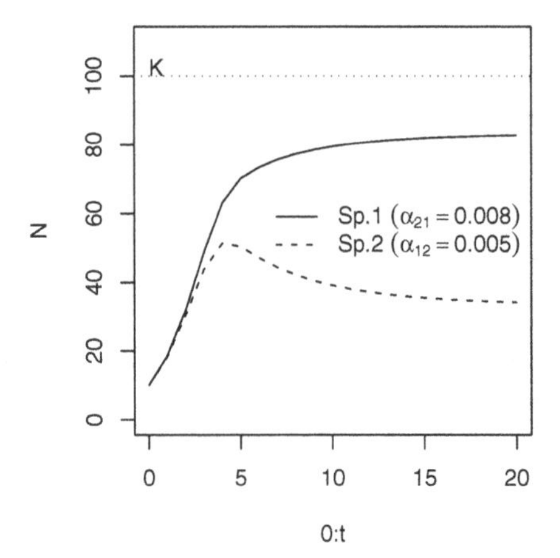

Fig. 5.2: Discrete population growth of two competing species. Both species have the same intraspecific competition coefficient, $\alpha_{ii} = 0.01$. In the absence of interspecific competition, both species would reach K. However, they both have negative effects on each other ($\alpha_{ij} > 0$), and species 1 (solid line) has a greater negative effect on species 2 ($\alpha_{21} > \alpha_{12}$).

Discrete logistic competition dynamics (Fig. 5.2)

First we specify the matrix of α's, the effects each species has on itself and each other, the initial population sizes, and the number of time steps.

```
> alphs <- matrix(c(0.01, 0.005, 0.008, 0.01), ncol = 2, byrow = TRUE)
> t <- 20
```

We then create a matrix to hold the results, put in the initial population sizes, and project the populations.

```
> N <- matrix(NA, nrow = t + 1, ncol = 2)
> N[1, ] <- c(10, 10)
> for (i in 1:t) N[i + 1, ] <- dlvcomp2(N[i, ], alphs)
```

At last, we can plot the populations, adding a reference line for the size of the populations, if there were only one species, at $K_i = 1/\alpha_{ii}$.

```
> matplot(0:t, N, type = "l", col = 1, ylim = c(0, 110))
> abline(h = 1/alphs[1, 1], lty = 3)
> text(0, 1/alphs[1, 1], "K", adj = c(0, 0))
> legend("right", c(expression("Sp.1 " * (alpha[21] == 0.008)),
+     expression("Sp.2 " * (alpha[12] == 0.005))), lty = 1:2,
+     bty = "n")
```

5.1.3 Continuous time model

Perhaps the classic model of competitive interactions is the continuous Lotka-Volterra model of interspecific competition [93]. Following directly the structure of the discrete version, we represent the two species as

$$\frac{dN_1}{dt} = r_1 N_1 (1 - \alpha_{11} N_1 - \alpha_{12} N_2) \qquad (5.5)$$

$$\frac{dN_2}{dt} = r_2 N_2 (1 - \alpha_{21} N_1 - \alpha_{22} N_2) \qquad (5.6)$$

where we interpret all the parameters as the instantaneous rates analogous to the parameters in the discrete version above, but with different units, because the effects are instantaneous, rather than effects over a given time interval (Table 5.1).

Table 5.1: Parameters of the continuous 2 species Lotka-Volterra competition model. The rest of the chapter explains the meaning and motivation.

Parameter	Description
r_i	Instantaneous rate of increase; intrinsic rate of growth; individuals produced per individual per unit time.
α_{ii}	Intraspecific density dependence; intraspecific competition coefficient; the negative effect of an individual of species i on its own growth rate.
α_{ij}	Interspecific density dependence; interspecific competition coefficient; the effect of interspecific competition; the negative effect that an individual of species j has on the growth rate of species i.
K_i	$1/\alpha_{ii}$; *carrying capacity* of species i; the population size obtainable by species i in the absence of species j.
α'_{ij}	α_{ij}/α_{ii}; the relative importance of interspecific competition.
β_{ij}	α_{ij}/α_{jj}; the *invasion criterion* for species i; the relative importance of interspecific competition; the importance of the effect of species j on species i relative to the effect of species j on itself; see sec. 5.3.5.

Continuous logistic competition

Here we simply write the code for 2-species Lotka-Volterra competition.

```
> lvcomp2 <- function(t, n, parms) {
+     with(as.list(parms), {
+         dn1dt <- r1 * n[1] * (1 - a11 * n[1] - a12 * n[2])
+         dn2dt <- r2 * n[2] * (1 - a22 * n[2] - a21 * n[1])
+         list(c(dn1dt, dn2dt))
+     })
+ }
```

We could then use this to numerically integrate the dynamics, using `ode` in the `deSolve` package, and plot it (graph not shown).

```
> library(deSolve)
> parms <- c(r1 = 1, r2 = 0.1, a11 = 0.2, a21 = 0.1, a22 = 0.02,
+     a12 = 0.01)
> initialN <- c(2, 1)
> out <- ode(y = initialN, times = 1:100, func = lvcomp2, parms = parms)
> matplot(out[, 1], out[, -1], type = "l")
```

These equations are also commonly represented using carrying capacity, K_i, to summarize intraspecific effects on abundance, and coefficients to modify the intraspecific effects quantified with $K_i = 1/\alpha_{ii}$. This representation looks like

$$\frac{\mathrm{d}N_1}{\mathrm{d}t} = r_1 N_1 \left(\frac{K_1 - N_1 - \alpha'_{12} N_2}{K_1} \right) \tag{5.7}$$

$$\frac{\mathrm{d}N_2}{\mathrm{d}t} = r_2 N_2 \left(\frac{K_2 - N_2 - \alpha'_{21} N_1}{K_2} \right). \tag{5.8}$$

In this version, $K_1 = 1/\alpha_{11}$, and *note that* α'_{12} *differs from* α_{12}. The α'_{12} in eq. 5.7 merely modifies the effect of $1/K_1$. It turns out the $\alpha'_{1,2}$ is equal to the *ratio* of the interspecific and intraspecific per capita effects, or

$$\alpha_{12} = \frac{\alpha'_{12}}{K_1}$$
$$\alpha'_{12} = \frac{\alpha_{12}}{\alpha_{11}}. \tag{5.9}$$

Another useful measure of the relative importance of interspecific competition is $\beta_{ij} = \alpha_{ij}/\alpha_{jj}$ (see *Invasion criteria*, below).

5.2 Equilbria

In this section, we describe how we find equilibria in a simple multispecies model, by solving the growth equations for zero, much the way we did in Chapters 3 and 4. We begin with *isoclines*, and move on to *boundary and internal equilibria* and *invasion criteria*.

5.2.1 Isoclines

An isocline is, in general, a line connecting points on a graph or map that have equal value. For instance, a topographic map has elevation isoclines, connecting points of equal elevation. Here, our isoclines will connect points in state space at which the growth rate for species i equals zero — every point on that line will represent $\dot{N}_i = 0$. We call these *zero net growth isoclines.*

A *zero net growth isocline*, typically referred to simply as an *isocline* or ZNGI, is the set of all points for which the growth of a population is zero, when all else (such as the population size of a competing species) is held constant. An *equilibrium* is one (or sometimes more than one) of those points, and in particular, it is a point at which the growth rates of *all* populations are zero. You saw in Chapter 3 that the carrying capacity of a single species logistic population is an equilibrium. With two species, it gets a tad trickier.

We find the isocline of a species by setting its growth rate equal to zero and solving the equation for that species in terms of the other species. As an example, let's focus on N_2 and solve for its zero net growth isocline. We find below that there is a straight line that describes all points for which $\mathrm{d}N_2/dt = 0$, if N_1 were held constant. We start by setting eq. 5.6 to zero, and solving for N_2.

$$\begin{aligned}\frac{\mathrm{d}N_2}{\mathrm{d}t} &= r_2N_2\left(1 - \alpha_{21}N_1 - \alpha_{22}N_2\right)\\ 0 &= r_2N_2\left(1 - \alpha_{21}N_1 - \alpha_{22}N_2\right)\\ 0 &= 1 - \alpha_{21}N_1 - \alpha_{22}N_2\\ N_2 &= \frac{1}{\alpha_{22}} - \frac{\alpha_{21}}{\alpha_{22}}N_1.\end{aligned} \tag{5.10}$$

Recall that the formula for a straight line is $y = mx + b$ where m is the slope and b is the intercept on the y axis. We can see that the expression for N_2 in eq. 5.10 is a straight line, where $y = N_2$, $m = \alpha_{21}/\alpha_{22}$, and $b = 1/\alpha_{22}$ (Fig. 5.3a). When N_1 is zero, $N_2 = 1/\alpha_{22}$. This is precisely what we saw in Chapter 3 (logistic growth), that a single species logistic population has an equilibrium at its carrying capacity, $K = 1/\alpha$.

The isocline (Fig. 5.3a) shows that as the competitor's population size, N_1, becomes larger, N_2 declines by α_{21}/α_{22} for each additional individual of the competing species 1, until finally $N_2 = 0$ when $N_1 = 1/\alpha_{21}$.

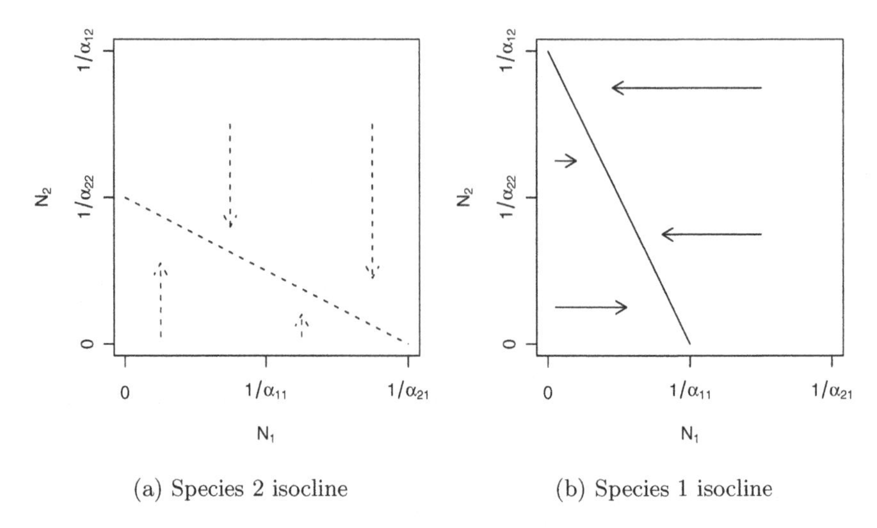

(a) Species 2 isocline

(b) Species 1 isocline

Fig. 5.3: Phase plane plots of each of the two competing species, with Lotka-Volterra zero growth isoclines. Arrows indicate population trajectories. Recall $K_i = 1/\alpha_{ii}$.

Graphing an Isocline

Here we graph something similar, but not identical, to Fig. 5.3a. First, we define a new matrix of competition coefficients, where $\alpha_{11} = \alpha_{22} > \alpha_{12} = \alpha_{21}$.

```
> a <- matrix(c(0.01, 0.005, 0.005, 0.01), ncol = 2, byrow = TRUE)
```

We create an expression to plot the N_2 isocline, as a function of possible values of N_1.

```
> N2iso <- expression(1/a[2, 2] - (a[2, 1]/a[2, 2]) * N1)
```

We then specify N_1, and then evaluate and plot N_2.

```
> N1 <- 0:200
> plot(N1, eval(N2iso), type = "l", ylim = c(0, 200), xlim = c(0,
+     200), ylab = expression("N"[2]))
```

We add arrows to remind us of what happens if N_2 is above or below the value on the isocline.

```
> arrows(x0 = 90, y0 = 150, x1 = 90, y1 = 80, length = 0.1)
> arrows(x0 = 75, y0 = 0, x1 = 75, y1 = 50, length = 0.1)
```

The isocline for N_2 (Fig. 5.3a) is the line at which $dN_2/dt = 0$ *for a fixed value of* N_1. Just as in the single species logistic growth model, if N_2 exceeds its equilibrium, it declines, and if N_2 is less than its equilibrium, it grows. The isocline (Fig. 5.3a) is the set of balance points between positive and negative growth. This is reflected in the arrows in Fig. 5.3a — if the N_2 is ever above this isocline, it declines and if it is ever below this isocline, it rises. This isocline shows that whether N_2 increases or decreases depends on N_1.

By analogy, the isocline for species 1 turns out to be

$$N_1 = \frac{1}{\alpha_{11}} - \frac{\alpha_{12}}{\alpha_{11}} N_2. \tag{5.11}$$

Note that these isoclines are merely equations for straight lines, and it is easy to do nonsensical things, such as specify coefficients that result in negative population sizes. Therefore, let us proceed with some thoughtfulness and care.

5.2.2 Finding equilibria

By themselves, the isoclines tell us that if species 2 becomes extinct ($N_2 = 0$), then species 1 reaches its carrying capacity ($N_1 = 1/\alpha_{11}$) (Fig. 5.3b). Similarly, if $N_1 = 0$, then $N_2 = 1/\alpha_{22}$. These are important equilibria, because they verify the internal consistency of our logical, and they provide end-points on our isoclines.

If the species coexist (N_1, $N_2 > 0$) it means that they must share one or more points on their isoclines — such an equilibrium is the point where the lines cross. We find these equilibria by solving the isoclines simultaneously. A simple way to do this is to substitute the right hand side of the N_2 isocline (eq. 5.10) in for N_2 in the N_1 isocline (eq. 5.11). That is, we substitute an isocline of one species in for that species' abundance in another species' isocline. Combining eqs. 5.10 and 5.11, we get

$$\begin{aligned} N_1 &= \frac{1}{\alpha_{11}} - \frac{\alpha_{12}}{\alpha_{11}} \left(\frac{1}{\alpha_{22}} - \frac{\alpha_{21}}{\alpha_{22}} N_1 \right) \\ N_1 &= \frac{1}{\alpha_{11}} - \frac{\alpha_{12}}{\alpha_{11}\alpha_{22}} + \frac{\alpha_{12}\alpha_{21}}{\alpha_{11}\alpha_{22}} N_1 \\ N_1 \left(1 - \frac{\alpha_{12}\alpha_{21}}{\alpha_{11}\alpha_{22}} \right) &= \frac{\alpha_{22} - \alpha_{12}}{\alpha_{11}\alpha_{22}} \\ N_1^* &= \frac{\alpha_{22} - \alpha_{12}}{\alpha_{11}\alpha_{22} - \alpha_{12}\alpha_{21}} \end{aligned} \tag{5.12}$$

When we do this for N_2, we get

$$N_2^* = \frac{\alpha_{11} - \alpha_{21}}{\alpha_{22}\alpha_{11} - \alpha_{12}\alpha_{21}} \tag{5.13}$$

We now have the values for N_1^* and N_2^* at the point at which their isoclines cross (Fig. 5.4a). These equilibria apply only when isoclines cross within feasible state space.

The expressions for N_1^* and N_2^* look pretty complicated, but can we use them to discern an intuitive understanding for species 1? First, we see that r_i is not in the expressions for the equilibria — they do not depend on r_i. It is important to remember that this intrinsic rate of increase is not germaine to the long term equilibria for the two species model. Second, we can confirm that as interspecific competition intensity falls to zero ($\alpha_{12} = \alpha_{21} = 0$), each species reaches its own carrying capacity. That is, when putative competitors occupy sufficiently different niches and no longer compete, then they both reach their own carrying capacities.

We can also say something a little less obvious about species 1. What happens when the negative effect of the competitor, N_2, starts to increase, that is, as α_{12} gets bigger? Or, put more obtusely but precisely, let's find

$$N_1^* = \lim_{\alpha_{12}\to\infty} \frac{\alpha_{22} - \alpha_{12}}{\alpha_{22}\alpha_{11} - \alpha_{12}\alpha_{21}} \tag{5.14}$$

that is, find the limit of the equilibrium (eq. 5.12) as α_{12} gets very large. Well, the α_{ii} become effectively zero because α_{12} gets so big. This leaves $-\alpha_{12}/(-\alpha_{12}\alpha_{21}) = 1/\alpha_{21}$. Huh? This means simply that as the negative effect of the competitor increases, the abundance of species 1 becomes increasingly dependent upon α_{21}, its negative effect on its competitor. Thus we have an arms race: as the negative effect of its competitor increases, the abundance of a species depends increasingly on its ability to suppress the competitor.

Summarizing, we see that in the absence of interspecific competition, species are attracted toward their carrying capacities. Second, if interspecific competition is intense, then a species' carrying capacity becomes less important, and its abundance is increasingly determined by its ability to suppress its competitor.

Coexistance — the invasion criterion

Based on the numerators in eqs. 5.12 and 5.13, it seems that N_1^* and N_2^* may both be greater than zero whenever $\alpha_{ii} - \alpha_{ji} > 0$. This is, indeed, the case. Below we step through the analysis of what we refer to as the "invasion criterion," which specifies the conditions for $N_i^* > 0$.

In general, the details of any model and its dynamics may be quite complicated, but as long as we know whether a species will always *increase when it is rare*, or *invade*,[1] then we know whether it can persist in the face of complex interactions. Thus we don't need to find its equilibrium, but merely its behavior near zero.

How do we determine whether a species can increase when rare? Let's explore that with the Lotka-Volterra competition model. We can start by reexamining species 1's growth equation (eq. 5.5)

$$\frac{dN_1}{dt} = r_1 N_1 (1 - \alpha_{11}N_1 - \alpha_{12}N_2).$$

From this we can see that in the absence of any density dependence ($\alpha = 0$) and assuming $r_1 > 0$, and $N_1 > 0$, the population grows exponentially. Further, we can see that $dN/dt > 0$ as long as $(1 - \alpha_{11}N_1 - \alpha_{12}N_2) > 0$. Therefore, let's examine this expression for density dependence and see what it looks like as N_1 gets very close to zero. We start by expressing dN_1/dt completely in terms of N_1 and α. We do this by substituting N_2's isocline (eq. 5.10) in place of N_2 in eq. 5.5. We then solve this for any growth greater than zero.

[1] Note that ecologists who study invasion of exotic species may use the word "invade" to mean the successful dispersal to, establishment and spread at a site. This incorporates a tremendous amount of species- or system-specific biology. Here we mean "invasion" in only a much more strict or narrow sense — to increase when rare.

$$0 < (1 - \alpha_{11}N_1 - \alpha_{12}N_2)$$

$$0 < \left(1 - \alpha_{11}N_1 - \alpha_{12}\left(\frac{1}{\alpha_{22}} - \frac{\alpha_{21}}{\alpha_{22}}N_1\right)\right) \tag{5.15}$$

Now — what is the value of eq. 5.15 as $N_1 \to 0$? We can substitute 0 for N_1, and eq. (5.15) becomes

$$0 < \left(1 - \alpha_{12}\left(\frac{1}{\alpha_{22}}\right)\right)$$

$$0 < 1 - \frac{\alpha_{12}}{\alpha_{22}}$$

$$\alpha_{12} < \alpha_{22}. \tag{5.16}$$

What is this saying? It is saying that as long as $\alpha_{12} < \alpha_{22}$, then our focal species can persist, increasing in abundance from near zero — *N_1 will increase when rare*, that is, it will successfully invade (Fig. 5.4a).

For two species to both persist, or coexist, it must be that case that

$$\alpha_{12} < \alpha_{22} \quad , \quad \alpha_{21} < \alpha_{11}.$$

Simply put, *for species to coexist stably, their effects on themselves must be greater than their effects on each other* (Fig. 5.4a).

Other equilibria

Given our isoclines and equilibria above, what other logical combinations might we see, other than coexistence? Here we list others, and provide graphical interpretations (Fig. 5.4).

Species 1 can invade when rare, but species 2 cannot (Fig. 5.4b).

$$\alpha_{12} < \alpha_{22} \quad , \quad \alpha_{21} > \alpha_{11}$$

This leads to competitive exclusion by species 1 — species 1 wins. This is referred to as a *boundary equilibrium*, because it is on the boundary of the state space for one species. Equilibria where all $N_i > 0$ are referred to as *internal equilibria*.

Species 1 cannot invade when rare, but species 2 can (Fig. 5.4c).

$$\alpha_{12} > \alpha_{22} \quad , \quad \alpha_{21} < \alpha_{11}$$

This leads to competitive exclusion by species 2 — species 2 wins. This is the other boundary equilibrium. Note that for both this and the previous boundary equilibrium, the equilibrium equations (eqs. 5.12, 5.13) can return N^* that are negative or too large ($> K$). Recall that these equations derive from simple equations of straight lines, and do not guarantee that they are used sensibly — equations aren't dangerous, theoreticians who use equations are dangerous.

Neither species can invade when rare (Fig. 5.4d).

$$\alpha_{12} > \alpha_{22} \quad , \quad \alpha_{21} > \alpha_{11}$$

This creates an unstable internal equilibrium — exclusion will occur, but either species could win. This condition is sometimes referred to as *founder control* [14] because the identity of the winner depends in part on the starting abundances. It creates a *saddle* in state space. What the heck is a saddle? More on that below. It suffices to say that from some directions, an saddle attracts the trajectories of the populations, while from other directions, it repels the trajectories.[2]

5.3 Dynamics at the Equilibria

Here we use eigenanalysis to analyze the properties of the equilibrium, whether they are attractors, repellers, or both, and whether the system oscillates around these equilibria.

In Chapter 3, we assessed stability with the partial derivative of the growth rate, with respect to population size. If it was negative the population was stable, and the more negative the value, the shorter the return time. Here we build on this, and present a general recipe for stability analysis [142]:

1. Determine the equilibrium abundances of each species by setting its growth equation to zero, and solving for N.
2. Create the Jacobian matrix. This matrix represents the response of each species to changes in its own population and to changes in each other's populations. The matrix elements are the partial derivatives of each species' growth rate with respect to each population.
3. Solve the Jacobian. Substitute the equilibrium abundances into the partial derivatives of the Jacobian matrix to put a real number into each element of the Jacobian matrix.
4. Use the Jacobian matrix to find the behavior of the system near the equilibria. The trace, determinant, and eigenvalues of the Jacobian can tell us how stable or unstable the system is, and whether and how it cycles.

5.3.1 Determine the equilibria

We just did this above. Given eqs. 5.12 and 5.13, we see that the $\boldsymbol{\alpha}$ determine completely N_1^* and N_2^*. *This is not true for Lotka-Volterra systems with more than two species*; such systems also depend on r_i.

[2] The topography of mountain ranges can include saddles, which are precisely the context in which we use "saddle" here. Search the web for Saddleback Mountain, New York, USA, Lat/Long: 44° 8' N; 73° 53' W. See also pictures of horse saddles — same shape.

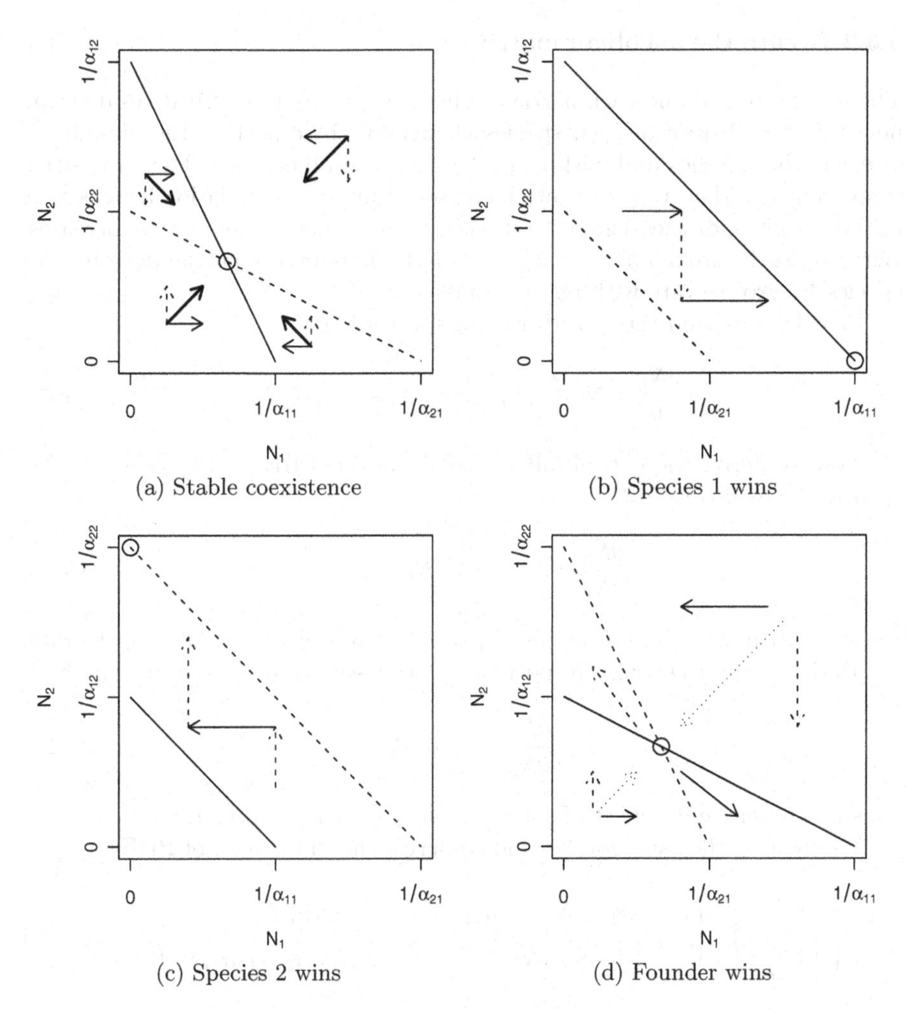

(a) Stable coexistence

(b) Species 1 wins

(c) Species 2 wins

(d) Founder wins

Fig. 5.4: Phase plane diagrams of Lotka-Volterra competitors under different invasion conditions. Horizontal and vertical arrows indicate directions of attraction and repulsion for each population (solid and dased arrows); diagonal arrows indicate combined trajectory. Circles indicate equilibria; additional boundary equilibria can occur whenever one species is zero.

Finding equilibria

We can create equations or *expressions* for the equilibria, N_1^* and N_2^*. These will be symbolic representations that we later evaluate.

```
> N1Star <- expression((a22 - a12)/(a22 * a11 - a12 * a21))
> N2Star <- expression((a11 - a21)/(a22 * a11 - a12 * a21))
```

Next we create the α and evaluate our expressions.

```
> a11 <- a22 <- 0.01; a12 <- 0.001; a21 <- 0.001
> N1 <- eval(N1Star); N2 <- eval(N2Star); N1

[1] 90.9
```

5.3.2 Create the Jacobian matrix

The next step is to find each partial derivative. The partial derivatives describe how the growth rate of each species changes with respect to the abundance of each other species and with respect to its own abundance. Thus a positive value indicates that a growth rate increases as another population increases. A negative value indicates a growth rate decreases as another population increases. Here, we work through an example, explicitly deriving the partial derivative of species 1's growth rate with respect to itself.

First let's expand the growth rate of species 1 (eq. 5.5)[3]

$$\frac{\mathrm{d}N_1}{\mathrm{d}t} = \dot{N}_1 = r_1 N_1 - r_1 \alpha_{11} N_1^2 - r_1 \alpha_{12} N_2 N_1. \tag{5.17}$$

Now we derive the partial differential equation (PDE)[4] with respect to N_1, *treating* N_2 *as a constant*[5]

$$\frac{\partial \dot{N}_1}{\partial N_1} = r_1 - 2 r_1 \alpha_{11} N_1 - r_1 \alpha_{12} N_2 \tag{5.18}$$

We should think of this as the per capita effect of species 1 on its growth rate.

To derive the PDE with respect to N_2, we treat N_1 as a constant, and find

$$\frac{\partial \dot{N}_1}{\partial N_2} = -r_1 \alpha_{12} N_1. \tag{5.19}$$

This is the per capita effect of species 2 on species 1's growth rate.

We then do the same for $\dot{N}_2$, and so derive the full matrix of PDE's,

$$\begin{pmatrix} \frac{\partial \dot{N}_1}{\partial N_1} & \frac{\partial \dot{N}_1}{\partial N_2} \\ \frac{\partial \dot{N}_2}{\partial N_1} & \frac{\partial \dot{N}_2}{\partial N_2} \end{pmatrix} = \begin{pmatrix} r_1 - 2 r_1 \alpha_{11} N_1 - r_1 \alpha_{12} N_2 & -r_1 \alpha_{12} N_1 \\ -r_2 \alpha_{21} N_2 & r_2 - 2 r_2 \alpha_{22} N_2 - r_2 \alpha_{21} N_1 \end{pmatrix}. \tag{5.20}$$

This matrix of PDE's is the Jacobian matrix, or simply the "Jacobian." As differential equations, they describe the slopes of curves (i.e. the slopes of tangents of curves) at a particular point. That is, they describe the straight line interpretations as that point. As *partial* differential equations, they describe how the growth rates change as population sizes change.

---

[3] Recall that the time derivative, $\mathrm{d}N/\mathrm{d}t$, can be symbolized with $\dot{N}$ ("n-dot").

[4] PDEs are typically written using a fancy symbol, delta, as in $\partial F/\partial N$. For most intents and purposes, these are equivalent to "d".

[5] Recall that when taking a derivative with respect to X, we treat all other variables as constants.

Finding partial differential equations and the Jacobian matrix

Here we create equations or *expressions* for the for the growth rates, $\dot{N}_1$ and $\dot{N}_2$, and use these to find the partial derivatives. First, expressions for the growth rates:

```
> dN1dt <- expression(r1 * N1 - r1 * a11 * N1^2 - r1 * a12 *
+     N1 * N2)
> dN2dt <- expression(r2 * N2 - r2 * a22 * N2^2 - r2 * a21 *
+     N1 * N2)
```

Next, we use each expression for $\dot{N}$ to get each the partial derivatives with respect to each population size. Here we use the R function `D()` (see also `?deriv`). We reveal here the result for the first one only, the partial derivative of $\dot{N}_1$ with respect to itself, and then get the others.

```
> ddN1dN1 <- D(dN1dt, "N1")
> ddN1dN1

r1 - r1 * a11 * (2 * N1) - r1 * a12 * N2
```

Here we find the remaining PDE's.

```
> ddN1dN2 <- D(dN1dt, "N2")
> ddN2dN1 <- D(dN2dt, "N1")
> ddN2dN2 <- D(dN2dt, "N2")
```

Last we put these together to create the Jacobian matrix, which is itself an expression that we can evaluate again and again.

```
> J <- expression(matrix(c(eval(ddN1dN1), eval(ddN1dN2), eval(ddN2dN1),
+     eval(ddN2dN2)), nrow = 2, byrow = TRUE))
```

5.3.3 Solve the Jacobian at an equilibrium

To *solve* the Jacobian at an equilibrium, we substitute the N_i^* (eqs. 5.12, 5.13) into the Jacobian matrix eq. (5.20). Refer to those equations now. What is the value of N_1 in terms of α_{ii} and α_{ij}? Take that value and stick it in each element of the Jacobian (eq. 5.21). Repeat for N_2. When we do this, and rearrange, we get,

$$\mathbf{J} = \begin{pmatrix} -r_1\alpha_{11}\left(\frac{\alpha_{22}-\alpha_{12}}{\alpha_{11}\alpha_{22}-\alpha_{12}\alpha_{21}}\right) & -r_1\alpha_{12}\left(\frac{\alpha_{22}-\alpha_{12}}{\alpha_{11}\alpha_{22}-\alpha_{12}\alpha_{21}}\right) \\ -r_2\alpha_{21}\left(\frac{\alpha_{11}-\alpha_{21}}{\alpha_{11}\alpha_{22}-\alpha_{12}\alpha_{21}}\right) & -r_2\alpha_{22}\left(\frac{\alpha_{11}-\alpha_{21}}{\alpha_{11}\alpha_{22}-\alpha_{12}\alpha_{21}}\right) \end{pmatrix}. \tag{5.21}$$

Yikes ...seems a little intimidating for such a small number of species. However, it is remarkable how each element can be expressed as a product of $-r_i\alpha_{ij}N_i^*$, where i refers to row, and j refers to column.

Evaluating the Jacobian matrix

Assuming that above we selected particular α, used these to determine N_1^* and N_2^*, found the PDEs and created an expression for the Jacobian matrix, and labeled everything appropriately, we can then evaluate the Jacobian at an equilibrium. For $\alpha_{ii} = 0.01$ and $\alpha_{ij} = 0.001$ (see above) we find

```
> r1 <- r2 <- 1
> J1 <- eval(J)
> J1

         [,1]     [,2]
[1,] -0.90909 -0.09091
[2,] -0.09091 -0.90909
```

Note that all of these PDEs are negative for this equilibrium. This indicates a stable equilibrium, because it means that each population's growth rate slows in response to an increase in any other.

5.3.4 Use the Jacobian matrix

Just the way we used eigenanalysis to understand long term asymptotic behavior of demographic matrices, we can use eigenanalysis of the Jacobian to assess the long-term asymptotic behavior of these competing Lotka-Volterra populations. We can again focus on its dominant, or leading, eigenvalue (λ_1). *The dominant eigenvalue will be the eigenvalue with the greatest real part*, and not necessarily the eigenvalue with the greatest magnitude.[6] In particular, the dominant eigenvalue, λ_1, may have a real part for which the magnitude, or absolute value is smaller, but which is less negative or more positive (e.g., $\lambda_1 = -.01$, $\lambda_2 = -1.0$). For continuous models, the dominant eigenvalue, λ_1, is approximately the rate of change of a perturbation, x, from an equilibrium,

$$x_t = x_0 e^{\lambda_1 t}. \tag{5.22}$$

Thus, the more negative the value, the faster the exponential decline back toward the equilibrium (i.e., toward $x = 0$). We can think of the dominant eigenvalue as a "perturbation growth rate": negative values mean negative growth (i.e. a decline of the perturbation), and positive values indicate positive growth of the perturbation, causing the system to diverge or be repelled away from the equilibrium.

In addition to the dominant eigenvalue, we need to consider the other eigenvalues. Table 5.2 provides a summary for interpreting eigenvalues with respect to the dynamics of the system. The eigenvalues depend upon elements of the Jacobian, and values calculated from the elements, notably the determinant, the trace, and the discriminant; a similar set of rules of system behavior can be based upon these values [181]. For instance, the *Routh-Hurwitz criterion* for

[6] This criterion is different than for demographic, discrete time projection matrices.

stability tells us that a two-species equilibrium will be locally stable, only if $\mathbf{J}_{11} + \mathbf{J}_{22} < 0$ and if $\mathbf{J}_{11}\mathbf{J}_{22} - \mathbf{J}_{12}\mathbf{J}_{21} > 0$. The biological interpretation of this criterion will be posed as a problem at the end of the chapter. For now, Table 5.2 will suffice.

Table 5.2: Interpretation of eigenvalues of Jacobian matrices.

Eigenvalues	Interpretation
All real parts < 0	Globally Stable Point (Point Attractor)
Some real parts < 0	Saddle (Attractor-Repellor)
No real parts < 0	Globally Unstable Point (Point Repellor)
Real parts = 0	Neutral
Imaginary parts absent	No oscillations
Imaginary parts present ($\pm\omega i$)	Oscillations with period $2\pi/\omega$

Eigenanalysis of the Jacobian matrix

Now that we have evaluated the Jacobian matrix (previous box), we simply perform eigenanalysis on the matrix (from previous boxes: $\alpha_{11} = \alpha_{22} = 0.01$, $\alpha_{12} = \alpha_{21} = 0.001$, $r = 1$).

```
> eigStable <- eigen(J1)
> eigStable[["values"]]
```

```
[1] -0.8182 -1.0000
```

The dominant eigenvalue is negative (the larger of the two: $\lambda_1 = -0.818$) indicating a globally stable equilibrium (Table 5.2). Both eigenvalues are real, not complex, indicating that there would be no oscillations (Table 5.2).

5.3.5 Three interesting equilbria

Here we examine the dynamical properties of three particularly interesting internal equilibria that are, respectively, stable, unstable, and neutral. In each case, our examples use $\alpha_{11} = \alpha_{22} = 0.01$ and $r_1 = r_2 = 1$. What is most important, however, is not the particular eigenvalues, but rather their sign, and how they vary with α_{12} and α_{21}, and the resulting stability properties and trajectories.

Given our stability criteria above, let us next examine the dominant eigenvalue of the Jacobian for each equilibrium ... but which values of α_{ij}, α_{ji} should we choose? We can describe our invasion criterion for species i as

$$\beta_{ij} = \alpha_{ij}/\alpha_{jj} \tag{5.23}$$

where, if $\beta_{ij} < 1$, species i can invade. This ratio is the relative strength of inter- *vs.* intraspecific competitive effect. It turns out to be useful to calculate λ_1 ("perturbation growth rate") for combinations of β_{ij}, β_{ji}.

Stable equilibrium – β_{ij}, $\beta_{ji} < 1$

These criteria correspond to $\alpha_{12} < \alpha_{22}$, $\alpha_{21} < \alpha_{11}$. As the relative strength of interspecific effects increases toward 1.0, λ_1 approaches zero, at which point the system would no longer have a single global point attractor.

When β_{ij}, $\beta_{ji} < 1$, then both species can invade each other. We find that all of the eigenvalues of the Jacobian are negative and real (Fig. 5.5), demonstrating that these populations will reach a stable equilibrium (Table 5.2). When we plot these eigenvalues for these combinations of β, we see that the dominant eigenvalue increases from negative values toward zero as either β_{12} or β_{21} approaches 1 (Fig. 5.5).

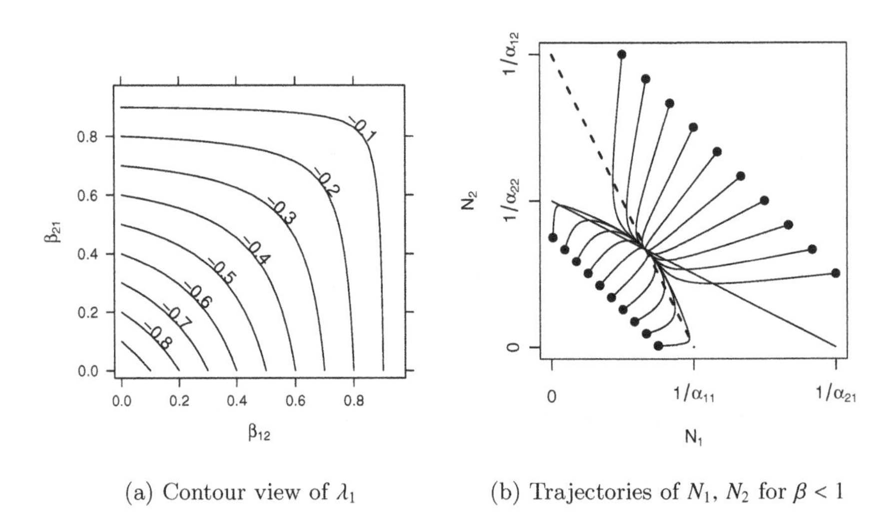

(a) Contour view of λ_1 (b) Trajectories of N_1, N_2 for $\beta < 1$

Fig. 5.5: Stable equilibria: as the relative strength of interspecific competition increases ($\beta_{ij} \to 1$), instability increases ($\lambda_1 \to 0$). (a) $\lambda_1 \to 0$ as $\beta \to 1$, (b) a globally stable equilibrium attracts trajectories from all locations (solid dots represent initial abundances).

Unstable equilibria – β_{ij}, $\beta_{ji} > 1$

These criteria correspond to $\alpha_{12} > \alpha_{22}$, $\alpha_{21} > \alpha_{11}$ (Fig. 5.6). As we saw above, the Lotka-Volterra competition model has not only stable equilibria, but also unstable equilibria, when both populations are greater than zero. Although an unstable equilibrium cannot persist, β_{ij}, $\beta_{ji} > 1$ creates interesting and probably important dynamics [74]. One of the results is referred to as *founder control*, where either species can colonize a patch, and whichever species gets there first (i.e. the founder) can resist any invader [14].

Another interesting phenomenon is the *saddle* itself; this unstable equilibrium is an *attractor-repeller*, that is, it attracts from some directions and repels from others (Fig. 5.6). This implies that the final outcome of dynamics may be difficult to predict from initial trajectories.

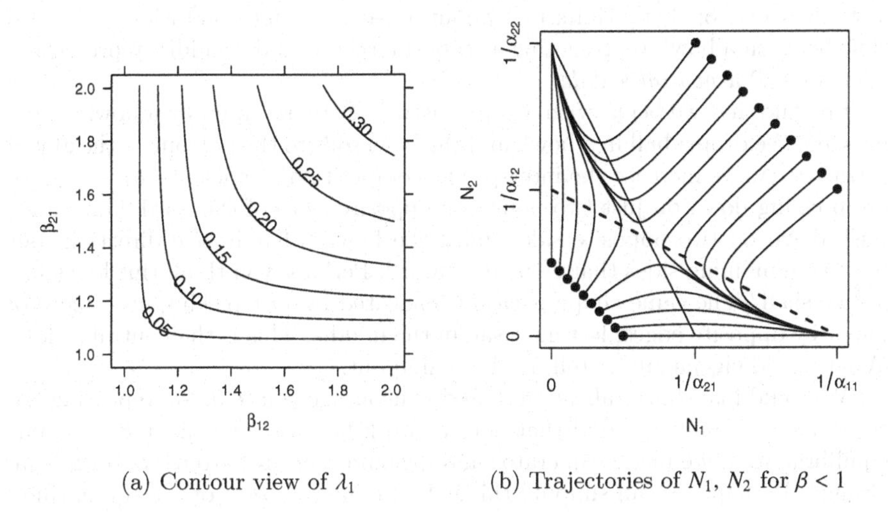

(a) Contour view of λ_1

(b) Trajectories of N_1, N_2 for $\beta < 1$

Fig. 5.6: Unstable equilibria: as the relative strength of interspecific competition increases ($\beta_{ij} > 1$), instability increases ($\lambda_1 > 0$). (a) λ_1 increases as β increases, (b) the unstable equilibrium may attract trajectories from some initial states, but repel from others (solid dots represent initial abundances).

Recall the geometric interpretation of this unstable equilibrium — a saddle. The trajectory of a ball rolling across a saddle can depend to a very large degree on where the ball starts. Place it on the crown of the saddle, and it will tend to roll in a very deterministic fashion directly toward the unstable equilibrium, even if it eventually rolls off the side.

Eigenanalysis of the Jacobian where $\beta_{ij}, \beta_{ji} > 1$

Here we create values for α that create an unstable equilbrium.

```
> a11 <- a22 <- 0.01
> a12 <- a21 <- 0.011
> N1 <- eval(N1Star)
> N2 <- eval(N2Star)
> eigen(eval(J))[["values"]]
```

```
[1]  0.04762 -1.00000
```

The dominant eigenvalue is now positive, while the other is negative, indicating a saddle (Table 5.2).

Neutral equilibria — $\beta_{ij} = \beta_{ji} = 1$

What happens when the inter- and intraspecific effects of each species are equal? This puts the populations on a knife's edge, between an unstable saddle and a stable attractor. Let's think first about a geometric interpretation, where we shift between a bowl, representing a stable attractor, and a saddle, representing what we call a *neutral saddle.*

Imagine that we begin with a stable attractor, represented by a bowl, where $\alpha_{ij} < \alpha_{ii}$. We drop a ball in a bowl, and the bowl rolls to the bottom — the global attractor. As we increase the interspecific competition coefficients, $\alpha_{ij} \rightarrow \alpha_{ii}$, we are pressing down on just two points on opposite sides of the bowl. Our hands push down on two opposite sides, until the bowl is flat in one direction, but has two remaining sides that slope downward. Perhaps you think this looks like a taco shell? The same shape is easily replicated by just picking up a piece of paper by opposite edges, letting it sag in the middle. This is the neutral saddle. What would eigenanalysis tell us? Let's find out.

We could just charge ahead in R, and I encourage you to do so, repeating the steps above. You would find that doesn't work because when $\beta_{ij} = \beta_{ji} = 1$, our equilibria are undefined (numerator and denominator are zero in eq. 5.12, 5.13. Hmmm. Perhaps we can simplify things by taking the *limit* of the equilibrium, as $\alpha_{ij} \rightarrow \alpha_{jj}$. Let $\alpha_{12} = a$ and $\alpha_{22} = a + h$, and let $\alpha_{21} = b$ and $\alpha_{11} = b + h$. Then we want the limit of the equilibrium as h goes to zero.

$$\lim_{h \to 0} \frac{(a + h) - a}{(a + h)(b + h) - ab} = \frac{1}{a + b} \tag{5.24}$$

Thus, $N_1^* = 1/(\alpha_{11} + \alpha_{22})$, assuming $\alpha_{12} = \alpha_{22}$ and $\alpha_{21} = \alpha_{11}$. Therefore, the equilibrium population size is simply the inverse of the sum of these coefficients.

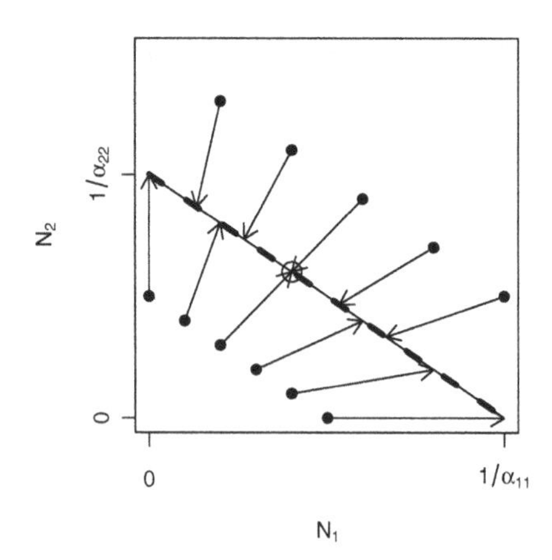

Fig. 5.7: Trajectories of N_1, N_2 for $\beta_{12}, \beta_{21} = 1$. The entire isocline is an attractor, a neutral saddle, and the final abundances depend on the initial abundances and the ratio of $\alpha_{11} : \alpha_{22}$. The circle represents our one derived equilibrium (eq. 5.24).

Eigenanalysis of the Jacobian where $\beta_{ij} = \beta_{ji} = 1$

Here we create values for α that create a neutral equilbrium.

```
> a11 <- a21 <- 0.01
> a22 <- a12 <- 0.015
```

We determine N^* differently (eq. 4.17) because the usual expression fails when the denominator equals 0.

```
> N1 <- N2 <- 1/(a11 + a22)
> eigen(eval(J))[["values"]]

[1] -1  0
```

The dominant eigenvalue is now zero, indicating a neutral equilibrium (Table 5.2). The neutral nature of this equilibrium results in more than one equilibrium. Let's try a different one, also on the isocline.

```
> N1 <- 1/(a11)
> N2 <- 0
> eigen(eval(J))[["values"]]

[1] -1  0
```

Again $\lambda_1 = 0$ so this equilibrium is also neutral.

When we perform eigenanalysis, we find that the largest of the two eigenvalues is zero, while the other is negative. This reveals that we have neither a bowl nor an unstable saddle, but rather, a taco shell, with a level bottom — a neutral saddle.

For example, if the populations start at low abundances, both populations will tend to increase at constant rates until they run into the isocline. Thus, both populations can increase when rare, but the relative abundances will never change, regardless of initial abundances.

Recall the Lotka-Volterra isoclines, and what we originally stated about them. We stated that *the equilibrium will be the point where the isoclines cross.* When all $\beta_{ij} = \beta_{ji} = 1$, *the isoclines completely overlap*, so we have an infinite number of equilibria—all the points along the line

$$N_2 = \frac{1}{\alpha_{22}} - \frac{\alpha_{11}}{\alpha_{22}} N_1 \tag{5.25}$$

and the initial abundances determine the trajectory and the equilibrium (Fig. 5.7).

5.4 Return Time and the Effect of *r*

Above, we referred to λ_1 as the perturbation growth rate. More commonly, people refer to another quantity known as *characteristic return time* (see Chapter

3). Return time is commonly calculated as the negative inverse of the largest real part of the eigenvalues,

$$RT = -\frac{1}{\lambda_1}. \tag{5.26}$$

It is the time required to return a fraction of the distance[7] back toward an equilibrium. Negative return times ($\lambda_1 > 0$) refer to "backward time," or time into the past when this population would have been this far away (Fig. 5.8).

If we envision the populations sitting at an equilibrium, we can then envision a small perturbation that shifts them away from that point in state space (see Chap. 3). Let's call this displacement x_0. The rate of change of in x is approximately the exponential rate,

$$\frac{dx}{dt} \approx c\lambda_1 t. \tag{5.27}$$

where c is a constant, so the distance traveled, x, is given by (eq. 5.22). Therefore, a negative λ_1 indicates an exponential decline in the disturbance, back toward the equilibrium (Fig. 5.8). The units of return time are the same as for r. Recall that all of this depends on the linearization of the curved surface around an equilibrium; it therefore applies exactly to only an infinitesimally small region around the equilibrium. It also usually provides the average, regardless of whether the perturbation is a population decline or a population increase.

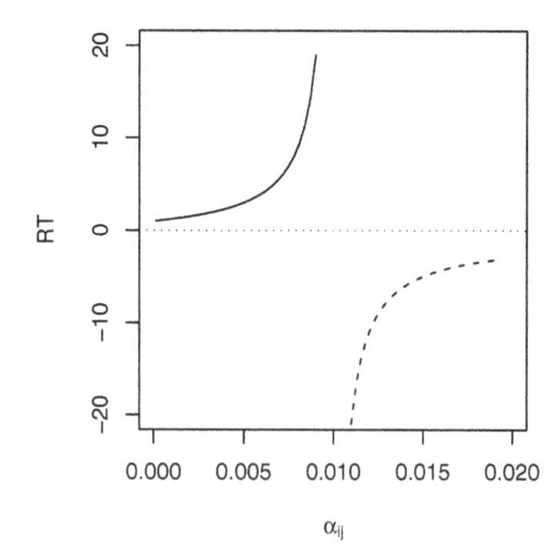

Fig. 5.8: For small β_{ij} ($\alpha_{ij} < \alpha_{jj}$), return time is positive because some time will lapse before the system returns toward to its equilibrium. For large β_{ij} ($\alpha_{ij} > \alpha_{jj}$), return time is negative, because it would have been some time in the past that this system was closer to its (unstable) equilibrium. ($\alpha_{ii} = 0.01$)

[7] This "fraction" happens to be about 63% or $1/e$; thus the hypothetical initial perturbation x_0 shrinks to $0.37x_0$.

Effect of r on stability and return time

Consider the Jacobian matrix (eq. 5.21), and note that $-r_i$ appears in each Jacobian element. Therefore, the larger the r, the greater the magnitude of the Jacobian elements. This causes λ_1 to increase in magnitude, reflecting greater responsiveness to perturbation at the equilibrium (Fig. 5.9).

If we consider return time for continuous models where $\beta_{12}, \beta_{21} < 1$, greater r shortens return time, increasing stability (Fig. 5.9). For continuous models where $\beta_{12}, \beta_{21} > 1$, greater r increases perturbation growth rate, decreasing stability (Fig. 5.9). For discrete models, which we have not discussed in this context, recall that increasing r_d of discrete logistic growth can destabilize the population because of the built-in lag. The same is true for discrete competition models — increasing r_d too much destabilizes the interaction.

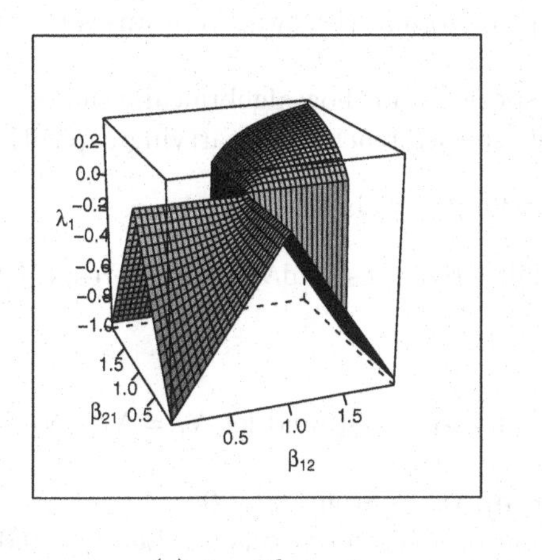

(a) λ_1 with $r = 1$

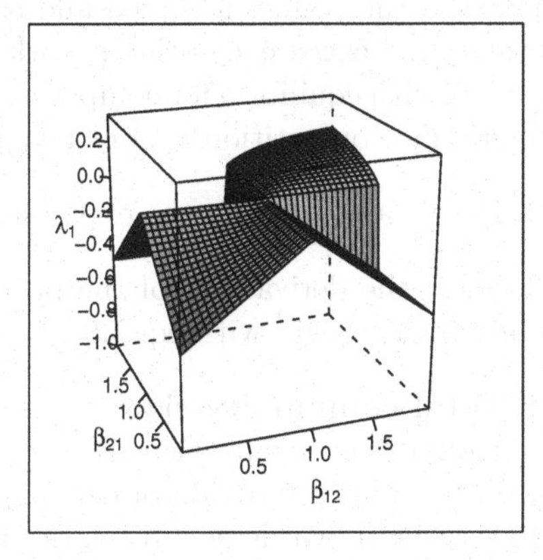

(b) λ_1 with $r = 0.5$

Fig. 5.9: The dominant eigenvalue of the Jacobian matrix varies with r as well as with β — higher r causes greater responsiveness to perturbations around an internal equilibrium. (a) $r = 1$, (b) $r = 0.5$.

5.5 Summary

This chapter has provided several useful results.

- We can represent species effects on each other in precisely the same way we represented their effects on themselves.
- Considering only two species, species i can invade species j when the effect of species j on species i is less than its effect of species j on itself.
- Two species coexist stably when their effects on each other are smaller than their effects on themselves.

- The dominant eigenvalue of the Jacobian matrix (perturbation growth rate), and its negative inverse, return time, are useful mathematical definitions of stability.
- Perturbation growth rate decreases as β_{ij}, β_{ji} decrease, and are either both less than one or both greater than 1 ($\beta_{ij} = \alpha_{ij}/\alpha_{jj}$).
- The magnitude of perturbation growth rate increases with r.

Problems

5.1. Basics
Let $\alpha_{11} = \alpha_{22} = 0.1$, $\alpha_{12} = 0.05$, $\alpha_{21} = 0.01$.
(a) Determine N_1^*, N_2^*, K_1, K_2.
(b) Draw (by hand, or in R) the ZNGIs (zero net growth isoclines); include arrows that indicate trajectories, and label the axes.
(c) Select other values for the α and repeat (a) and (b); swap your answers with a friend, and check each other's work.
(d) Start with equilibria for competing species, and show algebraically that when interspecific competition is nonexistent, species reach their carrying capacities.

5.2. Derive and simplify the expression for N_1^*/N_2^* in terms of the α.

5.3. Show the derivations of the partial derivatives of $\mathrm{d}N_2/\mathrm{d}t$, with respect to N_2 and to N_1; begin with eq. 5.6.

5.4. Total community size
Assume for convenience that $\alpha_{11} = \alpha_{22}$ and $\alpha_{12} = \alpha_{21}$, and let $N_T = N_1 + N_2$.
(a) Write N_T^* as a function of α_{11}, α_{22}, α_{12}, α_{21}.
(b) Describe in words how N_T varies as α_{ij} varies from $\alpha_{ii} \to 0$.
(c) Graph (by hand, or in R) the relation between N_T versus α_{ij}. Let $\alpha_{ii} = 0.01$.

5.5. Interpret the Routh-Hurwitz criterion in terms of species relative inter- and intraspecific competitive abilities.

5.6. The Jacobian matrix
Here we turn words into math. Note that this is one way of making our assumptions very precise and clear. In each case below (a.–d.), (i) use algebraic inequalities between the βs and between the αs to show what the assumptions imply for the equalities and inequalities with respect to all αs, (ii) use these inequalities to simplify the Jacobian matrix (eq. (5.21) as much as possible, (iii) show algebraically how these (in)equalities determine the sign of each element of the Jacobian, and (iv) explain in words how the magnitudes of the Jacobian elements determine stability properties.
(a) Assume that both species are functionally equivalent, and intraspecific competition is more intense than interspecific competition.
(b) Assume species are functionally equivalent and that each species has a greater impact on each other than they do on themselves.
(c) Assume species are functionally equivalent and interspecific competition is

precisely equal to intraspecific competition.
(d) Assume species have the same carrying capacity, and can coexist, but that species 1 is dominant.
(e) Assume species 1 is the better competitor (note: this may have multiple interpretations).
(f) Assume species 1 and 2 are equivalent (note: this may have multiple interpretations).

6

Enemy–Victim Interactions

Enemy–victim interactions, a.k.a. consumer–resource, or exploitative interactions are among the most dramatic interactions we can witness, whether that interaction is a cheetah chasing down a gazelle, or an osprey diving for a fish. Humans have always had a fascination with predators and death, and ecologists are humans, for the most part. In addition, plants are consumers too, but watching grass take up nitrate and CO_2 (i.e., grow) is somewhat less scintillating than tracking wolves across the tundra. Nonetheless, these are both examples of consumer–resource interactions; most competition, by the way, is thought to operate through uptake of shared, limiting resources [120, 200].

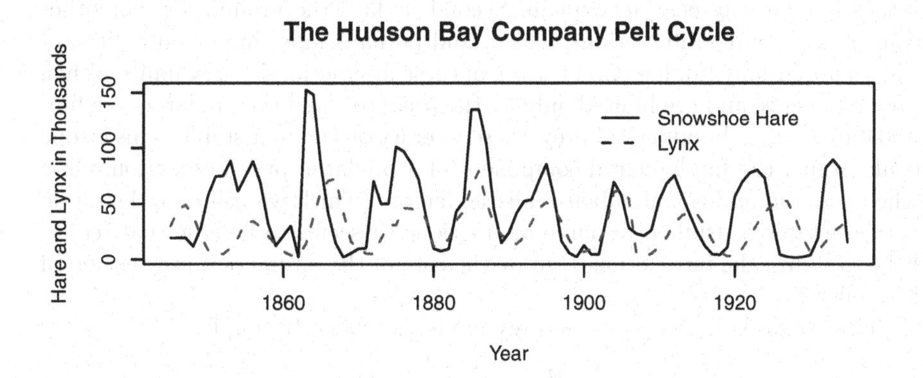

Fig. 6.1: Lynx–snowshoe hare cycles.

One of the most famous examples of species interactions in all of ecology is the lynx–snowshoe hare cycle, based on data from the Hudson Bay Trading

Co. trapping records (Fig. 6.1).[1] For decades, the lynx–hare cycle was used as a possible example of a predator-prey interaction, until a lot of hard work by a lot of people [94,191] showed an asymmetric dynamic — while the lynx depends quite heavily on the hare, and seems to track hare abundance, the hare cycles seem to be caused by more than just lynx.

In this chapter, we will do a few things. First, we will cover various flavors of consumer–resource models of different levels of complexity, and style. Also known as enemy–victim relations, or exploitative interactions, we represent cases in which one species has a negative effect and one a positive effect on the other. In doing so, we will illustrate some fundamental concepts about consumer–resource dynamics, such as how predators respond to prey abundances in both a *numerical* and a *functional* manner. We will try to develop an understanding of the range of dynamics for both continuous and discrete time systems.

6.1 Predators and Prey

This section covers a brief introduction to the classic predator–prey models, the Lotka–Volterra model, and the Rosenzweig-MacArthur extension.

6.1.1 Lotka–Volterra model

The Lotka–Volterra predator–prey model [117] is the simplest consumer–resource model we will cover. It is useful for a variety of reasons. First, its simplicity makes it relatively easy to explain. Second, it lays the groundwork for other consumer–resource interactions. Third, and perhaps more importantly, it captures a potentially fundamental feature of these interactions — instability. When prey reproduce and are limited only by the predator, and the predators are limited only by the abundance of prey, these interactions are not stable. This is, one could argue, the fundamental component of a predator–prey interaction. Only when you add additional, albeit realistic, factors (which we will cover later) do you get stable consumer–resource interactions. It seems to be true that reality helps stabilize the interactions, but at the core of the interaction is a tension of instability.

The Lotka–Volterra predator–prey model is relatively simple.

$$\frac{\mathrm{d}H}{\mathrm{d}t} = bH - aPH \tag{6.1}$$

$$\frac{\mathrm{d}P}{\mathrm{d}t} = eaPH - sP \tag{6.2}$$

Here the prey is H (herbivore, perhaps), the predator is P, and b, a, e, and s are parameters that we will discuss shortly. Let's break these equations down.

[1] These data are actually collected from a number of different regions and embedded with a complex food web, so it probably doesn't make much sense to think of this as only one predator-prey pair of populations.

First, we examine the model with regard to its terms that increase growth (+ terms) and those that reduce growth rate (– terms). Second, we find the terms where death of prey (in the prey equation) relates to growth of predators (in the predator equation).

The prey equation (eq. 6.1) has only two terms, bH and aPH. Therefore the prey population is growing at an instantaneous rate of bH. What does this mean? It means that its per capita rate is density independent, that is, it is growing exponentially, in the absence of the predator. This, as you know, frequently makes sense over very short time periods, but not over the long term. The units of b are similar to those for r — number of herbivores produced per herbivore.

What about the loss term, aPH? This term is known as *mass action*, a term borrowed from chemistry, where the rate of a reaction of two substances (e.g., P and H) is merely a linear function, a, of their masses. That is, given the abundances of each population, they encounter each other *with the outcome of prey death* at instantaneous rate a. Thus a is frequently known as the kill rate or "attack" rate.[2] The units of a are number of herbivores killed per predator per herbivore. When multiplied by PH, the units of the term become number of herbivores.

Lotka–Volterra predator–prey model

Here we create a function for the Lotka–Volterra predator–prey model that we use in `ode`.

```
> predpreyLV <- function(t, y, params) {
+     H <- y[1]
+     P <- y[2]
+     with(as.list(params), {
+         dH.dt <- b * H - a * P * H
+         dP.dt <- e * a * P * H - s * P
+         return(list(c(dH.dt, dP.dt)))
+     })
+ }
```

See the Appendix, B.10 for more information about using `ode`.

Functional response

What we have just described is the functional response of the predator [77]. The functional response is the rate at which a single predator kills prey. We can represent this a graph of the relation between the number of prey killed per predator per unit time (y-axis) *vs.* the prey available (x-axis) — the predators kill prey at rate aPH, so a single predator kills prey at rate aH. This simple relation is linear, and is known as a *type I* functional response (Fig. 6.2a). What does this assume about predator behavior? For one, it assumes that no

[2] In more detailed models, we may distinguish between encounter rate, attack rate, kill rate, and handling time, adding greatly to the detail of the model.

matter how abundant prey become, a single predator can and will always kill a fixed proportion, whether there are 0.1 or 100 prey$\cdot m^2$ available. Other types of functional responses are often discussed, including types II and III; these saturate (reach an asymptote) at high prey densities, and we will discuss these later in the chapter.

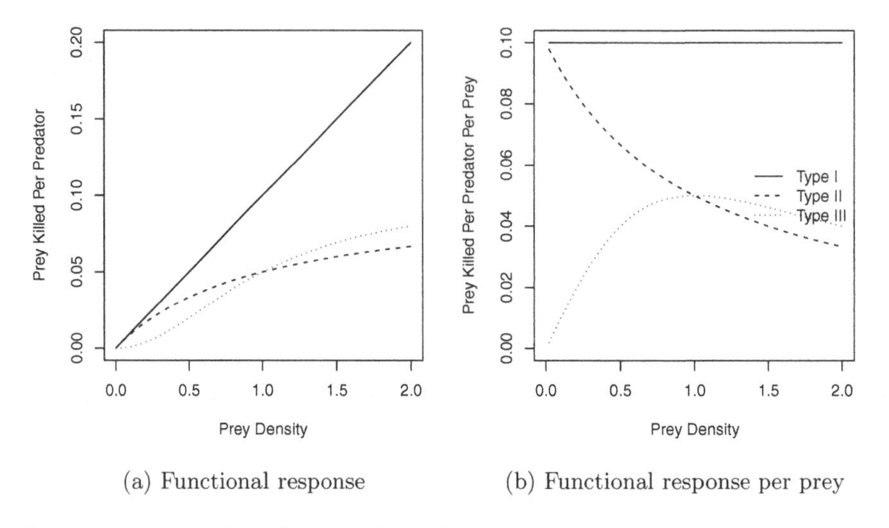

(a) Functional response (b) Functional response per prey

Fig. 6.2: Types I, II, and III predator functional responses; these are the rates at which predators kill prey across different prey densities. (a) The original functional responses; (b) Functional responses on a per prey basis. The Lotka–Volterra model assumes a type I functional response.

Ofttimes, especially with messy data, it is very difficult to distinguish among functional response types, especially at low prey densities [87]. It is sometimes easier to distinguish among them if we examine the functional responses *on a per prey basis*. This results in a much more stark contrast among the functional responses (Fig. 6.2b), at low prey densities.

Functional responses

This will graph types I, II, and III predator functional responses. Parameter a is the attack rate for mass action (type I). Parameter w is the maximum rate and D is the half saturation constant for the Michaelis-Menten/Holling disc equation in types II and III.

```
> a <- 0.1
> w <- 0.1
> D <- w/a
> curve(a * x, 0, 2, xlab = "Prey Density",
+     ylab = "Prey Killed Per Predator")
> curve(w * x/(D + x), 0, 2, add = TRUE, lty = 2)
> curve(w * x^2/(D^2 + x^2), 0, 2, add = TRUE, lty = 3)
```

It is sometimes easier to distinguish among these if we examine the per attack rate per prey, as a function of prey density.

```
> curve(w * x^2/(D^2 + x^2)/x, 0, 2, ylim = c(0, a), lty = 3,
+     xlab = "Prey Density", ylab = "Prey Killed Per Predator Per Prey")
> curve(w * x/(D + x)/x, 0, 2, lty = 2, add = TRUE)
> curve(a * x/x, 0, 2, add = TRUE, lty = 1)
> legend("right", c("Type I", "Type II", "Type III"), lty = 1:3,
+     bty = "n")
```

Numerical response

The *numerical response* of the predators is merely the population-level response of predators to the prey, that derives from both the growth and death terms.

In the Lotka–Volterra model, predators attack, kill or capture prey at rate a, but for the population to grow, they need to convert that prey into new predator biomass or offspring. They assimilate the nutrients in prey and convert the prey into predator body mass or offspring at rate e. Thus e is the efficiency (assimilation or conversion efficiency) of converting dead prey into new predator mass or predator numbers. The units of e are derived easily. Recall from above the units of a. The units of e must take the expression aPH with its units of numbers of herbivores, and give $eaPH$ units of numbers of predators. Therefore the units of e are numbers of predators per number of herbivores killed.

In this model, we pretend that predators die at a constant proportional rate, that is, at a constant per capita rate, s, in the absence of prey. At first blush, this may seem quite unrealistic. If, however, we consider that (i) not all predators are the same age, size, body condition, or occupying the same habitat, and (ii) they are all different genotypes, then we might expect many to die quickly, and a few to hang on for much longer. The exponential decay described by this simple formula of constant proportional loss may be a very good place to start when considering loss from a complex system such as a population. The numerical response is the combined response of this loss and the growth term.

Lotka–Volterra isoclines

What does this model tell us about the forces that govern the dynamics of predator–prey populations? Are there attractors or repellors? What can we learn?

As we saw in previous chapters, a good place to start learning about species interactions is with the ZNGIs or zero net growth isoclines. These tell us when populations tend to decrease or increase. How do we determine these ZNGIs for predator–prey interactions? We do so in the same fashion as for single species models, and for Lotka–Volterra competition — by setting the growth equations equal to zero and solving for the relevant state variable. Here we set the "herbivore" population growth rate equal to zero and solve for H^* — this is the value of H when $\dot{H} = 0$. As with L-V competition, it may be an isocline, a line along which all points in P and H space (the combinations of P and H) $\dot{H} = 0$. Solving for H,

$$0 = bH - aPH \tag{6.3}$$

$$0 = H(b - aP) \tag{6.4}$$

$$H = 0 \tag{6.5}$$

It seems at first that the only solution is for extinction of herbivores. What other condition, however, allows $\dot{H} = 0$? Further rearrangement shows us that this will hold if $(b - aP) = 0$, that is,

$$P = \frac{b}{a} \tag{6.6}$$

then this provides the isocline for which $\dot{H} = 0$ (Fig. 6.3). Note that it depends only on the traits of the predator and prey, or the ratio of the intrinsic growth rate, b, and the attack rate, a. The faster the prey grows, the more predators you need to keep them in check. The more vulnerable the prey are to predators, or the better the hunters are, the fewer predators are needed to keep the prey in check.

Now let's solve for the predator isocline, the set of points at which predator growth rate stops. We set $\dot{P} = 0$ and solve

$$0 = eaPH - sP$$

$$0 = P(eaH - s)$$

$$P = 0, \quad H = \frac{s}{ea} \tag{6.7}$$

What conditions cause the predator growth rate to equal zero? One equilibrium occurs at $P^* = 0$, and the other occurs at $H = s/(ea)$. As we saw for the herbivore, the state at which $\dot{P} = 0$ is independent of the predator population itself, but rather depends upon species' traits only. It depends upon the density-independent per capita death rate s, conversion efficiency, e, and the attack rate, a. The higher the predator's death rate, s, the more herbivores are required to maintain predator growth. The more efficient the predator is at assimilating

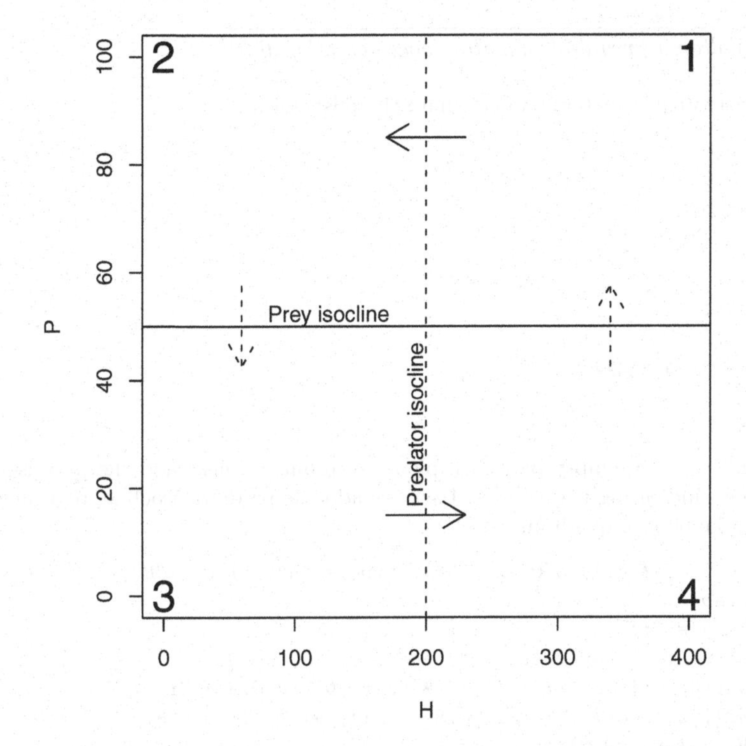

Fig. 6.3: Lotka–Volterra predator–prey isoclines. The isoclines (solid and dashed lines) are the set of all points for which the predator growth rate or the herbivore growth rate are zero. Increases and decreases in abundance are indicated by arrows (solid - prey, dashed - predator). Prey abundance, H, decreases in quadrants 1 and 2 because predator abundance, P, is high; prey abundance increases in quadrants 3 and 4 because predator abundance is low. In contrast, predator abundance, P, increases in quadrants 4 and 1 because prey abundance is high, whereas predator abundance decreases in quandrants 2 and 3 because prey abundance is low. These reciprocal changes in each species abundance results in counterclockwise dynamics between the two populations.

prey and converting them to biomass or progeny, e, the fewer prey are needed to maintain predator growth. The more efficient they are at encountering, attacking and killing prey, a, the fewer prey are required to maintain predator growth. We could also flip that around to focus on the prey, and state that the more nutritious the prey, or the more vulnerable they are to attack, the fewer are needed to maintain the predator population.

Fig. 6.3 illustrates the concepts we described above — predators increase when they have lots of food, and die back when food is scarce.

Lotka–Volterra prey and predator isoclines (Fig. 6.3)

We first select parameters and calculate these isoclines.

```
> b <- 0.5
> a <- 0.01
> (Hi <- b/a)
```

```
[1] 50
```

```
> e <- 0.1
> s <- 0.2
> (Pi <- s/(e * a))
```

```
[1] 200
```

We then set up an empty plot with plenty of room; we next add the prey isocline and arrows indicating trajectories. Last, we add the predator isocline, text, arrows, and then label each quadrant.

```
> plot(c(0, 2 * Pi), c(0, 2 * Hi), type = "n", xlab = "H",
+     ylab = "P")
> abline(h = Hi)
> text(Pi, Hi, "Prey isocline", adj = c(1.3, -0.3))
> arrows(x0 = c(0.85 * Pi, 1.15 * Pi), y0 = c(0.3 * Hi, 1.7 *
+     Hi), x1 = c(1.15 * Pi, 0.85 * Pi), y1 = c(0.3 * Hi, 1.7 *
+     Hi), len = 0.2)
> abline(v = Pi, lty = 2)
> text(Pi, Hi, "Predator isocline", adj = c(1.1, -0.2), srt = 90)
> arrows(x0 = c(0.3 * Pi, 1.7 * Pi), y0 = c(1.15 * Hi, 0.85 *
+     Hi), x1 = c(0.3 * Pi, 1.7 * Pi), y1 = c(0.85 * Hi, 1.15 *
+     Hi), lty = 2, len = 0.2)
> text(x = c(2 * Pi, 0, 0, 2 * Pi), y = c(2 * Hi, 2 * Hi, 0,
+     0), 1:4, cex = 2)
```

What do the dynamics in Fig. 6.3 tell us? Follow the path of the arrows. First, note that they cycle — they go around and around in a counterclockwise fashion, as each population responds to the changing abundances of the other species. We don't know much more than that yet, but we will later. The counter-clock wise direction illustrates a negative feedback loop between predators and prey.

Next we use linear stability analysis to learn more about the long-term behavior of this interaction. We will use this analysis to compare several different predator–prey models.

6.1.2 Stability analysis for Lotka–Volterra

In this section, we will perform the necessary analytical work to understand the dynamics of Lokta–Volterra predator–prey dynamics, and we follow this up with a peek at the time series dynamics to confirm our understanding based on the analytical work.

As before (e.g., Chapter 5), we can follow four steps: determine equilibria, create the Jacobian matrix, and solve and use the Jacobian.

Lotka–Volterra equilibrium

As you recall from Chapter 5, all we have to do is to solve the isoclines for where they cross. Thus we could set these equations equal to each other. It turned out, however, that the isoclines were so simple that we find that the prey and predator will come to rest at the (x, y) coordinates, $(b/a, s/(ea))$.

Creating, solving and using the Jacobian matrix

Take a look at the growth equations again (eqs. 6.1, 6.2). Here we take the partial derivatives of these because we want to know how each population growth rate changes in response to changes in the abundance each of the other population. The partial derivatives of the herbivore growth equation, with respect to itself and to the predator, go into the first row of the matrix, whereas the partial derivatives of the predator growth rate, with respect to the herbivore and itself go into the second row.[3]

$$\begin{pmatrix} \frac{\partial \dot{H}}{\partial H} & \frac{\partial \dot{H}}{\partial P} \\ \frac{\partial \dot{P}}{\partial H} & \frac{\partial \dot{P}}{\partial P} \end{pmatrix} = \begin{pmatrix} b - aP & -aH \\ eaP & eaH - s \end{pmatrix} \tag{6.8}$$

We can replace the P and H in the Jacobian with the equibria found above. When we do this, we get

$$\begin{pmatrix} b - a(b/a) & -a(s/(ae)) \\ ea(b/a) & ea(s/(ae)) - s \end{pmatrix} = \begin{pmatrix} 0 & -s/e \\ eb & 0 \end{pmatrix}. \tag{6.9}$$

Typically a system will be more stable if the diagonal elements are more negative — that would mean that each population is self regulating, and it corresponds to the Routh-Hurwitz criterion,[4]

$$\mathbf{J}_{11} + \mathbf{J}_{22} < 0. \tag{6.10}$$

We notice that in eq. 6.9 these diagonal elements are both zero; these zeroes reveal that there is no negative density dependence within each population; that is no self-regulation.

The other part of the Routh-Hurwitz criteria is the condition,

$$\mathbf{J}_{11}\mathbf{J}_{22} - \mathbf{J}_{12}\mathbf{J}_{21} > 0. \tag{6.11}$$

---

[3] Recall that a partial derivative is just a derivative of a derivative, with respect to another state variable. In this case, it is not "the second derivative" *per se*, because that would be with respect to time, not with respect to one of the populations.

[4] See Chapter 5 for an earlier use of the Routh-Hurwitz criteria

In the predator–prey context, this suggests that the herbivore *declines* due to the predator ($\mathbf{J}_{12} < 0$) and the predator *increases* due to the herbivore ($\mathbf{J}_{21} > 0$). The signs of these elements make their product negative, and help make the above condition true. Note that because $\mathbf{J}_{11}\mathbf{J}_{22}$, this condition reduces to $bs > 0$. Thus it seems that this will be true as along as both b and s are positive (which is always the case).

If we performed eigenanalysis on the above Jacobian matrix, we would find that the eigenvalues are complex (see next box). Because they are complex, this means that the populations will oscillate or cycle, with period $2\pi/\omega$ (Table 5.2). Because the real parts are zero, this means that the Lotka–Volterra predator–prey exhibits neutral stability (Table 5.2). Recall that neutral stability is the "in-between" case where perturbations at the equilibrium neither grow nor decline over time.

Lotka–Volterra predator–prey eigenanalysis

We can perform eigenanalysis given the parameters above.

```
> Jac <- matrix(c(0, -s/e, e * b, 0), byrow = TRUE, nr = 2)
> eigen(Jac)[["values"]]

[1] 0+0.3162i 0-0.3162i
```

Lotka–Volterra Dynamics

What do predator–prey cycles look like (Figs. 6.4a, 6.4b)? Typically, prey achieve higher abundances than predators — this makes sense if the "predators" are not pathogens (see Disease, below). It also makes sense when we assume that predators are not perfectly efficient at converting prey to offspring — that is, they have to metabolize a lot of energy per offspring ($e \ll 1$). Another characteristic we note is that the predator populations lag behind the prey — for instance, the prey peak occurs before the predator peak (Fig. 6.4a).

Lotka–Volterra predator–prey dynamics (Fig. 6.4a)

Here we set parameters and integrate the populations, with initial abundances of $H_0 = 25$, $P_0 = 5$.

```
> params1 <- c(b = b, a = a, s = s, e = e)
> Time <- seq(0, 100, by = 0.1)
> LV.out <- ode(c(H0 = 25, P0 = 5), Time, predpreyLV, params1)
```

Next we graph the populations over time.

```
> matplot(Time, (LV.out[, 2:3]), type = "l", ylab = "Population Size")
```

What do *neutral cycles* look like? Both populations oscillate indefinitely, going neither toward extinction, nor toward a stable *node* or point (Fig. 6.4b). An odd characteristic is that we could choose arbitrarially any new initial abundance, and the populations would continue to cycle on a new trajectory, passing through these new abundances every period (Fig. 6.4b).

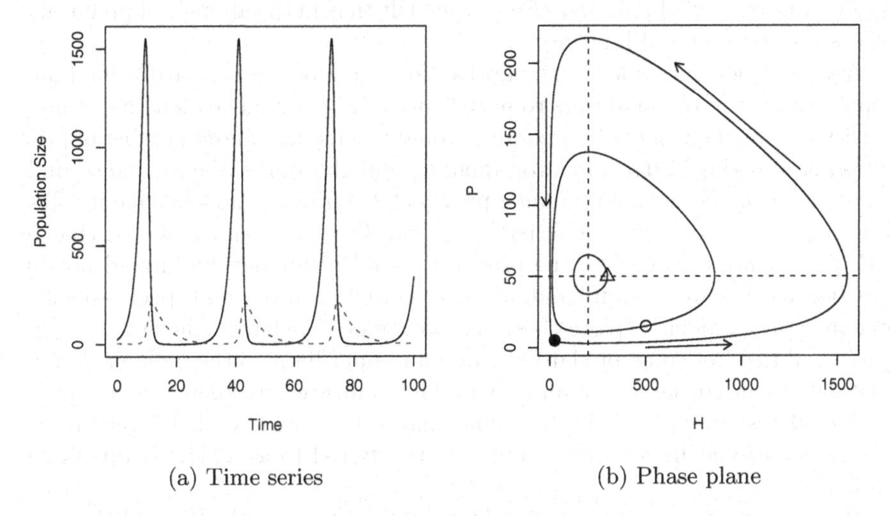

(a) Time series

(b) Phase plane

Fig. 6.4: Dynamics of the Lotka–Volterra predator–prey model. Both figures result from the same set of model parameters. (a) The times series shows the population sizes through time; these dynamics correspond to the largest oscillations in (b). (b) The phase plane plot includes three different starting abundances, indicated by symbols; the largest cycle (through solid dot) (a).

Lotka–Volterra predator–prey dynamics (Fig. 6.4b)

We integrate the same model as above twice more, but with arbitrarily different starting abundances — everything else is the same.

```
> LV.out2 <- ode(c(H0 = 500, P0 = 15), Time, predpreyLV, params1)
> LV.out3 <- ode(c(H0 = 300, P0 = 50), Time, predpreyLV, params1)
```

Now we plot the *phase plane portrait* of the first predator–prey pair, add trajectories associated with different starting points, and finally add the isoclines.

```
> plot(LV.out[, 2], LV.out[, 3], type = "l", ylab = "P", xlab = "H")
> points(25, 5, cex = 1.5, pch = 19)
> arrows(x0 = c(1300, -20, 500), y0 = c(125, 175, 0), x1 = c(650,
+     -20, 950), y1 = c(200, 100, 2), length = 0.1)
> lines(LV.out2[, 2], LV.out2[, 3])
> points(500, 15, cex = 1.5)
> lines(LV.out3[, 2], LV.out3[, 3])
> points(300, 50, cex = 1.5, pch = 2)
> abline(h = b/a, lty = 2)
> abline(v = s/(e * a), lty = 2)
```

6.1.3 Rosenzweig–MacArthur model

Michael Rosenzweig, a graduate student at the time, and his adviser, Robert MacArthur, proposed a predator–prey model that added two components to the

Lotka–Volterra model [177,180]. First, they felt that in the absence of predators, prey would become self-limiting.

A second element added to the Lotka–Volterra model was a saturating functional response of the predators to prey density (Fig. 6.2). They felt that if prey density became high enough, predators would reach a maximum number of prey killed per unit time. First, predator appetite could become satiated, and second, many predators need time to *handle* prey (catch, subdue, and consume it). The time required to do this is referred to as *handling time* and may not change with prey density. Therefore the predator would ultimately be limited not by prey density, but by its own handling time. Foraging theory and species-specific predator–prey models explore these components of predator behavior [72,123]. These realities (for some predators) cause an upper limit to the number of prey a predator can consume per unit time. This limitation is known as a type II functional response [77] (Fig. 6.2), and may take a variety of different forms — any monotonically saturating function is referred to as a type II functional response.

Here we add prey self-limitation by using logistic growth. We add the type II functional response using the Michaelis-Menten or Monod function.[5]

$$\frac{\mathrm{d}H}{\mathrm{d}t} = bH(1-\alpha H) - w\frac{H}{D+H}P \tag{6.12}$$

$$\frac{\mathrm{d}P}{\mathrm{d}t} = ew\frac{H}{D+H}P - sP \tag{6.13}$$

$$\tag{6.14}$$

where w and D are new constants.

Here we focus on the functional response, $w\frac{H}{D+H}$ (Fig. 6.2). How can we interpret this fraction? First let's consider what happens when prey density is very high. We do that by examining the limit of the functional response as prey density goes to infinity. As H gets really big, $D+H$ gets closer to H because D stays relatively small. In contrast, the numerator wH will continue to grow as a multiple of w, so w will always be important. Therefore, the functional response reduces to $wH/H = w$, or we state

$$\lim_{H\to\infty} w\frac{H}{D+H} = \frac{wH}{H} = w. \tag{6.15}$$

That is a pretty convenient interpretation, that w is the maximum capture rate.

What is the shape of the function at low prey abundance, or as $H \to 0$? H becomes smaller and smaller, and D becomes relatively more and more important. Therefore, the functional response reduces to $(w/D)H$, or

$$\lim_{H\to 0} w\frac{H}{D+H} = \frac{w}{D}H \tag{6.16}$$

Note that this is a linear functional response where $w/D = a$, where a was the attack rate in the Lotka–Volterra type I functional response aH. Cool!

[5] An alternative parameterization is the Holling disc equation $aHP/(1+bH)$.

Rosenzweig-MacArthur predator–prey function

Before we move on, let's make an R function of the Rosenzweig-MacArthur model.

```
> predpreyRM <- function(t, y, p) {
+     H <- y[1]
+     P <- y[2]
+     with(as.list(p), {
+         dH.dt <- b * H * (1 - alpha * H) - w * P * H/(D +
+             H)
+         dP.dt <- e * w * P * H/(D + H) - s * P
+         return(list(c(dH.dt, dP.dt)))
+     })
+ }
```

Rosenzweig-MacArthur isoclines

As usual, our first step is to find the zero net growth isoclines — we will first derive them, and then discuss them. We can begin with the prey,[6] setting $\dot{H} = 0$.

$$0 = bH(1 - \alpha H) - w\frac{PH}{D + H}$$
$$P = \frac{(D + H)}{w}b(1 - \alpha H)$$
$$P = \frac{b}{w}\left(D + (1 - \alpha D)H - \alpha H^2\right) \quad (6.17)$$

You will notice that it is a quadratic equation, and we can see that it is concave down ($-\alpha H^2$, when $\alpha > 0$), and the peak typically occurs at some $H > 0$.[7]

Next we find the predator isocline.

$$0 = ew\frac{PH}{D + H} - sP$$
$$0 = P\left(\frac{ewH}{D + H} - s\right)$$

One equilibrium is $P = 0$; let's focus on the other, where $\frac{ewH}{D+H} - s = 0$. This is telling us that the population growth rate of the predator will be zero when. . . ? Right — when H takes on a value such that this expression is true. To find that value, we solve this part of the expression for H.

$$0 = \frac{ewH}{D + H} - s$$
$$ewH - sH = sD$$
$$H = \frac{sD}{ew - s} \quad (6.18)$$

Thus $\dot{P} = 0$ when $H = sD/(ew - s)$. It is thus the straight line and independent of P — it depends entirely on the parameters.

[6] or as Rosenzweig liked to say, the "victim"

[7] Recall the rules you learned long ago, regarding the quadratic equation. . . .

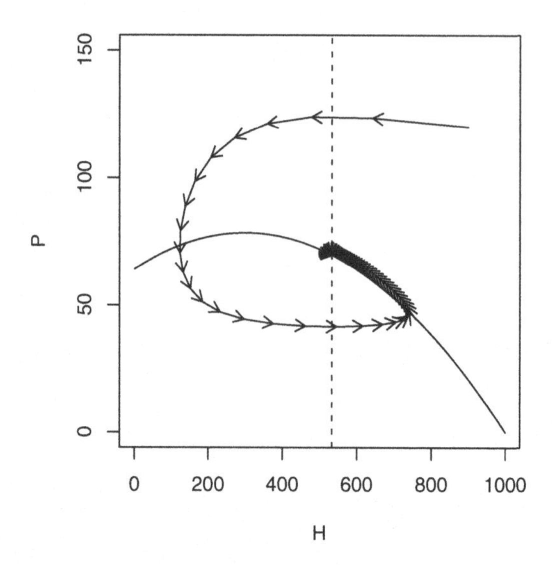

Fig. 6.5: Rosenzweig-MacArthur predator–prey isoclines for the predator (dashed) and the prey (solid). The isoclines are the set of all points for which the predator growth rate (dashed) and the herbivore growth rate (solid) are zero. The arrows get shorter, and the arrowheads closer together because the populations change more slowly as we approach the steady state equilibrium. Note that the x-intercept of the prey isocline is K.

Consider the dynamics around these isoclines (Fig. 6.5). Whenever the prey are abundant (right of dashed predator isocline), the predators increase, and when the prey are rare, the predators decrease. In contrast, whenever the predators are abundant (above solid prey isocline), then prey decline, and when predators are rare, then prey increase. In this case (Fig. 6.5), we see the classic counter-clockwise cycling through state space leading to a stable point equilibrium.

Rosenzweig-MacArthur isoclines (Fig. 6.5)

Let's graph the zero net growth isoclines for this model. First we set parameters, and make an expression for the prey isocline.

```
> b <- 0.8
> e <- 0.07
> s <- 0.2
> w <- 5
> D <- 400
> alpha <- 0.001
> H <- 0:(1/alpha)
> Hiso <- expression(b/w * (D + (1 - alpha * D) * H - alpha *
+     H^2))
> HisoStable <- eval(Hiso)
```

We also want to add a single trajectory, so we integrate using `ode`.

```
> p.RM <- c(b = b, alpha = alpha, e = e, s = s, w = w, D = D)
> Time <- 150
> RM1 <- ode(c(900, 120), 1:Time, predpreyRM, p.RM)
```

Finally, we plot everything, starting with the isoclines, and adding the trajectory using `arrows`.

```
> plot(H, HisoStable, type = "l", ylab = "P", xlab = "H", ylim = c(0,
+     150))
> abline(v = s * D/(e * w - s), lty = 2)
> arrows(RM1[-Time, 2], RM1[-Time, 3], RM1[-1, 2], RM1[-1,
+     3], length = 0.1)
```

Note that an arrow illustrates the change per one unit of time because we chose to have `ode` return H, P at every integer time step.

Creating and using the Jacobian

As we did for the Lotka–Volterra system, here we demonstrate stability by analyzing the Jacobian matrix of partial derivatives of each populations growth rate with respect to each other. We put all these together in a matrix, and it looks like this.

$$\begin{pmatrix} \frac{\partial \dot{H}}{\partial H} & \frac{\partial \dot{H}}{\partial P} \\ \frac{\partial \dot{P}}{\partial H} & \frac{\partial \dot{P}}{\partial P} \end{pmatrix} = \begin{pmatrix} b - 2\alpha bH - \left(\frac{wP}{D+H} - \frac{wPH}{(D+H)^2}\right) & -w\frac{H}{D+H} \\ \frac{ewP}{D+H} - \frac{ewPH}{(D+H)^2} & ew\frac{H}{D+H} - s \end{pmatrix} \tag{6.19}$$

You could review your rules of calculus, and prove to yourself that these *are* the correct partial derivatives.

Analysis of the Jacobian for Rosenzweig-MacArthur

Here we are going to write *expressions* for the two time derivatives, make a *list* of all the partial derivatives, find the stable equilibrium point, evaluate the partial derivatives as we stick them into the Jacobian matrix, and then perform eigenanalysis on the Jacobian. First, the time derivatives.

```
> dhdt <- expression(b * H * (1 - alpha * H) - w * P * H/(D +
+     H))
> dpdt <- expression(e * w * P * H/(D + H) - s * P)
```

Next we create a *list* of the partial derivatives, where their order will allow us, below, to fill columns of the Jacobian matrix.

```
> RMjac1 <- list(D(dhdt, "H"), D(dpdt, "H"), D(dhdt, "P"),
+     D(dpdt, "P"))
```

We need the stable equilibria, H^*, P^*. We know the value of H^* right away, because the predator isocline is a constant. Predator abundance exhibits zero growth when $H = sD/(ew - s)$, or

```
> H <- s * D/(e * w - s)
```

Now all we have to do is substitute that into the herbivore isocline, and we are finished.

```
> P <- eval(Hiso)
```

Now we "apply" the `eval` function to each *component* of the *list* of PD's, using the values of H and P we just created. We stick these values into a matrix and perform eigenanalysis.

```
> (RM.jac2 <- matrix(sapply(RMjac1, function(pd) eval(pd)),
+     nrow = 2))

        [,1]   [,2]
[1,] -0.2133 -2.857
[2,]  0.0112  0.000
```

```
> eigen(RM.jac2)[["values"]]

[1] -0.1067+0.1436i -0.1067-0.1436i
```

If we substitute in all of the relevant parameters and equilibria (see previous box), we would find that at the equilibrium, the Jacobian evaluates to

```
> RM.jac2

        [,1]   [,2]
[1,] -0.2133 -2.857
[2,]  0.0112  0.000
```

where we note[8] that the effect of both the prey itself and predators on the prey population growth rate (at the equilibrium) are negative, whereas the effect

[8] Recall that species 1 is in column 1 and row 1, and we interpret this as the effect of [column] on [row]

of prey on predators is positive, and the predators have no effect at all on themselves. How does this compare to the Jacobian matrix for Lotka–Volterra? There, the effect of prey on itself was zero.

If we examine the eigenvalues of this matrix,

```
> eigen(RM.jac2)[["values"]]

[1] -0.1067+0.1436i -0.1067-0.1436i
```

We find that for *these paramater values* the dominant eigenvalue is negative, indicating that the equilibrium is a stable attractor (Table 5.2). The presence of the imaginary part shows a cyclical approach toward the equilibrium — we could calculate the period of that cycle, if we wanted to (Table 5.2).

6.1.4 The paradox of enrichment

Rosenzweig and MacArthur [177, 180] realized that the type II functional response of the predator actually could *destablize* the interaction. They focused primarily on traits (or parameters) that bring the predator isocline closer to the y-axis, including predator assimilation efficiency —"overly" efficient predators (e.g., large e) can drive prey extinct.

In 1971, Rosenzweig described the "paradox of enrichment" where simply increasing the carrying capacity of the prey could destabilize the system [179]. In this case, the saturating nature of the functional response allows the prey to escape control when all individual predators become saturated — this allows prey to achieve higher peak abundances, until they reach carrying capacity. At that point, the predators can eventually achieve high abundance and drive the prey back down. In addition, when predators have a low half-saturation constant (small D) this allows them to kill a higher proportion of the prey when the prey are rare, thus driving the prey to even lower levels.

When would predator–prey cycles be stable, and when would they explode and destabilize? It all has to do with the position of the predator isocline relative to the hump of the prey isocline. When the hump or peak of the prey isocline is inside the predator isocline, the system has a stable point attractor (Fig. 6.5). When the carrying capacity of the prey increases, it draws the hump to the outside the predator isocline (Fig. 6.6a). This eliminates the stable point attractor (Fig. 6.6b), the cycles get larger, never returning to the equilibrium.

If we focus primarily on "enrichment" of the prey population, that means increasing carrying capacity or decreasing the self-limitation (decreasing α). If we look at the Jacobian matrix (eq. 6.19), we notice that α occurs only in the prey's negative effect on itself. It is common sense that this negative feedback of the prey upon itself is an important mechanism enhancing stability[9], preventing large oscillations. What we see in the Jacobian matrix is that *decreasing α reduces self regulation in the system.*

The phase plane portrait (Fig. 6.6a) reveals that the equilibrium, where the isoclines cross, is a repeller, an unstable equilibrium — the populations spiral away from the equilbrium.

[9] Recall the Routh-Hurwitz criteria, $\mathbf{J}_{11} + \mathbf{J}_{22} < 0$, $\mathbf{J}_{11}\mathbf{J}_{22} - \mathbf{J}_{12}\mathbf{J}_{21} > 0$

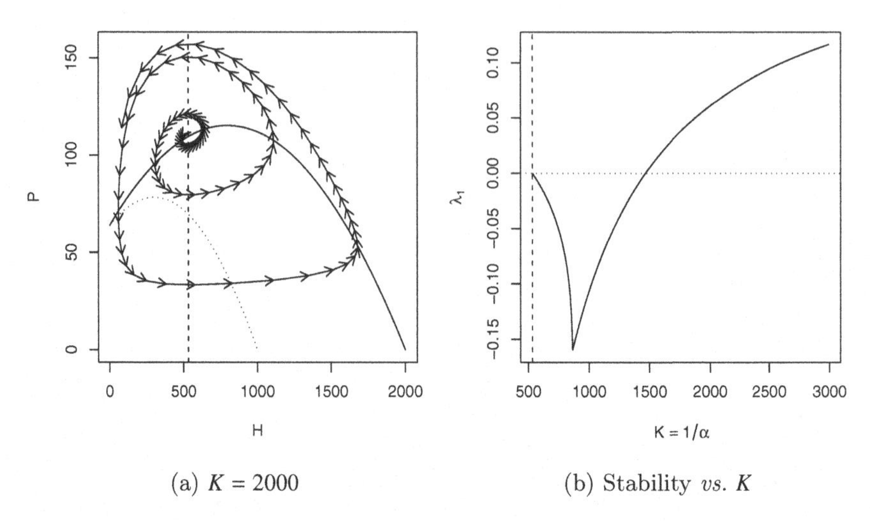

(a) $K = 2000$ (b) Stability *vs.* K

Fig. 6.6: Illustrating the paradox of enrichment. (a) One example of the paradox of enrichment, where large carrying capacity causes cycles instead of a stable attractor (compare to Fig. 6.5). (b) Stability declines when the prey are too strongly self-limiting (very small K) or especially when they have the capacity to achieve very high abundances (large K).

Phase plane portrait for the paradox of enrichment (Fig. 6.6a)

Let's graph the zero net growth isoclines for this model. We use previous parameter values, but change α, and reevaluate range of H we need, and the new trajectory.

```
> p.RM["alpha"] <- alpha <- 5e-04
> Time <- 100
> H <- 0:(1/alpha)
> RM2 <- ode(c(500, 110), 1:Time, predpreyRM, p.RM)
```

Next, we plot everything, starting with the prey isoclines with large K (solid line) and then the small K (dotted line). Last, we add the trajectory for the system with large K, small α.

```
> plot(H, eval(Hiso), type = "l", ylab = "P", xlab = "H", ylim = c(0,
+     max(RM2[, 3])))
> lines(0:1000, HisoStable, lty = 3)
> abline(v = s * D/(e * w - s), lty = 2)
> arrows(RM2[-Time, 2], RM2[-Time, 3], RM2[-1, 2], RM2[-1,
+     3], length = 0.1)
```

The Jacobian for the paradox of enrichment (Fig. 6.6b)

Using the expressions created above, we vary α and examine the effect on λ_1, the dominant eigenvalue. We select αs so that K ranges very small ($K = H^*$) to very large.

```
> H <- with(as.list(p.RM), s * D/(e * w - s))
> alphas <- 1/seq(H, 3000, by = 10)
```

For each α_i in our sequence, we calculate P^*, then evaluate the Jacobian, arranging it in a matrix. The result is a *list* of evaluated Jacobian matrices.

```
> RM.jacList <- lapply(1:length(alphas), function(i) {
+     alpha <- alphas[i]
+     P <- eval(Hiso)
+     matrix(sapply(RMjac1, function(pd) eval(pd)), nrow = 2)
+ })
```

For each evaluated Jacobian matrix, we perform eigenanalysis, and retain the maximum real part of the eigen values.

```
> L1 <- sapply(RM.jacList, function(J) max(Re(eigen(J)[["values"]])))
```

Finally, we plot these dominant eigenvalues *vs.* the carrying capacities that generated them.

```
> plot(1/alphas, L1, type = "l", xlab = quote(italic(K) ==
+     1/alpha), ylab = quote(lambda[1]))
> abline(h = 0, lty = 3)
> abline(v = H, lty = 2)
```

6.2 Space, Hosts, and Parasitoids

A parasitoid is a ghastly thing. These animals, frequently small wasps, characteristically lay an egg on their hosts, often a caterpillar, the young hatch out, and then slowly consume the juicy bits of the host. Neither the adult wasp nor their larvae immediately kill their host. Many eggs can be laid on a host, and many larvae can live (at least briefly) in the host. Ultimately, however, the host is gradually consumed and one or a few larvae per host metamorphosizes into an adult. Thus parasitoid larvae always consume and kill their hapless host. In this sense, their natural history is intermediate between that of predator and parasite — they are parasite-like, or a parasitoid. Thank the gods for parasitoids, as they often help suppress other animals that we like even less, such as herbivores that eat our crops.

There are a variety of characteristics about host–parasitoid relations that might make them different from the kind of predator–prey relations that we have been thinking about up until now. In particular, the life cycles of host and prey are so intimately connected that there is a one-to-one correspondence between dead hosts and the number of parasitoids in the next generation. If we know how many hosts are killed, then we know approximately how many parasitoids are born.

6.2.1 Independent and random attacks

In keeping with convention of parasitoid models, let us describe dynamics of hosts and their parasitoid enemies with a discrete time model [131]. This makes sense, as these organisms, typically insects, frequently complete their life cycles together and within a year. Let us pretend that the hosts, H, grow geometrically in the absence of parasitoids, such that $H_{t+1} = RH_t$. If we assume that some individuals are attacked and killed by parasitoids, H_a, then this rate becomes

$$H_{t+1} = R(H_t - H_a) \tag{6.20}$$

Further, we assume a couple things about parasitoid natural history. First, we pretend the parasitoids *search randomly and independently* for prey, and are able to search area, a (the "area of discovery"), in their life time. The total number of *attack events* then is $E_t = aH_tP_t$. "Attack events" is sort of a strange way to put it but it makes perfect sense, given the natural history. Adults may lay an egg on any host it encounters, but this event does not kill the host. Therefore, hosts can receive more than one egg, thereby making the number of attacked hosts lower than the number of attack events, E_t. Each attack occurs as a random event, and for now we assume that each attack is independent of all others.

For many species of parasitoids, only one adult can emerge from a single host, *regardless of how many eggs were laid on that host.* Therefore, the number of emerging adult parasitoids, P_{t+1}, is simply the number of hosts that were attacked at all, whether the host received one egg or many. Therefore, the number of parasitoids at time $t + 1$ is merely the number of hosts that were attacked, and we can represent this as

$$P_{t+1} = H_a. \tag{6.21}$$

The assumption of one dead host = one new parasitoid can be easily relaxed if we introduce a constant q, such that $P_{t+1} = qH_a$.

Nicholson and Bailey [149] took advantage of a well known discrete probability distribution, the *Poisson distribution*,[10] to create a simple discrete analytical model of host–parasitoid dynamics.

$$H_{t+1} = RH_te^{-aP_t} \tag{6.22}$$

$$P_{t+1} = H_t\left(1 - e^{-aP_t}\right) \tag{6.23}$$

[10] This distribution was discovered by a French mathematician (Siméon-Denis Poisson (1781–1840), so we pronounce the name "pwah-sohn," like the "fois" of fois gras, and the "sohn" with a quiet "n," like, well, like the French would say it. It is the series

$$\frac{1}{e^{\mu}}, \frac{\mu}{1!e^{\mu}}, \frac{\mu^2}{2!e^{\mu}}, \ldots \frac{\mu^r}{r!e^{\mu}}$$

representing the probabilities of occurrence of 0, 1, 2, $\ldots r$.

Here aP_t is the mean number of attacks per larva (λ of the Poisson distribution) and results from P_t parasitoids searching a particular area, a. Parameter a is known as the "area of discovery," the area searched during a parasitoid's lifetime — more area searched (i.e., larger a) means more hosts killed, and fewer to survive and contribute to the next generation. The Poisson distribution tells us that if attacks are random and independent, then e^{-aP_t} is the probability of a host not being attacked. Therefore, $H_t e^{-aP_t}$ is the expected number of hosts which are not attacked and survive to reproduce.

We can find the equilibrium for this discrete model. Recall that, in models of continuous differential equations, this meant letting $dN/dt = 0$. In the discrete case, it means letting $N_{t+1} - N_t = 0$. The equilibrium, N^*, is the value of N when this is true, so $N^* = N_{t+1} = N_t$. We use this to find the equilibrium of the parasitoids, starting with eq. 6.22,

$$H^* = RH^* e^{-aP_t}$$

$$1 = Re^{-aP_t}$$

$$P^* = P_t = \frac{\log R}{a} \tag{6.24}$$

If we recognize that in eq. 6.23, $P_{t+1} = H_t - H_{t+1}/R$, and that $H^* = H_t = H_{t+1}$, then we find the host equilibrium

$$H^* = P^* \frac{R}{R-1}. \tag{6.25}$$

The next section motivates the Nicholson-Bailey model with a simulation of these dynamics. It may provide a more intuitive understanding than simply throwing a probability distribution at the problem. In the end, we wind up at the same place, and this should be reassuring.

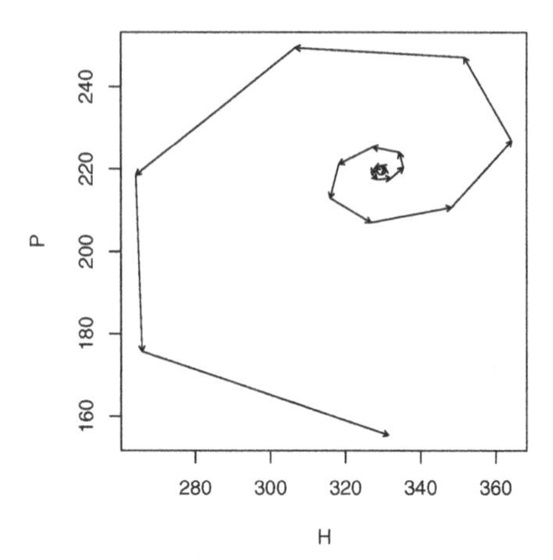

Fig. 6.7: Dynamics around the unstable equilibrium of the Nicholson-Bailey host–parasitoid model ($R = 3$, $a = 0.005$). Arrows indicate a single time step; the point is the equilibrium at the center.

Dynamics of the Nicholson-Bailey host–parasitoid model (Fig. 6.7)

This model assumes that parasitoids search a fixed area, and each attacks hosts randomly and independently. We set the duration of the time series, the model parameters, and create an empty data frame to hold the results.

```
> time <- 20
> R <- 3
> a <- 0.005
> HPs <- data.frame(Hs <- numeric(time), Ps <- numeric(time))
```

Next we calculate the equilibrium, and use a nearby point for the initial abundances.

```
> Pst <- log(R)/a
> Hst <- Pst * R/(R - 1)
> HPs[1, ] <- c(Hst + 1, Pst)
```

We then project the dynamics, one generation at a time, for each time step.

```
> for (t in 1:(time - 1)) HPs[t + 1, ] <- {
+     H <- R * HPs[t, 1] * exp(-a * HPs[t, 2])
+     P <- HPs[t, 1] * (1 - exp(-a * HPs[t, 2]))
+     c(H, P)
+ }
```

Last we plot the trajectory, placing a point at the (unstable) equilibrium, and using arrows to highlight the direction and magnitude of the increment of change per time step.

```
> plot(HPs[, 1], HPs[, 2], type = "n", xlab = "H", ylab = "P")
> points(Hst, Pst)
> arrows(HPs[-time, 1], HPs[-time, 2], HPs[-1, 1], HPs[-1,
+     2], length = 0.05)
```

Simulating Random Attacks

This section relies on code and could be skipped if necessary.

This model makes strong assumptions about the proportion of hosts that escape attacks, that depends on parasitoid attacks being random and independent. Therefore let us simulate parasitoid attacks that are random and independent and compare those data to field data. Let's start with some field data collected by Mark Hassell [131], which are the number of parasitoid larvae per host, either 0, 1, 2, 3, or 4 larvae.

```
> totals <- c(1066, 176, 48, 8, 5)
```

Here, 1066 hosts have no larvae. We can recreate the data set, where we have one observation per host: the number of larvae in that host.

```
> obs <- rep(0:4, totals)
```

To simulate random attacks, let's use the same number of hosts, and the same number of attacks, and let the computer attack hosts at random. We calculate the total number of hosts, H, and the total and mean number of attacks experienced by hosts.

```
> H <- sum(totals)
> No.attacks <- sum(obs)
> mean.attacks <- mean(obs)
```

Next, the predators "sample" the hosts at random and with equal probability. To code this, we identify the hosts by numbering them 1–H. We then attack these hosts independently, that is, we *replace* each attacked host back into the pool of possible prey.

```
> attack.events <- sample(x=1:H, size=No.attacks, replace=TRUE)
```

We then count the number times different hosts were attacked.

```
> N.attacks.per.host <- table(attack.events)
```

and then find count the number of hosts that were attacked once, twice, or more.

```
> (tmp <- table(N.attacks.per.host))

N.attacks.per.host
  1   2   3
236  34   4
```

We see, for instance, that 34 hosts were attacked twice. This also allows us to know how many hosts were *not* attacked,

```
> not.att <- H - sum(tmp)
```

Let's compare the observed data to the simulated data.

```
> obs.sim <- rep(0:max(N.attacks.per.host), c(not.att, tmp))
> table(obs)
```

```
obs
   0    1    2    3    4
1066  176   48    8    5
```

```
> table(obs.sim)
```

```
obs.sim
   0    1    2    3
1029  236   34    4
```

There seem to be fewer unattacked hosts and more attacked hosts in the random-attack simulation than in the observed data. Is this simply a consequence of the observed values being one of many random possibilities? Or is it a systematic difference? How do we check this?

One way to check whether the observed data differ from our model of random and independent attacks is to simulate many such observations, and compare the observed data (1066 unattacked hosts) to the *distribution* of the simulated results. Here we do this by performing n simulations of the same steps we used above.

```
> n <- 1000
> unatt.sim <- sapply(1:n, function(j) {
+   host.sim.att <- sample(x=1:H, size=No.attacks, replace=TRUE)
+   attacks.per <- table(host.sim.att)
+   n.attd <- length(attacks.per)
+   H - n.attd
+ })
```

Next we just make a histogram (Fig. 6.8) of the number of unattacked hosts, adding a dotted line to show where the true data lie, and a dashed line for the prediction, under the assumption of random and independent attacks, based on the Poisson distribution.

```
> hist(unatt.sim, xlab = "Simulated # of Unattacked Hosts",
+      prob = TRUE, xlim = c(1000, 1070))
> abline(v = 1066, lty = 3)
> abline(v = exp(-mean.attacks), lty = 2)
```

Our simulation results (Fig. 6.8) indicate that the observed data include far more unattacked hosts than expected by chance.

Another, quicker way to evaluate the assumption of independent and random attacks is to compare the ratio of the variance of the observed larvae per host to the mean. If the data follow a Poisson distribution, this ratio is expected to equal to one.

```
> (I <- var(obs)/mean(obs))
```

```
[1] 1.404
```

This ratio is greater than one, but we do not know if this could be due to chance. We can test it, because under the null hypothesis, we expect that the product of the ratio and the number of hosts, H, follows a χ^2 distribution, with $H - 1$ degrees of freedom. We can ask how likely this ratio, or a more extreme

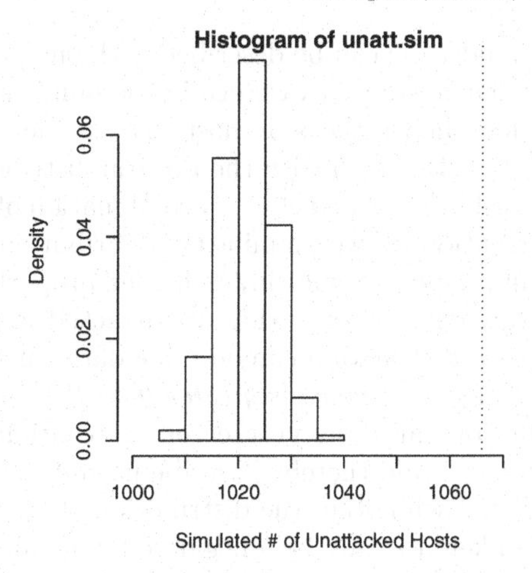

Fig. 6.8: Histogram of simulated host populations, attacked at a rate of 0.24 mean attacks per host, assuming a attacks on hosts are random and independent of each other.

ratio, would be, if attacks are random and independent. We compare it to the cumulative probability density function for the χ^2 distribution.

```
> 1 - pchisq(I * H, df = H - 1)

[1] 0
```

We find that, given this large number of observations (1303 hosts), it is exceedingly unlikely to observe a ratio this large or larger. It is nice that this agrees with our simulation! We feel more confident when a parameteric test agrees with a simulation of our ideas; perhaps both are not necessary, but working through both helps us to understand what we are doing.

6.2.2 Aggregation leads to coexistence

The above model (eqs. 6.22, 6.23) has two problems. The first is that it doesn't reflect the biology — parasitoids tend to aggregate on particular hosts or on local populations of the hosts. Some hosts are more likely to get attacked than others, resulting in more unattacked hosts and more hosts receiving multiple attacks, than predicted under random and independent attacks. This may be related to their searching behavior, or to spatial distributions of hosts in the landscape. The second problem is that the above model is not stable, and predicts that the parasitoid or host becomes extinct. We know that in nature they don't become extinct! Perhaps we can kill two birds with one stone, and fix both problems with one step. That is what Robert May [131] and others have done, by assuming that parasitoids *aggregate*.

May proposed the following logic for one reason we might observe more unattacked hosts than expected. Imagine that the distribution of parasitoids *among* patches in the landscape can be described with one probability distribution, with some particular mean and variance in the number of parasitoids per patch. Imagine next that their attacks on hosts *within* each patch are random and independent and thus described with the Poisson distribution. This will result in a compound distribution of attacks — a combination of the among-patch, and within-patch distributions. If we examine the distribution of larvae per host, for all hosts in the landscape, we will find a higher proportion of unattacked hosts, and a higher proportion of hosts that are attacked many times.

The *negative binomial distribution* can describe data, such as the number of larvae per host, in which *the variance is greater than the mean.* Thus the negative binomial distribution can describe greater aggregation than the Poisson distribution (where $\mu = \sigma^2$), and thereby describe nature somewhat more accurately. May suggested that while the true distribution of larvae in hosts, in any particular case, was unlikely to truly be a negative binomial, it was nonetheless a useful approximation.

Ecologists frequently use the negative binomial distribution for data where the variance is greater than the mean. There are different derivations of the distribution, and therefore different parameterizations [13]. In ecology, we typically use the mean, μ, and the *overdispersion* parameter k. The variance, σ^2, is a function of these, where $\sigma^2 = \mu + \mu^2/k$; by overdispersion we mean that $\sigma^2 > \mu$. Large values of k ($k > 10$) indicate randomness, and the distribution becomes indistinguishable from the Poisson. Small values ($k < 2$) indicate aggregation.[11] Fig. 6.9 shows examples.

Showing the negative binomial distribution (Fig. 6.9)

R has the negative binomial distribution built in, but does not use `k` as one of its arguments; rather, `size` = k. Here we generate a graph showing the distribution with different values of k.

```
> nb.dat <- cbind(Random = dnbinom(0:5, mu = 0.5, size = 10^10),
+      `k=10` = dnbinom(0:5, mu = 0.5, size = 10), `k=1` = dnbinom(0:5,
+          mu = 0.5, size = 1), `k=0.01` = dnbinom(0:5, mu = 0.5,
+          size = 0.01))
> matplot(0:5, nb.dat, type = "b", pch = 1:4, col = 1, ylim = c(0,
+      1), xlab = "Attacks per Hosts", ylab = "Probability")
> legend("topright", rev(colnames(nb.dat)), pch = 4:1, lty = 4:1,
+      bty = "n")
> mtext(quote(mu == 0.5), padj = 2)
```

The proportion of unattacked hosts expected under the negative binomial is $(1 + aP_t/k)^{-k}$. Therefore, we can write the analytical expressions for the population dynamics that are very similar to those of the Nicholson-Bailey model, but using the negative binomial distribution.

[11] Although k is often referred to as a "clumping" or "aggregation" parameter, we might think of it as a *randomness* parameter, because larger values result in more random, Poisson-like distributions.

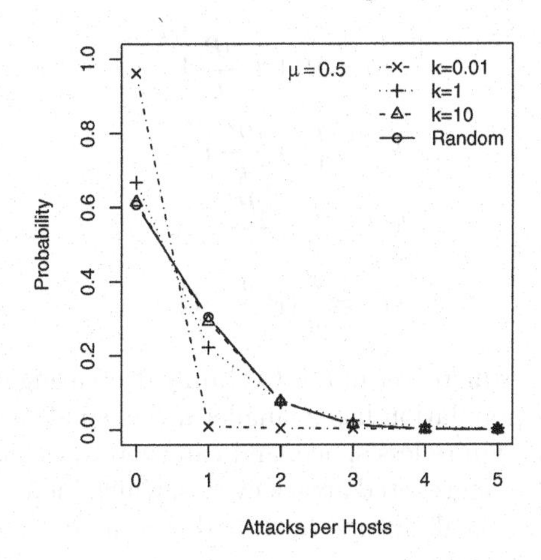

Fig. 6.9: The negative binomial distribution, where the dispersion parameter k controls variance or breadth the distribution. For a given mean, smaller k causes a greater variance, and results in a higher proportion of zeroes. These k are representative of values from Pacala *et al.* (1990).

$$H_{t+1} = RH_t\left(1 + \frac{aP_t}{k}\right)^{-k} \tag{6.26}$$

$$P_{t+1} = H_t - H_t\left(1 + \frac{aP_t}{k}\right)^{-k} \tag{6.27}$$

May [131] referred to this as a phenomenlogical model (as opposed to mechanistic) because the negative binomial merely approximates the true, perhaps compound, distribution.

Equilibria for a discrete-time model

The equilibria of the host and parasitoid populations in May's model (eqs. 6.26, 6.27) are derived simply, once we decide upon how to describe the condition for an equilibrium (a.k.a. steady state). In models of differential equations, this meant letting $\mathrm{d}N/\mathrm{d}t = 0$. In the discrete case it means letting $N_{t+1} - N_t = 0$. The equilibrium is the value of N when this is true, so $N^* = N_{t+1} = N_t$. Following this train of thought, we have $H^* = H_{t+1} = H_t$ and $P^* = P_{t+1} = P_t$.

To solve for equilbria, we begin with the expression for hosts (eq. 6.26), and solve for equilibrium parasitoid density. In this case, we can divide eq. 6.26 both sides by H^*.

$$
\begin{aligned}
1 &= R\left(1+\frac{aP_t}{k}\right)^{-k}\\
R^{-1} &= \left(1+\frac{aP_t}{k}\right)^{-k}\\
R^{1/k} &= 1+\frac{aP_t}{k}\\
P^* &= \frac{k}{a}\left(R^{1/k}-1\right)
\end{aligned}
\tag{6.28}
$$

Given this, what causes increases in P^*? Certainly decreasing a leads to increases in P^*. If a parasitoid population has a smaller a (i.e. smaller area of discovery), this means that they require less space, and can thereby achieve higher *density*. Increasing k means less aggregated attack events, which increases the probability that a parasitoid larvae is alone on a host, and this increases parasitoid density, but only up to a limit.

If we want H^* as well, we can solve for that too. One of many ways is to realize that, eq. 6.27 can be rearranged to show

$$P_{t+1} = H_t - \frac{H_{t+1}}{R}$$

and given contant H and P,

$$H^* = P^*\left(\frac{R}{R-1}\right) \tag{6.29}$$

where P^* is eq. 6.26.

6.2.3 Stability of host–parasitoid dynamics

As with differential equation models, we can analyze the stability in the immediate vicinity of equilibria in discrete time models using eigenanalysis of the Jacobian matrix. Although analogous, there are some important differences between continuous and discrete models.

- The Jacobian matrix is comprised of partial derivatives of growth increments (e.g., $\Delta N = N_{t+1} - N_t$), rather than of time derivatives (e.g., $\dot{N}$).
- Interpretation of eigenvalues of the Jacobian reflects the difference between finite rate of increase (λ) and the intrinsic rate of increase (r); e.g., for discrete models, $\lambda = 1$ indicates zero growth.
- Populations with discrete generations can grow too quickly to be stable (e.g., chaos in discrete logistic growth).

In the discrete time case, the Jacobian matrix is the set of *partial derivatives of the discrete growth increment* rather than of the time derivatives used for continuous growth[12]. The growth increments are the increments of change over an entire time step, or $\Delta H = H_{t+1} - H_t$ and $\Delta P = P_{t+1} - P_t$. Taking those increments from eqs. 6.26, 6.27, the Jacobian matrix of partial differential equations is

[12] Note $\mathrm{d}N/\mathrm{d}t$ is also an increment — it is the increment ΔN as $\Delta t \to 0$, whereas in the discrete case, $\Delta t = 1$.

$$\begin{pmatrix} \frac{\partial \Delta H}{\partial H} & \frac{\partial \Delta H}{\partial P} \\ \frac{\partial \Delta P}{\partial H} & \frac{\partial \Delta P}{\partial P} \end{pmatrix} = \begin{pmatrix} R(1+aP)^{-k} - 1 & -akHR(1+aP)^{-(k+1)} \\ 1-(1+aP)^{-k} & akH(1+aP)^{-(k+1)} \end{pmatrix} \tag{6.30}$$

To analyze this at the equilibrium, we substitute H^*, P^* for H, P, and perform eigenanalysis. The resulting eigenvalues, λ_1, λ_2, then reflect *perturbation growth increments*, the discrete time analogy of the instantaneous perturbation growth rates of continuous models.

Recall that for continuous time, λ_1 is the instantaneous rate of change in a peturbation at equilibrium,

$$\dot{x} = \lambda_1 x, \quad x_t = x_0 e^{\lambda_1 t} \tag{6.31}$$

where x is the perturbation at the equilibrium; if $\lambda_1 < 0$, the perturbation would decline. For discrete growth, we can think of λ_1 as the discrete growth factor of a perturbation at equilibrium,

$$x_{t+1} = \lambda_1 x_t \tag{6.32}$$

where x is the perturbation. Here, we see that x will decline as long as $0 < \lambda_1 < 1$. Second, if $\lambda_1 < 0$, then x_{t+1} changes sign with each time step, and the perturbation oscillates around 0. That is alright, as the magnitude also decreases with each time step, so the perturbation still declines toward zero. However, if $\lambda < -1$, those oscillations grow and the perturbation oscillates permanently. This is directly analogous to the oscillations, or stable limit cycles, we saw in the discrete logistic growth model. Thus, a criterion for stability in discrete growth models is that for all eigenvalues, $-1 < \lambda < 1$.

For discrete models, we also need to be a little more concerned about the imaginary part of the eigenvalues, because they contribute to the magnitude of the oscillations and eigenvalues. We therefore add that the magnitude of $\lambda = a + bi$ is $|\lambda| = \sqrt{a^2 + b^2}$. Thus, the system is stable when $|\lambda| \leq 1$. The magnitude of a complex number is known as its *modulus.* The moduli (*plural* of modulus) of the host–parasite model therefore includes the complex plane, were the real part of each eigenvalue is on the x-axis, and the imaginary part is on the y-axis (Fig. 6.10a). The magnitude or modulus of an eigenvalue is the length of the vector in this plane.

We can illustrate the stability criterion for discrete models in a few ways. The phase plane portrait or time series would illustrate the dynamics directly. However, it is also useful to show the relation between a measure of stability and parameters influencing stability (e.g., Fig. 6.6b). Since aggregation seems so important in the host–parasitoid model, we can show how stability (λ) varies with k. We can thus investigate *how much* aggregation is required for stability [154]. We would anticipate that stability declines ($|\lambda| \rightarrow 1$) as k increases.

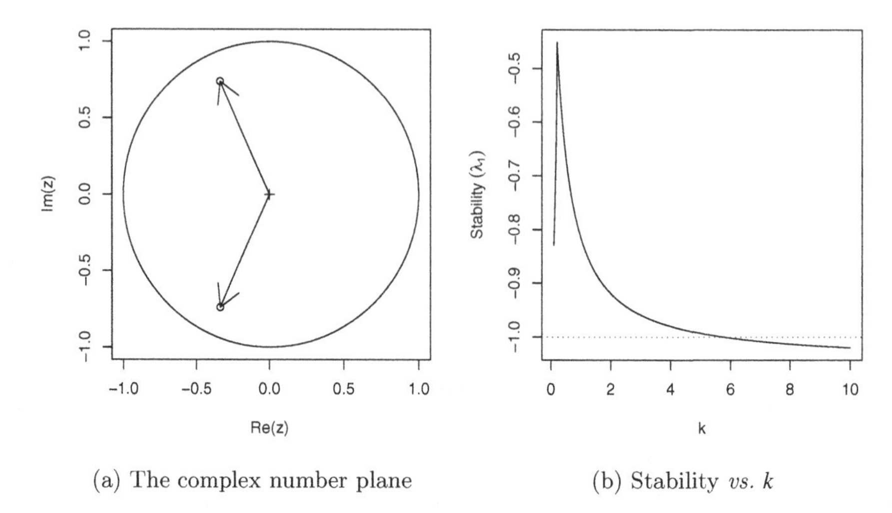

(a) The complex number plane

(b) Stability *vs.* k

Fig. 6.10: Dynamical stability of a discrete host–parasitoid model with aggregation (a) Region of stability for the rate of change following a small perturbation away from equilibrium. The plus sign "+" is (0, 0) and the two small circles are the complex eigenvalues. The length of the vector is the modulus.

Stability of the host–parasitoid model with aggregation (Fig. 6.10b)

We proceed largely as we did for continuous models, first with expressions for, and partial derivatives of, the relevant functions — for discrete models we use $F(N_t)$, where of $N_{t+1} = F(N_t)$.

```
> F.H <- expression(R * H * (1 + a * P/k)^-k - H)
> F.P <- expression(H - H * (1 + a * P/k)^-k - P)
> F.H.H <- D(F.H, "H")
> F.H.P <- D(F.H, "P")
> F.P.H <- D(F.P, "H")
> F.P.P <- D(F.P, "P")
```

We next specify a sequence of k's, and for each k, find the equilibria, evaluate the Jacobian, and return the eigenvalues of the Jacobian.

```
> k <- 10^seq(-1, 1, by = 0.01)
> R <- 3
> a <- 0.005
> HPeigs <- sapply(k, function(ki) {
+     k <- ki
+     P <- k * (R^(1/k) - 1)/a
+     H <- P * R/(R - 1)
+     jac <- matrix(c(eval(F.H.H), eval(F.H.P), eval(F.P.H),
+         eval(F.P.P)), nrow = 2, byrow = TRUE)
+     eigen(jac)[["values"]]
+ })
```

Last, we plot the eigenvalue with the greatest absolute magnitude, and retain the sign of the real part, λ *vs.* k.

```
> modmaxs <- apply(HPeigs, 2, function(lambdas) {
+     i <- which.max(Mod(lambdas))
+     sign(Re(lambdas[i])) * Mod(lambdas[i])
+ })
> plot(k, modmaxs, type = "l", ylab = quote("Stability " *
+     (lambda[1])))
> abline(h = -1, lty = 3)
```

Graphing eigenvalues in the complex number plane (Fig. 6.10a)

It is typically important to evaluate the modulus, or magnitude, of the eigenvalues of a Jacobian matrix for a discrete model. First we set up the unit circle which will define the stability region in the complex number plane.

```
> th <- seq(-pi, pi, len = 100)
> z <- exp((0+1i) * th)
```

We then plot the circle and add the eigenvalues for our smallest k;

```
> par(pty = "s")
> plot(z, type = "l")
> points(0, 0, pch = 3)
> points(HPeigs[, 100])
> arrows(x0 = c(0, 0), y0 = c(0, 0), x1 = Re(HPeigs[, 100]),
+     y1 = Im(HPeigs[, 100]))
```

The length of the arrows are the moduli, $|\lambda|$.

Dynamics of the May host–parasitoid model

Here we simply play with May's model. The following generates a phase plane diagram of the dynamics, although it is not shown. We set the duration of the time series, the model parameters, and create an empty data frame to hold the results.

```
> time <- 20
> R <- 3
> a <- 0.005
> k <- 0.6
> HP2s <- data.frame(Hs <- numeric(time), Ps <- numeric(time))
```

Next we calculate the equilibrium, and use a nearby point for the initial abundances.

```
> P2st <- k * (R^(1/k) - 1)/a
> H2st <- P2st * R/(R - 1)
> P2st

[1] 628.8

> H2st

[1] 943.2

> HP2s[1, ] <- c(1000, 500)
```

We then project the dynamics, one generation at a time, for each time step.

```
> for (t in 1:(time - 1)) HP2s[t + 1, ] <- {
+     H <- R * HP2s[t, 1] * (1 + a * HP2s[t, 2]/k)^(-k)
+     P <- HP2s[t, 1] - HP2s[t, 1] * (1 + a * HP2s[t, 2]/k)^(-k)
+     c(H, P)
+ }
```

Last we plot the trajectory, placing a point at the equilibrium, and using arrows to highlight the direction and magnitude of the increment of change per time step.

```
> plot(HP2s[, 1], HP2s[, 2], type = "l", xlab = "H", ylab = "P")
> points(H2st, P2st)
> arrows(HP2s[-time, 1], HP2s[-time, 2], HP2s[-1, 1], HP2s[-1,
+     2], length = 0.05)
```

6.3 Disease

Here we discuss epidemiological disease models. Pathogens cause diseases, and are typically defined as microorganisms (fungi, bacteria, and viruses) with some host specificity, and which undergo population growth within the host.

Our simplest models of disease are funny, in that they don't model pathogens (the enemy) at all. These famous models, by Kermack and McCormick [91], keep track of different types of hosts, primarily those with and without disease symptoms. That makes them *epidemiological models.* Specifically, these are SIR models [91] that model all N hosts by keeping track of

Susceptible hosts Individuals which are not infected, but could become infected,

Infected hosts Individuals which are already infected, and

Resistant hosts Individuals which are resistant to the disease, typically assumed to have built up an immunity through past exposure,

where $N = S + I + R$. It is important to note that N, S, I, and R are *densities.* That is, we track numbers of individuals in a fixed area. This is important because it has direct consequences for the spread, or transmission, of disease [136].

Disease spreads from infected individuals to susceptible individuals. The rate depends to some extent on the number or alternatively, on the fraction of the population that is infected. Resistant individuals are not typically considered vectors of the pathogens, and so increased abundance of resistant individuals slow the transmission rate by diluting the other two groups.

A good starting place is a simple SIR model for a population of constant size, with no immigration or emigration [48, 91].

$$\frac{dS}{dt} = -\beta IS \tag{6.33}$$

$$\frac{dI}{dt} = \beta IS - \gamma I \tag{6.34}$$

$$\frac{dR}{dt} = \gamma I \tag{6.35}$$

Density-dependent SIR *model*

Here we create the function for the system of ODE's in eq. 6.33.

```
> SIR <- function(t, y, p) {
+     {
+         S <- y[1]
+         I <- y[2]
+         R <- y[3]
+     }
+     with(as.list(p), {
+         dS.dt <- -B * I * S
+         dI.dt <- B * I * S - g * I
+         dR.dt <- g * I
+         return(list(c(dS.dt, dI.dt, dR.dt)))
+     })
+ }
```

In this model, the *transmission coefficient*, β, describes the instantaneous rate at which the number of infected hosts increases per infected individual. It is directly analogous to the attack rate of type I Lotka–Volterra perdator–prey models. Recall that it is based on the law of mass action, borrowed from physics and chemistry. It assumes that the rate at which events occur (new infections) is due to complete and random mixing of the reactants (S, I), and the rate at which the reactants collide and react can be described by a single constant, β. As density of either type of molecule increases, so too does the rate of interaction. In Lotka–Volterra predation, we referred to βS as a linear functional response; here we refer to βS as the *transmission function* and in particular we call it a *mass action* or *density-dependent* transmission function. The *transmission rate* is the instantaneous rate for the number of *new infections* or cases per unit time [136].

Resistant individuals might be resistant for one of two reasons. They may die, or they may develop immunities. In either case, we assume they cannot catch the disease again, nor spread the disease. As this model assumses a constant population size, we continue to count all R individuals, regardless of whether they become immune or die.

The individuals become resistant to this disease at the constant per capita rate, γ. The rate γ is also the inverse of the mean residence time, or *duration*, of the disease[13].

Disease *incidence* is the number of new infections or cases occurring over a defined time interval. This definition makes incidence a discrete-time version of transmission rate. *Prevalence* is the fraction of the population that is infected I/N.

A common question in disease ecology is to ask under what conditions will an outbreak occur. Another way of asking that is to ask what conditions cause

[13] This is an interesting phenomenon of exponential processes — the mean time associated with the process is equal to the inverse of the rate. This is analogous to turnover time or residence time for a molecule of water in a lake.

$\dot{I} > 0$. We can set $\mathrm{d}I/\mathrm{d}t > 0$ and solve for something interesting about what is required for an outbreak to occur.

$$\begin{aligned} 0 &< \beta IS - \gamma I \\ \frac{\gamma}{\beta} &< S \end{aligned} \tag{6.36}$$

What does this tell us? First, because we could divide through by I, it means that if no one is infected, then an outbreak can't happen — it is the usual, but typically unstable equilibrium at 0. Second, it tells us that an outbreak will occur if the absolute density of susceptibles[14] is greater than γ/β. If we consider the pre-outbreak situation where $S \approx N$, then simply making the population size (and density) low enough can halt the spread of disease. This is why outbreaks tend to occur in high density populations, such as agricultural hosts (e.g., cattle), or historically in urban human populations, or in schools.

Vaccinations are a way to reduce S without reducing N. If a sufficient number of individuals in the population are vaccinated to reduce S below γ/β, this tends to protect the unvaccinated individuals as well.

Another common representation of this is called the "force of infection" or "basic reproductive rate of the disease." If we assume that in a large population $S \approx N$, then rearranging eq. 6.36 gives us

$$R_0 = \frac{\beta N}{\gamma} \tag{6.37}$$

where R_0 is the basic reproductive rate of the disease. If $R_0 > 1$, then an outbreak (i.e., disease growth) is plausible. This is analogous to the finite rate of increase of a population where $\lambda > 1$.

Simple SIR *dynamics (Fig. 6.11)*

Here we model the outbreak of a nonlethal disease (e.g., a new cold virus in winter on a university campus). We assume that the disease is neither live-threatening, and nor is anyone resistant, thus $R_{t=0} = 0$. We can investigate the SIR model by pretending that, as is often the case, we begin with a largely uninfected population and $t = 0$, so $I_0 = 1$ and $S_0 \approx N$. We first set parameters.

```
> N <- 10^4
> I <- R <- 1
> S <- N - I - R
> parms <- c(B = 0.01, g = 4)
```

We next integrate for three months.

```
> months <- seq(0, 3, by = 0.01)
> require(deSolve)
> SIR.out <- data.frame(ode(c(S, I, R), months, SIR, parms))

> matplot(months, SIR.out[, -1], type = "l", lty = 1:3, col = 1)
> legend("right", c("R", "I", "S"), lty = 3:1, col = 3:1, bty = "n")
```

[14] S is the absolute density, whereas S/N is the relative density.

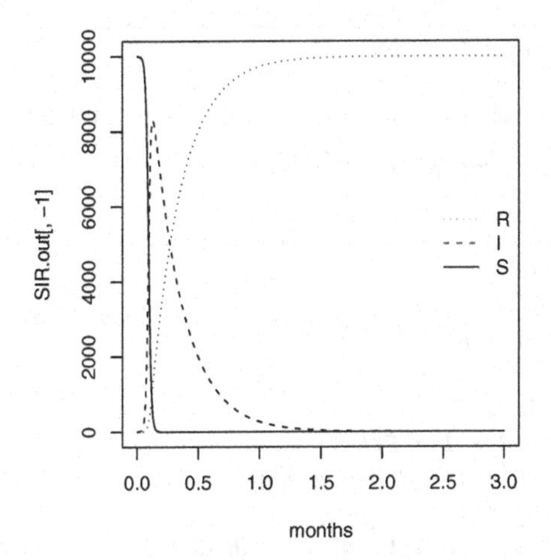

Fig. 6.11: Epidemic with the simplest SIR model. Assumes constant population size.

6.3.1 SIR with frequency–dependent transmission

It is important at this point to reiterate a point we made above — *these conclusions apply when S, I, and R are densities* [136]. If you increase population size but also the area associated with the population, then you have not changed density. If population size only increases, but density is constant, then interaction frequency does not increase. Some populations may increase in density as they increase is size, but some may not. Mass action dynamics are the same as type I functional response as predators — there is a constant linear increase in per capita infection rate as the number of susceptible hosts increases.

In addition to mass action (a.k.a. density dependent) transmission, investigators have used other forms of density dependence. One the most common is typically known as *frequency–dependent* transmission, where

$$\frac{\mathrm{d}S}{\mathrm{d}t} = -\beta\frac{SI}{N} \tag{6.38}$$

$$\frac{\mathrm{d}I}{\mathrm{d}t} = \beta\frac{SI}{N} - \gamma I \tag{6.39}$$

$$\frac{\mathrm{d}I}{\mathrm{d}t} = \gamma I. \tag{6.40}$$

Frequency–dependent SIR *model*

Here we create the function for the system of ODEs in eq. 6.33.

```
> SIRf <- function(t, y, p) {
+     {
+         S <- y[1]
+         I <- y[2]
+         R <- y[3]
+         N <- S + I + R
+     }
+     with(as.list(p), {
+         dS.dt <- -B * I * S/N
+         dI.dt <- B * I * S/N - g * I
+         dR.dt <- g * I
+         return(list(c(dS.dt, dI.dt, dR.dt)))
+     })
+ }
```

The proper form of the transmission function depends on the mode of transmission [136]. Imagine two people are on an elevator, one sick (infected), and one healthy but susceptible, and then the sick individual sneezes [48]. This results in a particular probability, β, that the susceptible individual gets infected. Now imagine resistant individuals get on the elevator — should adding resistant individuals change the probability that the susceptible individual gets infected? Note what has and has not changed. First, with the addition of a resistant individual, N has increased, and prevalence, I/N, has decreased. However, the densities of I and S remain the same (1 per elevator). What might happen? There are at least two possible outcomes:

1. If sufficient amounts of the virus spread evenly throughout the elevator, adding a resistant individual does *not* change the probability of the susceptible becoming sick, and the rate of spread will remain dependent on the densities of I and S — the rate will not vary with declining prevalence.
2. If only the immediate neighbor gets exposed to the pathogen, then the probability that the neighbor is susceptible declines with increasing R, and thus the rate of spread *will* decline with declining prevalence.

It is fairly easy to imagine different scenarios, and it is very important to justify the form of the function.

Density–dependent transmission (Fig. 6.12) is independent of the number of resistant individuals; having higher density of infected individuals or more susceptible individuals always enhances transmission rate, assuming both I, $S > 0$. In contrast, frequency–dependent transmission does depend on the density of resistant (living) individuals because they can "get in the way" of transmission (consider our elevator example above). That is, they reduce to probability of infected individuals coming into contact with susceptible hosts[15]. That is, there is a greater probability that the immediate neighbor of an infected host is already infected, and so cannot become a new infection or case. Similarly, having more

[15] Note $N = S + I + R$, and that increasing R cause decreasing SI/N.

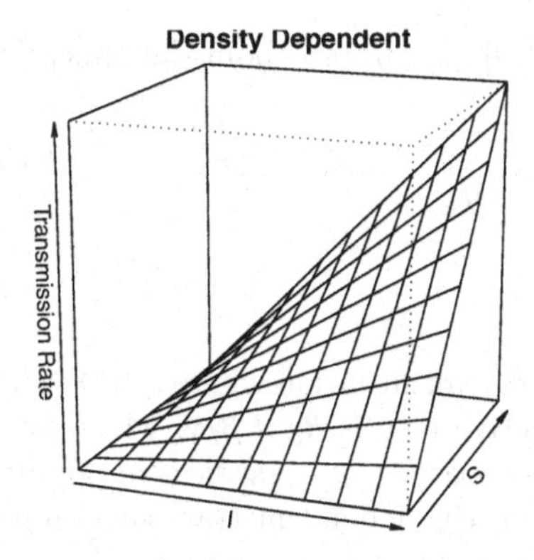

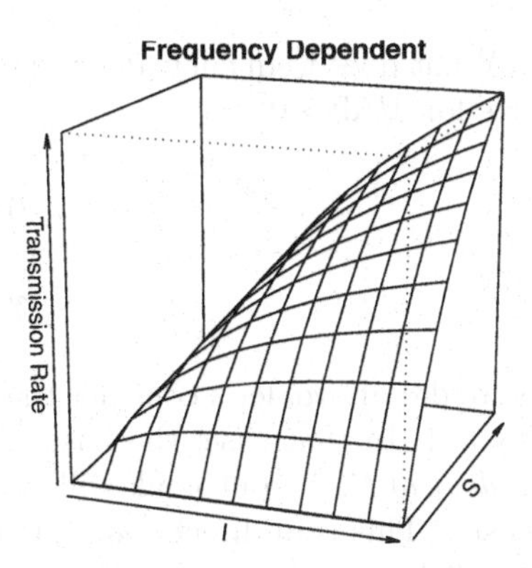

Fig. 6.12: The rate of transmission may depend only on the linear dependence of mass action, or may depend curvilinearly on the prevalence, the frequency of infected individuals in the population.

susceptible hosts makes it more likely the immediate neighbor of a susceptible host is another susceptible host and not a source of infection.

Transmission models (Fig. 6.12)

Here we plot density–dependent and frequency–dependent transmission rates, as functions of S and I. We rescale the transmission coefficient appropriately ($\beta_F = N_{max}\beta_D$) [136].

```
> R <- 0
> S <- I <- 1000
> Ss <- Is <- seq(1, S, length = 11)
> N <- S + I + R
> betaD <- 0.1
> betaF <- betaD * N
```

We use `sapply` to calculate the transmission functions for each combination of the values of I and S.

```
> mat1 <- sapply(Is, function(i) betaD * i * Ss)
> mat2 <- sapply(Is, function(i) betaF * i * Ss/(i + Ss + R))
```

Now we plot these matrices.

```
> layout(matrix(1:2, nr = 1))
> persp(mat1, theta = 20, phi = 15, r = 10, zlim = c(0, betaD *
+     S * I), main = "Density Dependent", xlab = "I", ylab = "S",
+     zlab = "Transmission Rate")
> persp(mat2, theta = 20, phi = 15, r = 10, zlim = c(0, betaF *
+     S * I/N), main = "Frequency Dependent", xlab = "I", ylab = "S",
+     zlab = "Transmission Rate")
```

What does frequency–dependent transmission imply about dynamics? Let's solve for $\mathrm{d}I/\mathrm{d}t > 0$.

$$\begin{aligned} 0 &< \beta\frac{SI}{N} - \gamma I \\ \gamma &< \beta\frac{S}{N}. \end{aligned} \tag{6.41}$$

As we did above, let's consider the pre-outbreak situation where $S \approx N$, so that $S/N \approx 1$. In that case, the basic reproductive rate is $R_0 = \beta/\gamma$, which is *independent of* N. An outbreak will occur as long as $\beta > \gamma$, regardless of population density. This is in direct contrast to the density–dependent transmission model (eqs. 6.38, 6.40), where outbreak could be prevented if we simply reduce the population, N, to a sufficently low density. Both modes of transmission are observed in human and non-human populations, so it is important to understand how the disease is spread in order to predict its dynamics.

Another interesting phenomenon with frequency–dependent transmission is that prevalence (I/N) can decline with increasing population size (Fig. 6.13). Two phenomena contribute to this pattern. First, outbreak in a completely susceptible population typically begins with a single individual, and so initial prevalence is always $I/N = 1/N$. Second, as a consequence of this, the transmission rate is lower in larger populations because $\beta SI/N$ is small. As a consequence, prevalence remains low for a relatively long time. In a seasonal population, most individuals in large populations remain uninfected after four months. Depending upon the organism, this could be long enough to reproduce. In contrast, a density–dependent model typically shows the oppositive, pattern, with more rapid, extreme outbreaks and greater prevalence in more dense populations (Fig. 6.13).

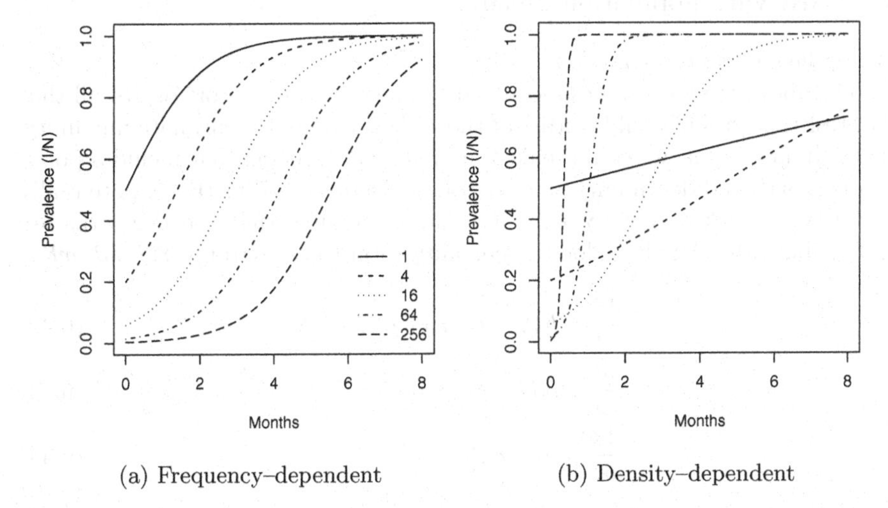

(a) Frequency–dependent (b) Density–dependent

Fig. 6.13: Prevalence (I/N) *vs.* population density. With frequency–dependent transmission, (a), prevalence may decrease with population density. In contrast, with density–dependent transmission, (b), prevalence may increase with density.

SIR *dynamics with frequency–dependent transmission (Fig. 6.13)*

Here we demonstrate that prevalence can decline with increasing population size in a frequency–dependent disease (e.g., a smut on plant [3]). Let us assume that resistance cannot be acquired, so $\gamma = 0$, and $R = 0$. We can investigate the SIR model by pretending that, as is often the case, we begin with a largely uninfected population and $t = 0$, so $I_0 = 1$ and $S_0 \approx N$. We first set parameters.

```
> S <- 4^(0:4); I <- 1
> parmsf <- c(B = 1, g = 0)
> parmsd <- c(B = 1/16, g = 0)
```

We next integrate for six months, letting $R = S/2$.

```
> Months <- seq(0, 8, by = 0.1)
> outd <- sapply(S, function(s) {
+     out <- ode(c(s, I, R), Months, SIR, parmsd)
+     out[, 3]/apply(out[, 2:4], 1, sum)
+ })
> outf <- sapply(S, function(s) {
+     out <- ode(c(s, I, R), Months, SIRf, parmsf)
+     out[, 3]/apply(out[, 2:4], 1, sum)
+ })
```

Last, we make the figures.

```
> matplot(Months, outd, type = "l", col = 1, ylab = "Prevalence (I/N)")
> matplot(Months, outf, type = "l", col = 1, ylab = "Prevalence (I/N)")
> legend("bottomright", legend = S, lty = 1:length(S), bty = "n")
```

6.3.2 SIR with population dynamics

(*The following sections rely on code.*)

The above model assumes a constant population size — sort of. Recall that the "resistant group" could consist of those that acquire the ultimate immunity, death. In any event, we could make a more complex model that includes population growth and death unrelated to disease. Here we add births, b, potentially by all types, sick or not $(S + I + R)$, and we assume that the newborns are susceptible only. We also added a mortality term to each group (mS, mI, mR).

$$\frac{\mathrm{d}S}{\mathrm{d}t} = b(S + I + R) - \beta S I - mS \quad (6.42)$$

$$\frac{\mathrm{d}I}{\mathrm{d}t} = \beta S I - \gamma I - mI \quad (6.43)$$

$$\frac{\mathrm{d}R}{\mathrm{d}t} = \gamma I - mR \quad (6.44)$$

Note that the births add only to the susceptible group, whereas density independent mortality subtracts from each group.

Disease model with population growth

Here we create the function for the system of ODE's in eq. 6.42.

```
> SIRbd <- function(t, y, p) {
+      S <- y[1]
+      I <- y[2]
+      R <- y[3]
+      with(as.list(p), {
+          dS.dt <- b * (S + I + R) - B * I * S - m * S
+          dI.dt <- B * I * S - g * I - m * I
+          dR.dt <- g * I - m * R
+          return(list(c(dS.dt, dI.dt, dR.dt)))
+      })
+ }
```

Let's start to work with this model — that frequently means making simplifying assumptions. We might start by assuming that if infected and resistant individuals can contribute to offspring, then the disease is relatively benign. Therefore, we can assume that mortality is the same for all groups ($m_i = m$). Last, let us assume (again) a constant population size. This means that birth rate equals mortality or $b = m$.

Now imagine a large city, with say, a million people. Let's then assume that we start of with a population of virtually all susceptible people, but we introduce a single infected person.

```
> N <- 10^6
> R <- 0
> I <- 1
> S <- N - I - R
```

Let us further pretend that the disease runs its course over about 10–14 days. Recall that γ ("gamma") is the inverse of the *duration* of the disease.

```
> g <- 1/(13/365)
```

Given a constant population size and exponential growth, then the average life span is the inverse of the birth rate. Let us pretend that the average life span is 50 years.

```
> b <- 1/50
```

For this model, the force of infection turns out to be $R_0 = 1 + 1/(b + \alpha)$, where α is the average age at onset of the disease [48]. We can therefore estimate β from all the other parameters, including population size, average life span, average age at onset, and the average duration of the disease. For instance, imagine that we have a disease of children, where the average onset of disease is 5 y, so we have

```
> age <- 5
> R0 <- 1 + 1/(b * age)
```

so β becomes

```
> B <- R0 * (g + b)/N
```

Finally, we can integrate the population and its states. We create a named vector of parameters, and decide on the time interval.

```
> parms <- c(B = B, g = g, b = b, m = b)
> years <- seq(0, 30, by = 0.1)
```

It turns out that because of the relatively extreme dynamics (Fig. 6.14), we want to tell the ODE solver to take baby steps, so as to properly capture the dynamics — we use the `hmax` argument to make sure that the maximum step it takes is relatively small.

```
> SIRbd.out <- data.frame(ode(c(S = S, I = I, R = R), years,
+     SIRbd, parms, hmax = 0.01))
```

```
> matplot(SIRbd.out[, 1], sqrt(SIRbd.out[, -1]), type = "l",
+     col = 1, lty = 1:3, ylab = "sqrt(No. of Individuals)",
+     xlab = "Years")
> legend("right", c("S", "I", "R"), lty = 1:3, bty = "n")
```

Note that the population quickly becomes resistant (Fig. 6.14). Note also that we have oscillations, albeit damped oscillations. An analytical treatment of the model, including eigenanalysis of the Jacobian matrix could show us precisely the predicted periodicity [48]. It depends on the the age at onset, and the duration of the disease.

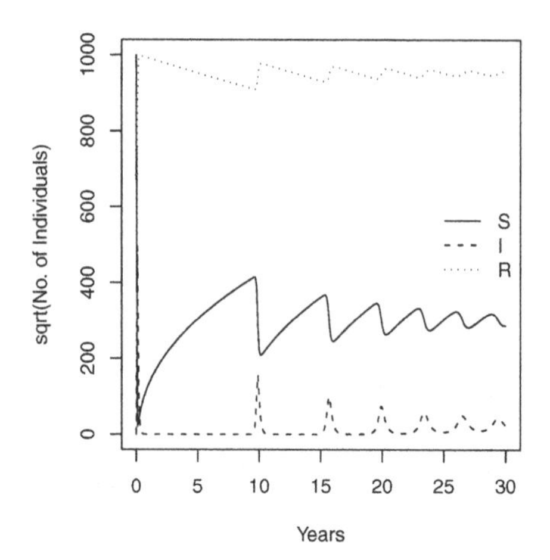

Fig. 6.14: Epidemic for a nonlethal disease, with an SIR model which includes births and deaths, and a constant population size.

6.3.3 Modeling data from Bombay

Here we try our hand at fitting the SIR model to some data. Kermack and McCormick [91] provided data on the number of plague deaths per week in Bombay[16] in 1905–06. We first enter them and look at them[17].

```
> data(ross)
> plot(CumulativeDeaths ~ Week, data = ross)
```

As with most such enterprises, we wish we knew far more than we do about the which depends on fleas, rodents, humans, and *Yersinia pestis* on which the dynamics depend. To squeeze this real-life scenario into a model with a small number of parameters requires a few assumptions.

A good starting place is a simple SIR model for a population of constant size (eq. 6.33) [48,91].

Optimization

We next want to let R find the most accurate estimates of our model parameters β, γ. The best and most accessible reference for this is Bolker [13]. Please read the Appendix B.11 for more explanation regarding optimization and objective functions.

Now we create the objective function. An objective function compares a model to data, and calculates a measure of fit (e.g., residual sum of squares, likelihood). Our objective function will calculate the likelihood[18] of particular

[16] Bombay is the coastal city now known as Mumbai, and is the capital of Maharashtra; it is one of the largest cities in the world.

[17] Data provided kindly by S.P. Ellner

[18] *Likelihood* is the probability of data, given a particular model and its parameters.

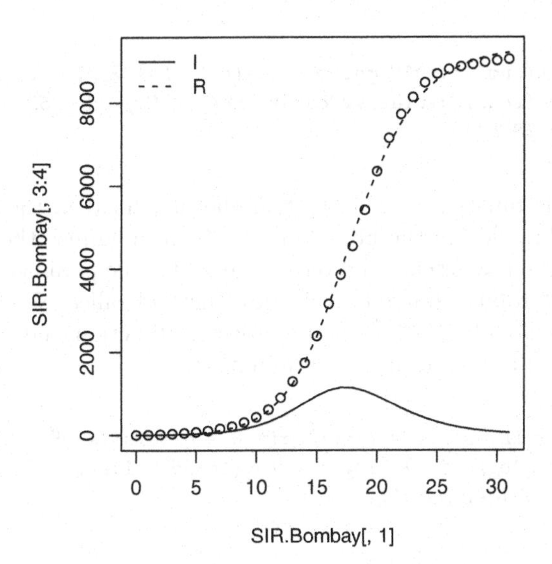

Fig. 6.15: Cumulative deaths for plague, in Bombay, India, 1905–1906 (raw data and fitted model, as described in this section).

values for SIR model parameters. These parameters include γ and β of the SIR model. The parameters will also include two other unknowns, (i) N, total relevant population size of Bombay at the time (1905–1906), and (ii) I_0, initial size of the infected population at our $t = 0$. A survey of easily accessed census data suggests the population at the time was in the neighborhood of $\sim 10^6$ individuals. We also might assume that I_0 would be a very small number, perhaps ~ 1 in principle.

Some details about our objective function:

1. Arguments are transformed parameters (this allows the optimizer to work on the logit[19] and log scales).
2. Transformed parameters are backtransformed to the scale of the model.
3. Parameters are used to integrate the ODEs using `ode`, retaining only the resistant (i.e. dead) group (the fourth column of the output); this provides the *predicted values* given the parameters and the time from onset, and the standard deviation of the residuals around the predicted values.
4. It returns the negative log-likelihood of the data, given the parameter values.

Here is our objective function. Note that its last value is the negative sum of the log-probabilities of the data (given a particular realization of the model).

```
> sirLL = function(logit.B, logit.g, log.N, log.I0) {
+     parms <- c(B = plogis(logit.B), g = plogis(logit.g))
+     x0 <- c(S = exp(log.N), I = exp(log.I0), R = 0)
+     Rs <- ode(y = x0, ross$Week, SIR, parms, hmax = 0.01)[,
```

[19] A logit is the transformation of a proportion which will linearize the logistic curve, logit (p) $= \log(p/(1-p))$.

```
+          4]
+      SD <- sqrt(sum((ross$CumulativeDeaths - Rs)^2)/length(ross$Week))
+      -sum(dnorm(ross$CumulativeDeaths, mean = Rs, sd = SD,
+          log = TRUE))
+ }
```

We then use this function, sirLL, to find the likelihood of the best parameters. The mle2 function in the bbmle library[20] will minimize the negative log-likelihood generated by sirLL, and return values for the parameters of interest.

We will use a robust, but relatively slow method called Nelder-Mead (it is the default). We supply mle2 with the objective function and a list of initial parameter values. This can take a few minutes.

```
> require(bbmle)
> fit <- mle2(sirLL, start = list(logit.B = qlogis(1e-05),
+     logit.g = qlogis(0.2), log.N = log(1e+06), log.I0 = log(1)),
+     method = "Nelder-Mead")

> summary(fit)

Maximum likelihood estimation

Call:
mle2(minuslogl = sirLL, start = list(logit.B = qlogis(1e-05),
    logit.g = qlogis(0.2), log.N = log(1e+06), log.I0 = log(1)),
    method = "Nelder-Mead")

Coefficients:
        Estimate Std. Error z value  Pr(z)
logit.B  -9.4499     0.0250 -377.39 <2e-16
logit.g   1.0180     0.0663   15.34 <2e-16
log.N     9.5998     0.0181  530.15 <2e-16
log.I0    1.2183     0.1386    8.79 <2e-16

-2 log L: 389.2
```

This gets us some parameter estimates, but subsequent attempts to actually get confidence intervals failed. This occurs frequently when we ask the computer to estimate too many, often correlated, parameters for a given data set. Therefore, we have to make assumptions regarding selected parameters. Let us assume for the time being that the two variable estimates are correct, that the population size of the vulnerable population was approximately exp(9.6) and the number of infections at the onset of the outbreak was 1.2. We will hold these constant and ask R to refit the model, using the default method.

```
> fit2 <-mle2(sirLL,start=as.list(coef(fit)),fixed=list(log.N=coef(fit)[3],
+     log.I0 = coef(fit)[4]), method = "Nelder-Mead")

> summary(fit2)
```

[20] You will need to load the bbmle package from a convenient mirror, unless someone has already done this for the computer you are using. See the Appendix for details about packages (A.3) and optimization in R (B.11).

```
Maximum likelihood estimation

Call:
mle2(minuslogl= sirLL, start= as.list(coef(fit)), method = "Nelder-Mead",
    fixed = list(log.N = coef(fit)[3], log.I0 = coef(fit)[4]))

Coefficients:
         Estimate Std. Error z value  Pr(z)
logit.B -9.44971    0.00632 -1495.4 <2e-16
logit.g  1.01840    0.03423    29.8 <2e-16

-2 log L: 389.2
```

Next we want to find confidence intervals for β and γ. This can take *several* minutes, but results in a likelihood profile for these parameters, which show the confidence regions for these parameters (Fig. 6.16).

```
> pr2 <- profile(fit2)

> par(mar = c(5, 4, 4, 1))
> plot(pr2)
```

We see that the confidence intervals for the transformed variables provide estimates of our confidence in these parameters.

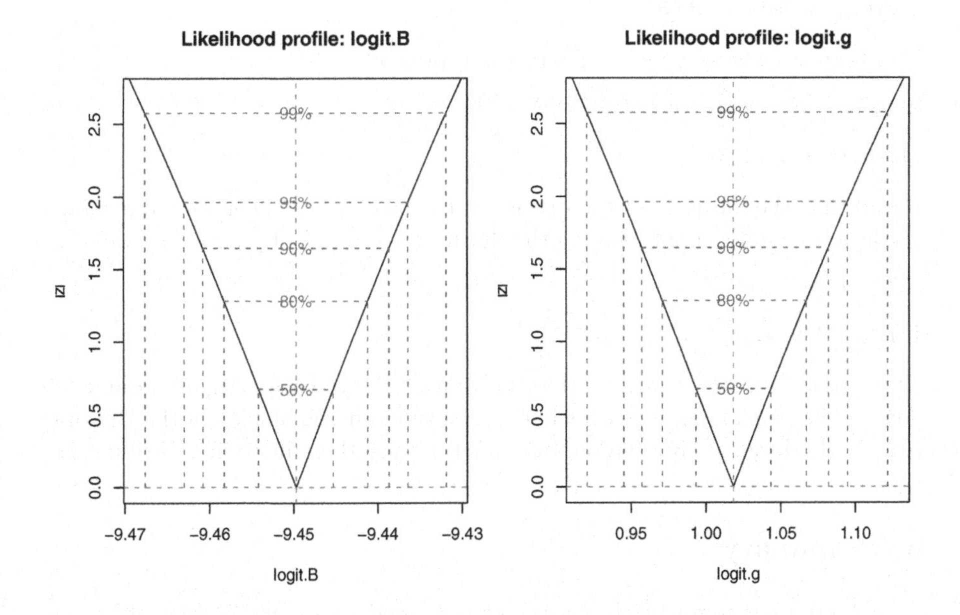

Fig. 6.16: Likelihood profile plots, indicating confidence intervals on transformed SIR model parameters.

Last we get to plot our curve with the data. We first backtransform the coefficients of the objective function.

```
> p <- as.numeric(c(plogis(coef(fit2)[1:2]), exp(coef(fit2)[3:4])))
> p
```

```
[1] 7.871e-05 7.347e-01 1.476e+04 3.381e+00
```

We then get ready to integrate the disease dynamics over this time period.

```
> inits <- c(S = p[3], I = p[4], R = 0)
> params <- c(B = p[1], g = p[2])
> SIR.Bombay <- data.frame(ode(inits, ross$Week, SIR, params))
```

Last, we plot the model and the data (Fig. 6.15).

```
> matplot(SIR.Bombay[, 1], SIR.Bombay[, 3:4], type = "l", col = 1)
> points(ross$Week, ross$CumulativeDeaths)
> legend("topleft", c("I", "R"), lty = 1:2, bty = "n")
```

So, what does this mean (Fig. 6.15)? We might check what these values mean, against what we know about the reality. Our model predicts that logit of γ was a confidence interval,

```
> (CIs <- confint(pr2))
```

```
          2.5 % 97.5 %
logit.B -9.4630 -9.436
logit.g  0.9452  1.095
```

This corresponds to a confidence interval for γ of

```
> (gs <- as.numeric(plogis(CIs[2, ])))
```

```
[1] 0.7201 0.7493
```

Recall that the duration of the disease in the host is $1/\gamma$. Therefore, our model predicts a confidence interval for the duration (in days) of

```
> 7 * 1/gs
```

```
[1] 9.720 9.342
```

Thus, based on this analysis, the duration of the disease is right around 9.5 days. This seems to agree with what we know about the biology of the Bubonic plague. Its duration, in a human host, is typically thought to last 4–10 days.

6.4 Summary

This chapter has skimmed the surface of a tremendous amount of territory. Some of the things we've learned include the notion that predator–prey relations can unstable (Lotka–Volterra predator–prey), and that additional biology stabilize or destabilize dynamics (negative density dependence, type II and III functional responses); increasing resources does not always help those we intend to help (paradox of enrichment). We learned that space, and the spatial distribution of hosts and their parasitoids matter (host–parasitoid dynamics) to population dynamics. We also learned that disease outbreaks are expected, under some conditions and not others, and that the mode of disease transmission matters, and can be modeled in a variety of ways.

Problems

6.1. Lotka–Volterra Predator–prey Model
(a) Write down the two species Lotka–Volterra predator–prey model.
(b) Describe how Fig. 6.4 illustrates neutral oscillatory dynamics.
(c) What are the units of the predator–prey model coefficients b, a, e, and s? How do we interpret them?

6.2. Rosenzweig-MacArthur Predator–prey Model
(a) Write down the two species Rosenzweig-MacArthur predator–prey model.
(b) How do we interpret b, K, w, D, e and s? What are their units?
(c) What is the value of the functional response when $H = D$? Explain how this result provides the meaning behind the name we use for D, the *half saturation* constant.
(d) For each point A–D in Fig. 6.5, determine whether the growth rate for the predator and the herbivore are zero, positive, or negative.
(e) In what way is the Rosenzweig-MacArthur predator isocline (Fig. 6.5) similar to the Lotka–Volterra model? It also differs from the Lotka–Volterra isocline–explain the ecological interpretation of D in the type II functional response and its consequence for this isocline.
(f) Explain the interpretation of real and imaginary parts of the eigenvalues for this paramterization of the Rosenzweig-MacArthur predator–prey model.
(g) In what ways does Fig. 6.6a match the interpretation of the eigenanalysis of this model?
(h) Examine the prey isoclines in Fig. 6.6a. How you can tell what the carrying capacities of the prey are?
(i) What do the above eigenanalyses tell us about how the stability of the predator–prey interaction varies with the carrying capacity of the prey?
(j) Consult Fig. 6.6a. What is the relation between the carrying capacity of the prey and the magnitude of the oscillations? What is the relation between the carrying capacity of the prey and the minimum population size? What does this interpretation imply about natural ecosystems?

6.3. Effects of dispersion on host–parasitoid dynamics
(a) Demonstrate the effects of aggregation on host–parasitoid dynamics. Specifically, vary the magnitude of k to find the effects on stability.
(b) Demonstrate the effects of a on stability.
(c) Demonstrate the effects of R on stability.

6.4. Effects of age at onset and disease duration on outbreak periodicity
(a) Create three simulations showing how diseases of different durations influence the periodicity of the outbreaks.
(b) Create three simulations showing how the age at onset for different diseases influence the periodicity of the outbreaks.
(c) Consider which factor is more important in influencing outbreak interval. How do you measure the interval? What criteria would you use to determine "importance"? How do the realities of age and duration influence your selection of criteria? Write a short essay that asserts a thesis, and then provides support based on this exercise.

Part III

Special Topics

7

An Introduction to Food Webs, and Lessons from Lotka–Volterra Models

A food web is a real or a model of a *set of feeding relations among species or functional groups.* This chapter has two foci, (i) a very brief introduction to multi-species webs as networks, and (ii) a re-examination of old lessons regarding the effect of food chain length on a web's dynamical properties.

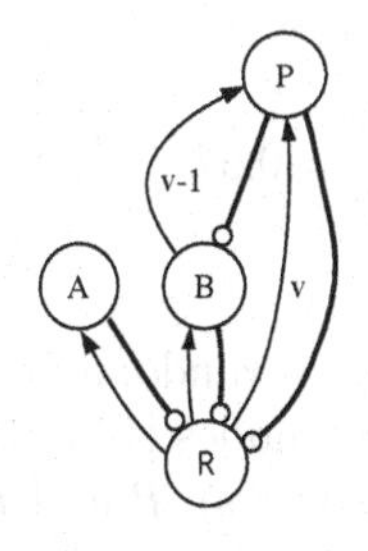

As prey	*As predators* R	B	A	P
R	0	–	–	–
B	+	0	0	–
A	+	0	0	0
P	+	+	0	0

Fig. 7.1: Two representations of a food web with 4 nodes and 8 directed links. The link label "v" indicates the proportion of *P*'s diet comprised of *R*.

7.1 Food Web Characteristics

We need language to describe the components of a food web, such as *links* and *nodes*, and we also need language to describe the properties of a web as a whole. Generally speaking, networks such as food webs have *emergent properties* [133], such as the number of nodes in the network. Emergent properties are typically considered to be nothing more than characteristics which do not exist at simpler levels of organization. For instance, one emergent property of a population is its density, because population density cannot be measured in an individual; that is why density is an emergent property.[1] While all food web models are

[1] If we assume no supernatural or spectral interference, then we can also assume that density arises mechanistically from complicated interactions among individuals and

based on simple pairwise interactions, the resulting emergent properties of multispecies webs quickly become complex due to indirect interactions and coupled oscillations [11, 208]. In addition, any extrinsic factor (e.g., seasonality) that might influence species interactions may also influence the emergent properties of food webs.

A few important network descriptors and emergent properties include,

Node A point of connection among links, a.k.a. trophospecies; each node in the web may be any set of organisms that share sufficiently similar feeding relations; in Fig. 7.1, P may be a single population of one species, or it may be a suite of species that all feed on both B and R.

Link A feeding relation; a connection between nodes or trophospecies; may be directed (one way) or undirected. A directed link is indicated by an arrow, and is the effect (+, −) of one species on another. An undirected link is indicated by a line (no arrow head), and is merely a connection, usually with positive and negative effects assumed, but not quantified.

Connectance The proportion of possible links realized. Connectance may be based on either directed, C_D, or undirected, C_U, links. For Fig. 7.1 these would be

$$C_D = \frac{L}{S^2} = \frac{8}{16} = 0.5$$
$$C_U = \frac{L}{(S^2 - S)/2} = \frac{4}{6} = 0.67$$

where S is the number of species or nodes.

Degree distribution, $\Pr(i)$ The probability that a randomly chosen node will have degree i, that is, be connected to i other nodes [133]. In Fig. 7.1, A is of degree 1 (i.e., is connected to one other species). P and B are of degree 2, and R is of degree 3. If we divide through by the number of nodes (4, in Fig. 7.1), then the *degree distribution* consists of the probabilities $\Pr(i) = \{0.25, 0.5, 0.25\}$. As webs increase in size, we can describe this distribution as we would a statistical distribution. For instance, for a web with randomly placed connections, the degree distribution is the binomial distribution [34].

Characteristic path length Sometimes defined as the average shortest path between any two nodes [47]. For instance, for Fig. 7.1, the shortest path between P and R is 1 link, and between A and P is 2 links. The average of all pairwise shortest paths is $(1+1+1+1+2+2)/6 = 1.\bar{3}$. It is also sometimes defined as the average of *all paths* between each pair of nodes.

Compartmentation, C_I The degree to which subsets of species are highly connected or independent of other species.

To calculate compartmentation in a food web, first assume each species interacts with itself. Next, calculate the proportion of shared interactions, p_{ij} for each pair of species, by comparing the lists of species with which each

their environments, rather than via magic. Other scientists will disagree and say that properties like density that appear to be simple additive aggregates do not qualify for the lofty title of "emergent property."

species in a pair interacts. The numerator of p_{ij} is the number of species with which both of the pair interact. The denominator is the total number of different species with which either species interacts.

As an example, let's calculate this for the above food web (Fig. 7.1). A interacts with A and R, B interacts with B, R, and P. Therefore, A and B both interact with only R, whereas, together, A and B interact with A, B, R, and P. The proportion, p_{ij}, therefore is $1/4 = 0.25$. We do this for each species pair.

Next we sum the proportions, and divide the sum by the maximum possible number of undirected links, C_U. To reiterate: For any pair of species, i and

Species	A	B	P
R	2/4	3/4	3/4
A		1/4	1/4
B			3/3

$$C_I = \frac{\sum_{i=1}^{S-1} \sum_{j=i+1}^{S} p_{ij}}{(S^2 - S)/2} = 3.5/6 = 0.58$$

j ($i \neq j$), p_{ij} is the proportion of shared interactions, calculated from the number of species that interact with *both* species i and j, divided by the number of species that interact with *either* species i or species j. As above, S is the number of species or nodes in the web.

Trophic Level Trophic *position* may simply be categorized as basal, intermediate or top trophic positions. Basal positions are those in which the trophospecies feed on no other species. The top positions are those in which the trophospecies are fed upon by nothing. One can also calculate a quantitative measure of trophic *level*. This is important in part because omnivory, something rampant in real food webs, complicates identification of a trophic level. We can calculate trophic level for the top predator, P (Fig. 7.1), and let us assume that P gets two-thirds of what it needs from B, and gets one-third from A. B itself is on the second trophic level, so given that, the trophic level of P is calculated as

$$T_i = 1 + \sum_{j=1}^{S} T_j p_{ij} = 1 + (2\,(0.67) + 1\,(0.33)) = 2.67$$

where T_i is the trophic level of species i, T_j is the trophic level of prey species j, and p_{ij} is the proportion of the diet of predator i consisting of prey j.

Omnivory Feeding on more than one *trophic* level ($v > 0$, Fig. 7.1); it is *not* merely feeding on different species or resources.

Intraguild predation A type of omnivory in which predation occurs between consumers that share a resource; in Fig. 7.1 P and B share prey R. When P gets most of its energy from B, we typically refer to that as omnivory ($v < 0.5$); when P gets most of its energy from R, we typically refer to that as intraguild predation ($v > 0.5$).

This list of food web descriptors is a fine start but is by no means exhaustive.

7.2 Food chain length — an emergent property

There are many interesting questions about food webs that we could address; let us address one that has a long history, and as yet, no complete answer: What determines the length of a food chain? Some have argued that chance plays a large role [34, 81, 221], and others have shown that area [122, 178] or ecosystem size [167] may play roles. The explanation that we will focus on here is *dynamical stability*. Communities with more species had been hypothesized to be less stable, and therefore less likely to persist and be observed in nature. Stuart Pimm and John Lawton [159] extended this work by testing whether *food chain length* could be limited by the instability of long food chains [167].

7.2.1 Multi-species Lotka–Volterra notation

A many-species Lotka–Volterra model can be represented in a very compact form,

$$\frac{\mathrm{d}X_i}{\mathrm{d}t} = X_i\left(b_i + \sum_{j=1}^{S} a_{ij}X_j\right) \tag{7.1}$$

where S is the number of species in the web, b_i is the intrinsic rate of increase of species i (i.e., r_i), and a_{ij} is a per capita effect of species j on species i.

When $i = j$, a_{ij} refers to an *intra*specific effect, which is typically negative. Recall that in our earlier chapters on competition, we used α_{ii} to represent intraspecific per capita effects. Here for notational convenience, we leave i and j in the equation, realizing that $i = j$ for intraspecific interactions. Further, we let a_{ij} be any sign, either positive or negative, and sum the effects. If we let $X = N$, $b = r$, and $a = r\alpha$, then the following are equivalent:

$$\dot{N}_1 = r_1 N_1 \left(1 - \alpha_{11}N_1 - \alpha_{12}N_2\right)$$
$$\dot{X} = X_1 \left(b_1 + a_{11}X_1 + a_{12}X_2\right)$$

The notation in eq. 7.1 is at once both simple and flexible. When a_{ij} is negative, it may represent competition or the effect of a predator, j, on its prey, i. When a_{ij} is positive, it may represent mutualism or the effect of prey j on a predator i.

7.2.2 Background

In the early and mid-1970's, Robert May and others demonstrated that important predictions could be made with relatively simple Lotka–Volterra models [127], and this work still comprises an important compendium of lessons for ecologists today [133]. May used simple Lotka–Volterra models to show that increasing the number of species in a food web tended to make the food web less stable [127, 134]. In species-rich webs, species were more likely to become extinct. Further, he showed that the more connections there were among species in the web (higher connectance), and the stronger those connections (higher interaction strength), the *less* stable the web. At the time, this ran counter to a

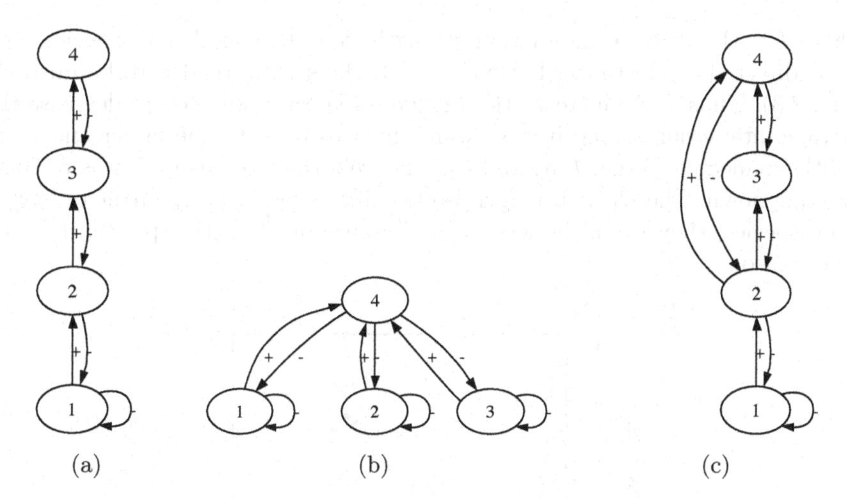

Fig. 7.2: (a), (b), and (c) correspond to Pimm and Lawton (1977) Figs. 1A, E, and B. Note that (b) differs from (a) in that (b) has only two trophic levels instead of four. Note also that (c) differs from (a) only in that species 4 has an additional omnivorous link. All basal species exhibit negative density dependence.

prevailing sentiment that more diverse ecosystems were more stable, and led to heated discussion.

May used specific quantitative definitions of all of his ideas. He defined connectance as the proportion of interactions in a web, given the total number of all possible directed interactions (i.e., directed connectance). Thus a linear food chain with four species (Fig. 7.2a), and intraspecific competition in the basal (bottom) species would have a connectance of $4/16 = 0.25$. May's definition of interaction strength was the square root of the average of all a_{ij}^2 $(i \neq j)$,

$$I = \sqrt{\frac{\sum_{i=1}^{S} \sum_{j=1, i \neq j}^{S} a_{ij}}{S^2 - S}}. \tag{7.2}$$

Squaring the a_{ij} focuses on magnitudes, putting negative and positive values on equal footing.

An important component of May's work explored the properties of *randomly* connected food webs. At first glance this might seem ludicrous, but upon consideration, we might wonder where else one could start. Often, simpler (in this case, random) might be better. The conclusions from the random connection models act as null hypotheses for how food webs might be structured; deviations from May's conclusions might be explained by deviations from his assumptions. Since this work, many ecologists have studied the particular ways in which webs in nature appear non-random.

One conclusion May derived was a threshold between stability and instability for random webs, defined by the relation

$$I (S C_D)^{1/2} = 1 \tag{7.3}$$

where I is the average interaction strength, S is the number of species, and C_D is directed connectance. If $I(SC)^{1/2} > 1$, the system tended to be unstable (Fig. 7.3). Thus, if we increase the number of species, we need to decrease the average interaction strength if we want them to persist. The larger and more tightly connected (larger I, S, and C_D) the web, the more likely it was to come crashing down. Therefore, if longer food chains were longer by virtue of having more species, they would be less stable because of the extra species, if for no other reason.

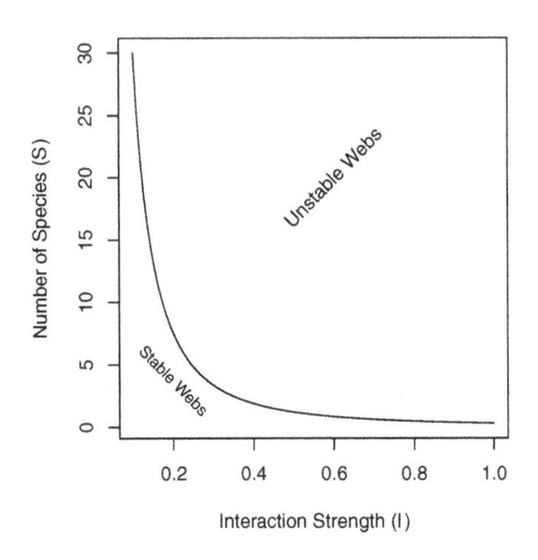

Fig. 7.3: Relation between the average interaction strength and the number of species able to coexist (here directed connectance is $C_D = 0.3$). The line represents the maximum number of species that are predicted to be able to coexist at equilibrium. Fewer species could coexist, but, on average, more species cannot coexist at equilibrium.

Pimm and Lawton felt that it seemed reasonable that long chains might be less stable also because predator-prey dynamics appear inherently unstable, and a connected series of unstable relations seemed less likely to persist than shorter chains. They tested whether food chain length *per se*, and not the number of species, influenced the stability of food chains. Another question they addressed concerned omnivory. At the time, surveys of naturally occurring food webs had indicated that omnivory was rare [160]. Pimm and Lawton tested whether omnivory stabilized or destabilized food webs [159, 160].

Like May, Pimm and Lawton [159] used Lotka–Volterra models to investigate their ideas. They designed six different food web configurations that varied food chain length, but held the number of species constant (Fig. 7.2). For each food web configuration, they varied randomly interaction strength and tested whether an otherwise randomly structured food web was stable. Their food webs included negative density dependence only in the basal species.

Pimm and Lawton concluded that (i) shorter chains were more stable than longer chains, and (ii) omnivory destabilized food webs (Fig. 7.5). While these conclusions have stood the test of time, Pimm and Lawton failed to highlight another aspect of their data — that omnivory shortened return times for those webs that were qualitatively stable (Fig. 7.7). Thus, omnivory could make more stable those webs that it didn't destroy. Subsequent work has elaborated on this, showing that weak omnivory is very likely to stabilize food webs [137,138].

7.3 Implementing Pimm and Lawton's Methods

Here we use R code to illustrate how one might replicate, and begin to extend, the work of Pimm and Lawton [159].

In previous chapters, we began with explicit time derivatives, found partial derivatives and solved them at their equilibria. Rather than do all this, Pimm and Lawton bypassed these steps and went straight to the evaluated Jacobian matrix. They inserted random estimates for the elements of the Jacobian into each non-zero element in the food web matrix. These estimates were constrained within (presumably) reasonable limits, given large less abundant predators and small more abundant prey.

Their methods followed this approach.

1. Specify a food web interaction matrix.[2]
2. Include negative density dependence for basal species only.
3. Set upper and lower bounds for the effects of predators on prey (0 to −10) and prey on predators(0 to +0.1); these are the Jacobian elements.
4. Generate a large number of random Jacobian matrices and perform linear stability analysis.
5. Determine qualitative stability (test $\lambda_1 < 0$), and return time for each random matrix. Use these to examine key features of the distributions of return times (e.g., average return time).
6. Compare the stability and return times among different food web configurations that varied systematically in food chain length and the presence of omnivory, but hold the number of species constant.

It is worth discussing briefly the Jacobian elements. May [127] defined interaction strength as the Jacobian element of a matrix, which represents the *total effect of one individual on the population growth rate of another species.* Think about how you calculate the Jacobian — as the partial derivative of one species' growth rate with respect to the size of the other population. It is the instantaneous change in the *population* growth rate per *unit change* in the population of another, at the equilibrium. The units chosen for the Jacobian elements thus mean that individual predators have relatively much larger effects on the population growth rates of prey than *vice versa.*

Let's build a function that does what Pimm and Lawton did. There are an infinite number of ways to do this, but this will suffice. First, we'll create a matrix that represents *qualitatively* the simplest longest food chain (Fig. 7.2a) where each species feeds only on one other species and where no prey are fed

[2] In the original publication, webs E and D seem to be represented incorrectly.

upon by more than one consumer. Initially, we will use the values of greatest magnitude used by Pimm and Lawton.

```
> Aq = matrix(c(-1, -10, 0, 0, 0.1, 0, -10, 0, 0, 0.1, 0, -10,
+     0, 0, 0.1, 0), nrow = 4, byrow = TRUE)
```

Note that this matrix indicates a negative effect of the basal species on itself, large negative effects (-10) of each consumer on its prey, and small positive effects of each prey on its consumer.

For subsequent calculations, it is convenient to to find out from the matrix itself how big the matrix is, that is, how many species, S, are in the web.

```
> S <- nrow(Aq)
```

Next, we create a random realization of this matrix by multiplying each element times a unique random number between zero and 1. For this matrix, that requires 4^2 unique numbers.

```
> M <- Aq * runif(S^2)
```

Next we perform eigenanalysis on it, retaining the eigenvalues.

```
> eM <- eigen(M)[["values"]]
```

Pimm and Lawton tested whether the dominant eigenvalue was greater than 0 (unstable) and if less than zero, they calculated return time. We will simply record the dominant eigenvalue (the maximum of the real parts of the eigenvalues).

```
> deM <- max(Re(eM))
```

Given the stabilizing effect of the intraspecific negative density dependence, we will hang on to that as well.

```
> intraNDD <- sqrt(sum(diag(M)^2)/S)
```

Given lessons from May's work [134], we might also want to calculate the average interaction strength, not including the intraspecific interactions. Here we set the diagonal interactions equal to zero, square the remaining elements, find their average, and take the square root.

```
> diag(M) <- 0
> IS <- sqrt(sum(M^2)/(S * (S - 1)))
```

Recall that weak omnivory is supposed to stabilize food webs [138]. For webs that include omnivory, we will calculate the interaction strength of omnivory in the same way we do for other interactions, as the square root of the average of the squared a_{ij} (eq. 7.2).

We can wrap all this up in a function where we specify the i, j of one of the directed omnivorous links.[3]

```
> args(pimmlawton)

function (mat, N = 1, omni.i = NA, omni.j = NA, omega = NULL)
```

[3] It does not matter which we specify, either the ij or the ji.

Now we can check this function for a single simulation for our first web,

```
> set.seed(1)
> pimmlawton(Aq)

    DomEig     Im IntraDD     I
1 -0.01593 0.4626  0.1328 2.304
```

Now let's do it 2000 times, as Pimm and Lawton did. Each row will be an independent randomization, and the columns will be the dominant eigenvalue, the intraspecific density dependence, and the average interaction strength.

```
> out.A <- pimmlawton(Aq, N = 2000)
```

We might like to look at basic summary statistics of the information we collected — what are their minima and maxima and mean?

```
> summary(out.A)

     DomEig                 Im            IntraDD                 I
 Min.   :-2.28e-01   Min.   :0.000   Min.   :0.000178   Min.   :0.265
 1st Qu.:-4.30e-02   1st Qu.:0.301   1st Qu.:0.117873   1st Qu.:2.241
 Median :-1.50e-02   Median :0.634   Median :0.245611   Median :2.851
 Mean   :-2.92e-02   Mean   :0.592   Mean   :0.246466   Mean   :2.766
 3rd Qu.:-3.53e-03   3rd Qu.:0.881   3rd Qu.:0.376127   3rd Qu.:3.369
 Max.   :-7.19e-08   Max.   :1.373   Max.   :0.499709   Max.   :4.762
```

We see that out of 2000 random food chains, the largest dominant eigenvalue is still less than zero ($\lambda_1 < 0$). What does that mean? It means that all of the chains are qualitatively stable, and that the return times are greater than zero ($-1/\lambda_1 > 0$).[4]

May's work showed that stability is related to interaction strength. Let's examine how the dominant eigenvalue is related to interaction strength.[5]

```
> pairs(out.A)
```

The results of our simulation (Fig. 7.4) show that the dominant eigenvalue can become more negative with greater intraspecific negative density dependence (IntraDD) and greater intersepcifiic interaction strength (I). Recall what this means — the dominant eigenvalue is akin to a perturbation growth rate at the equilibrium and is the negative inverse of return time. Therefore, stability can increase and return time decrease with increasing interaction strengths.

Note also (Fig. 7.4) that many eigenvalues seem very close to zero — what does this mean for return times? The inverse of a very small number is a very big number, so it appears that many return times will be very, very large, and rendering the webs effectively unstable. Let's calculate return times and examine a summary.

[4] Recall that a negative return time indicates that any "perturbation" at the equilibrium would have been closer to zero at some time in the past, i.e., that the perturbation is growing.

[5] Recall that if we think of stability analysis as the analysis of a small perturbation at the equilibrium, then the dominant eigenvalue is the growth rate of that perturbation.

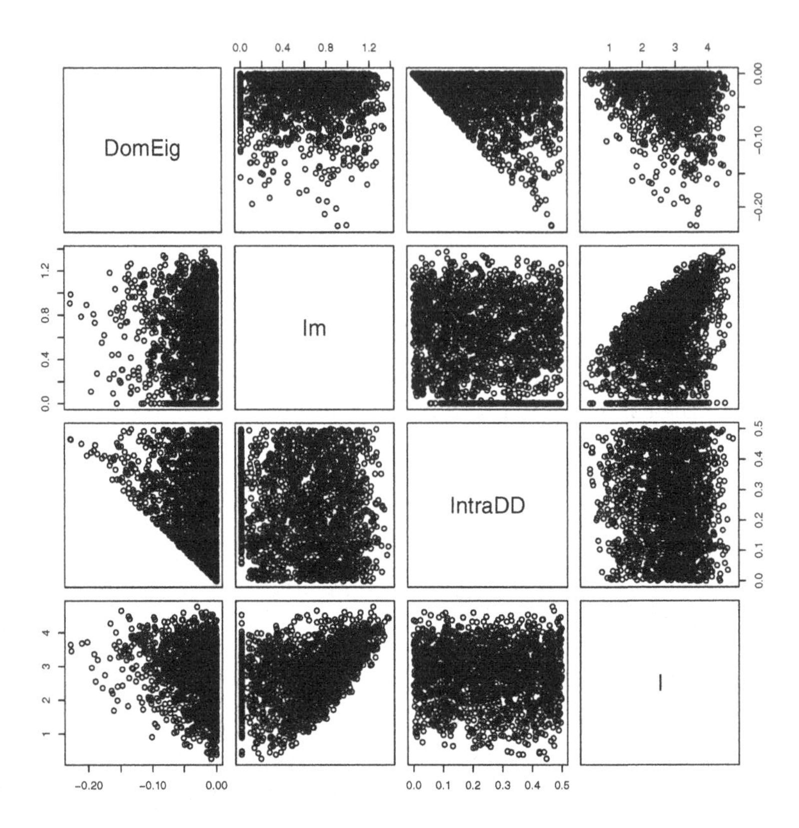

Fig. 7.4: Perturbations at the equilibrium tend to dissipate more rapidly (more negative dominant eigenvalues) with greater intraspecific negative density dependence (IntraDD) and greater interspecifiic interaction strength (I). This graph also demonstrates the independence of IntraDD and I in these simulations.

```
> RT.A <- -1/out.A[["DomEig"]]
> summary(RT.A)

    Min.  1st Qu.   Median     Mean  3rd Qu.     Max.
4.38e+00 2.32e+01 6.68e+01 1.17e+04 2.83e+02 1.39e+07
```

We find that the maximum return time is a very large number, and even the median is fairly large (67). In an ever-changing world, is there any meaningful difference between a return time of 1000 generations *vs.* neutral stability?

Pimm and Lawton addressed this by picking an arbitrarily large number (150) and recording the percentage of return times greater than that. This percentage will tell us the percentage of webs that are not effectively stable.

```
> sum(RT.A > 150)/2000

[1] 0.348
```

Now let's extract the return times that are less than or equal to 150 and make a histogram with the right number of divisions or bins to allow it to look like the one in the original [159].

```
> A.fast <- RT.A[RT.A < 150]
> histA <- hist(A.fast, breaks = seq(0, 150, by = 5), main = NULL)
```

This histogram (Fig. 7.5a) provides us with a picture of the stability for a food chain like that in Fig. 7.2a. Next, we will compare this to other webs.

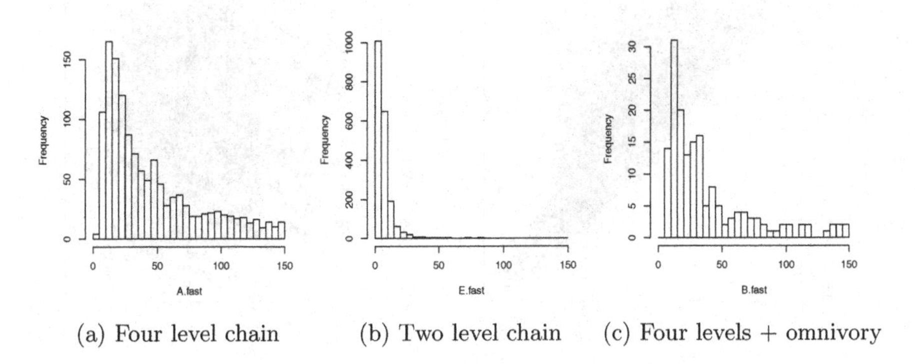

(a) Four level chain (b) Two level chain (c) Four levels + omnivory

Fig. 7.5: Histograms for three of the six food chains (A, E, and B) used by Pimm and Lawton.

7.4 Shortening the Chain

Now let's repeat all this (more quickly) for a shorter chain, but with the same number of species (Fig. 7.2b). So, we first make the web function.

```
> Eq = matrix(c(-1, 0, 0, -10, 0, -1, 0, -10, 0, 0, -1, -10,
+     0.1, 0.1, 0.1, 0), nrow = 4, byrow = TRUE)
```

Next we run the 2000 simulations, and check a quick summary.

```
> out.E <- pimmlawton(Eq, N = 2000)
> summary(out.E)
```

```
     DomEig                Im             IntraDD              I
 Min.   :-0.48631   Min.   :0.000   Min.   :0.0471   Min.   :0.206
 1st Qu.:-0.28429   1st Qu.:0.142   1st Qu.:0.3894   1st Qu.:2.301
 Median :-0.20151   Median :0.674   Median :0.4901   Median :2.861
 Mean   :-0.20865   Mean   :0.583   Mean   :0.4805   Mean   :2.799
 3rd Qu.:-0.12473   3rd Qu.:0.897   3rd Qu.:0.5814   3rd Qu.:3.372
 Max.   :-0.00269   Max.   :1.423   Max.   :0.8346   Max.   :4.772
```

The summary shows that, again, that all webs are stable ($\lambda_1 < 0$). A histogram of return times also shows very short return times (Fig. 7.5b). Plots of λ_1 *vs.*

interaction strengths show that with this short chain, and three basal species that the role of intraspecfic density dependence becomes even more important, and the predator-prey interactions less important in governing λ_1.

```
> layout(matrix(1:2, nr = 1))
> plot(DomEig ~ IntraDD, data = out.E)
> plot(DomEig ~ I, data = out.E)
```

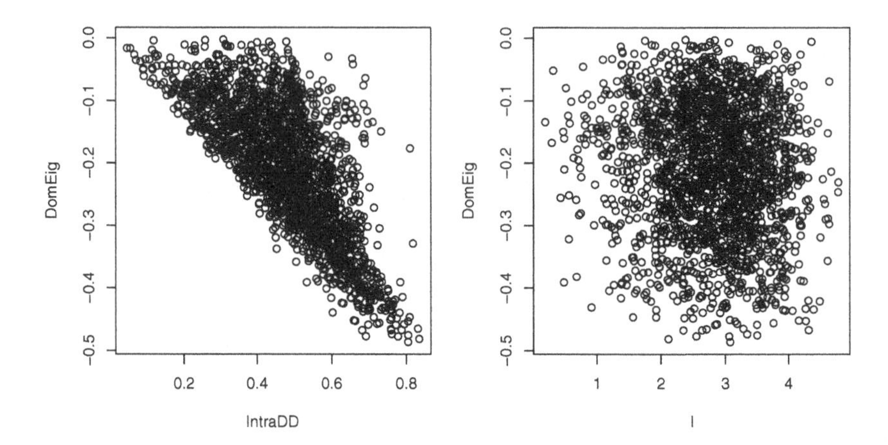

Fig. 7.6: For a food chain with two levels, and three basal species, perturbation growth rate (λ_1) declines with increasing intraspecific negative density dependence (IntraDD) and is unrelated to predator-prey interaction strengths.

Note that with the shorter food chain, a greater proportion of the λ_1 are more negative (farther away from zero) than in the four level food chain. Clearly then, shortening the web stabilizes it, in spite of still having the same *number* of species.

Let us again categorize these as having long and short return times, and graph the distribution of the short ones.

```
> RT.E <- -1/out.E[["DomEig"]]
> E.fast <- RT.E[RT.E < 150]
> histE <- hist(E.fast, breaks = seq(0, 150, by = 5), main = NULL)
```

7.5 Adding Omnivory

Real webs also have omnivory — feeding on more than one trophic level. A nagging question, then and now, concerns the effect of omnivory on food web dynamics. Pimm and Lawton compared food web dynamics with and without omnivory. Let's now create the web (Fig. 7.2c) that they used to compare directly with their linear food chain (Fig. 7.2a).

```
> Bq = matrix(c(-1, -10, 0, 0, 0.1, 0, -10, -10, 0, 0.1, 0,
+     -10, 0, 0.1, 0.1, 0), nrow = 4, byrow = TRUE)
```

Next we run the 2000 simulations, and check a quick summary.

```
> out.B <- pimmlawton(Bq, N = 2000, omni.i = 2, omni.j = 4)
> summary(out.B)
```

```
     DomEig            IntraDD                I              I.omni
 Min.   :-0.182   Min.   :0.000291   Min.   :0.568   Min.   :0.00643
 1st Qu.: 0.178   1st Qu.:0.125840   1st Qu.:2.707   1st Qu.:1.70230
 Median : 0.527   Median :0.245658   Median :3.275   Median :3.45300
 Mean   : 0.576   Mean   :0.248152   Mean   :3.217   Mean   :3.49159
 3rd Qu.: 0.913   3rd Qu.:0.371943   3rd Qu.:3.782   3rd Qu.:5.24184
 Max.   : 1.839   Max.   :0.499965   Max.   :5.389   Max.   :7.06030
```

With omnivory, we now see that most webs have $\lambda_1 > 0$, and thus are *unstable.* This was one of the main points made by Pimm and Lawton. Let's look at the data.

```
> pairs(out.B)
```

It means that most of the randomly constructed webs were not stable point equilibria. To be complete, let's graph what Pimm and Lawton did.

```
> RT.B <- -1/out.B[["DomEig"]]
> B.fast <- RT.B[RT.B < 150 & RT.B > 0]
> out.B.fast <- out.B[RT.B < 150 & RT.B > 0, ]
> out.B.stab <- out.B[RT.B > 0, ]
> histB <- hist(B.fast, breaks = seq(0, 150, by = 5), main = NULL)
```

7.5.1 Comparing Chain A versus B

Now let's compare the properties of the two chains, without, and with, omnivory, chains **A** and **B** (Figs. 7.2a, 7.2c). Because these are stochastic simulations, it means we have *distributions* of results. For instance, we have a distribution of return times for chain **A** and a distribution for return times for chain **B**. That is, we can plot histograms for each of them. Pimm and Lawton compared their webs in common sense ways. They compared simple summaries, including

- the proportion of random webs that were stable (positive return times),
- the proportion of stable random webs with return times greater than 150.

Now let's try graphical displays. Rather than simply chopping off the long return times, we use base 10 logarithms of return times because the distributions are so right-skewed. We create a histogram[6] of the return times for chain **A**, and nonparametric density functions for both chain **A** and **B**.[7]

[6] Note that now we use probabilities for the y-axis, rather than counts. The probability associated with any particular return time is the product of the height of the column and the width of the column (or bin).

[7] These density smoothers do a good job describing empirical distributions of continuous data, often better than histograms, which have to create discrete categories or "bins" for continuous data.

```
> hist(log(RT.A, 10), probability = T, ylim = c(0, 1), main = NULL,
+     xlab = expression(log[10]("Return Time")))
> lines(density(log(RT.A, 10)))
> lines(density(log(RT.B[RT.B > 0], 10)), lty = 2, lwd = 2)
> legend("topright", c("Chain A", "Chain B"), lty = 1:2, lwd = 1:2,
+     bty = "n")
```

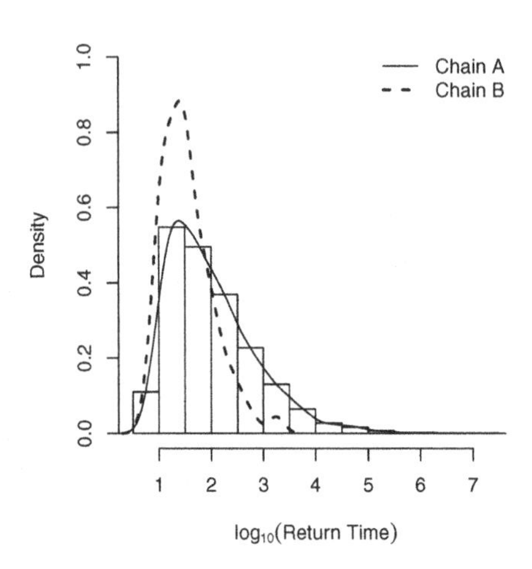

Fig. 7.7: Comparing the distributions of return times for chain **A** and **B**. "Density" is probability density. The distribution of return times for chain **A** is the solid line, and the distribution of return times for chain **B** is the dashed line.

By overlaying the density function of web B on top of web A return times (Fig. 7.7), we make an interesting observation. The omnivorous webs with positive return times (those plotted) actually tended to have shorter return times than the linear chain. Pimm and Lawton noted this, but did not emphasize it. Rather, they sensibly focused on the more obvious result, that over 90% of the omnivorous webs had negative return times, indicating an absence of a stable point equilibrium.

7.6 Re-evaluating Take-Home Messages

The primary messages made by Pimm and Lawton [159] were that

- shorter webs are more stable than long chains,
- omnivory destabilized webs.

These conclusions were a major part of the lively debate surrounding these issues. It was consistent with the apparent observation of the time, that empirical food webs revealed little omnivory [159,160], and that food chains in nature seemed to be much shorter than could occur, if primary productivity (green plants and algae) was channeled upwards into a linear chain.

Let's consider their assumptions.

First, Pimm and Lawton made the argument, as many others have (including us), that systems with stable point equilibria are more likely to persist than systems with oscillations, such as stable limit cycles. That is, we presume a strong correlation between the tendency to oscillate, and the tendency for species to become extinct (i.e., the system to collapse). It is easy to show that a system can be pushed from a stable equilibrium into oscillations which eventually become so big as to drive an element of the system to extinction. This is a very reasonable assumption, but one which is not challenged enough. Other measures of system stability could be used, such as the minimum that occurs in a long series of fluctuations [85, 138].

Second, Pimm and Lawton ignored the effects of self-regulated basal species. By exhibiting negative density dependence, the basal species stabilized the web. When Pimm and Lawton made a shorter web, they also added more self-regulated populations. Thus, they necessarily confounded chain length with the number of species with negative density dependence. Which change caused the observed differences among webs? We don't know.

Third, they assumed that the effect of web topology (i.e., short *vs.* long chain) was best evaluated with the *average* properties of the topology, rather than the maximum properties of the topology. By these criteria, webs without omnivory were clearly better. On average, webs without omnivory were more often stable than chains with omnivory, even if some of their return times tended to be quite long. Therefore, one might argue that if a web assembles in nature, it is more likely to persist (i.e., be stable) if it lacks omnivory.

However, let us consider this preceding argument further. The world is a messy place, with constant insults and disturbances, and resources and environmental conditions fluctuating constantly. In addition, there is a constant rain of propagules dropping into communities, and species abundances are changing all the time. In a sense then, communities are being constantly perturbed. The only webs that can persist in the face of this onslaught are the *most* stable ones, that is the ones with the shortest return times. We just showed that Pimm and Lawton's own analyses showed that *the most stable webs tended to be those with omnivory.* Subsequent work supports this claim that omnivory is rampant in nature [166], and this is supported by theory that shows weak interactions, including omnivory, stabilize food webs [137, 138].

Pimm and Lawton made very important contributions to this lengthy debate, and we are finally figuring out how to interpret their results.

7.7 Summary

Over 35 years ago, May started using simple, highly artificial dynamical descriptions of communities, using mathematical approaches that had been well established, if somewhat controversial, more than 50 years earlier by Alfred Lotka, Vito Volterra, and others [93]. Such simple abstractions are still useful today [14]. May's results, and those of Pimm and Lawton remain logical deductions that have resonance throughout community ecology. May, and Pimm and Lawton showed that under very simple assumptions, adding complexity usually destabilizes food webs. We have found that in practice, it is very difficult to build or restore structurally complex, speciose ecosystems [6, 19]. Further, we all now realize that omnivory is quite widespread [199], and additional theory indicates that omnivory can actually stabilize food webs by speeding a return to equilibrium and bounding systems farther from minima [137, 207]. Simple Lotka–Volterra webs will likely reveal more interesting generalizations in the years ahead.

Problems

7.1. General questions

(a) For each web, write out all four species' differential equations, using row, column subscripts for each parameter. Label species 1 X_1 or N_1, and the others accordingly.
(b) State the type of predator functional response used in these models, explain how you know, and explain the effect that this type of response typically has on dynamics.
(c) Describe the role played by intraspecific negative density dependence in these models — which species have it and what is it likely to do?
(d) Explain whether the results of this chapter support the contention that longer food chains are less stable.
(e) Explain whether the results of this chapter support the contention that omnivory destabilizes food chains.

7.2. More models

(a) Rewrite the above code to replicate the rest of Pimm and Lawton's results.
(b) Replicate the results of Pimm and Lawton's companion paper [160].
(c) Test the effects of intraspecific negative density dependence. Vary the average *magnitude* of negative density dependence.
(d) Design a new simulation experiment of your own.

8

Multiple Basins of Attraction

8.1 Introduction

Historically, simple models may have helped to lull some ecologists into thinking either that (i) models are useless because they do not reflect the natural world, or (ii) the natural world is highly predictable.[1] Here we investigate how simple models can create unpredictable outcomes, in models of Lotka–Volterra competition, resource competition, and intraguild predation. In all cases, we get totally different outcomes, or alternative stable states, depending on different, stochastic, initial conditions.

8.1.1 Alternative stable states

Alternative stable states, or alternative stable equilibria (ASE), *are a set of two or more possible stable attractors that can occur given a single set of external environmental conditions.* For a single list of species, there may exist more than one set of abundances that provide stable equilibria. One key to assessing alternative stable states is that *the stable states must occur with the same external environmental conditions.* If the external conditions differ, then the system is merely governed by different conditions.[2]

If stable attractors exist, how does the system shift from one attractor to another? The system can be shifted in a variety of ways, but the key is that the system (i.e., the set of species abundances) gets shifted into a different part of the state space, and then is attracted toward another state. System shifts may occur due to demographic stochasticity, the random variation in births and deaths that may be especially important when populations become small. System shifts may also occur due to different assembly sequences. For instance the outcome of succession may depend upon which species arrive first, second,

[1] One notable exception was May's work revealing that chaos erupts from a very simple model of discrete logistic growth.

[2] A very important complication is that if an abiotic factor is coupled dynamically to the biotic community, then it becomes by definition an internal part of the system and no longer external to it.

third, etc. System shifts might also arise via a physical disturbance that causes mortality. Different abundances may also arise from the gradual change and return of an environmental factor, and the resulting set of alternative equilibria is known as *hysteresis*, and below we examine a case related to resource competition.

The term *priority effects* refers to the situation in which initial abundances favor one species or group of species over others. We use "priority" to imply that if a species is given an early advantage ("priority") then it can suppress other species, whereas without that head start, it would not do as well. In contrast, for most stable models we considered previously, the location of the attractor was independent of any early advantage. We use the terms *effects of initial conditions* and priority effects to refer to situations in which starting conditions influence the outcome. So, in a 4 billion year old world, what are "starting conditions"? Starting conditions can refer to any point resulting from a system shift, as we described above (e.g., a disturbance).

Part of our thinking about ASEs relies on two assumptions that some find untenable, or at least not very useful [74]. First, the concept assumes that fixed stable states exist and that, if perturbed, populations can return to these states. Second, the concept assumes that communities move to these states with all due haste, and achieve these states over observable time periods. If these conditions are met, then ASEs can arise and be observed. We might recognize that ASEs might be a fuzzier concept, in which there exists more than one attractor, and where complex dynamics and long return times make the attractors difficult to observe. Further, we can imagine that short-term *evolution of species' traits* cause attractors to shift through time. Last, we might also want to acknowledge that saddles (i.e., attractor–repellers) also influence dynamics and species abundances. If we embrace all of these complications, then we might prefer a term other than ASE for this complex dynamical landscape — *multiple basins of attraction*.

8.1.2 Multiple basins of attraction

Multiple basins of attraction (MBA) is a phrase describing an entire landscape of "tendencies" in community dynamics. Imagine for a moment a two- or three-dimensional coordinate system in which each axis is the abundance of one species. An attractor is merely a place in the coordinate system (i.e., a particular set of species abundances) that exerts a pull on, or attracts, the dynamics of the populations. A stable equilibrium is an example of a global attractor; each local minimum or maximum in a stable limit cycle is also an attractor. The unstable equilibrium we observed in a Lotka–Volterra competition model is another example — it is a saddle, or an attractor–repellor, because it attracts from one direction, but repels in another.

MBAs can be visualized as a topographic landscape, or mountain range. We see lots of little valleys (attractors) and lots of peaks (repellers) (Fig. 8.1). If we think more broadly about multiple basins of attraction, then we begin to see that a strict definition of ASS lies at one end of a continuum, and it

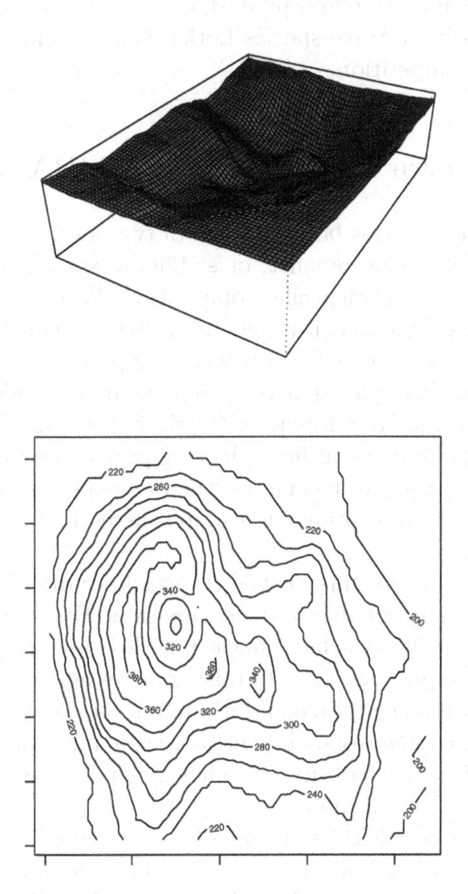

Fig. 8.1: Perspective and contour plots of a single complex dynamical landscape, containing multiple basins of attraction. Imagine putting a ball into this landscape, and jiggling the landscape around. The landscape represents the possible states of the community, the ball represents the actual community structure at any one point in time, and the jiggling represents stochasticity, either demographic, or environmental.

is matched at the other end by a system with one global stable equilibrium. The entire continuum, and much more besides, can be conceived of in terms of a landscape with some number of both basins of attraction (attractors) and peaks (repellors) in the landscape; each of these may act with some unique force or strength.

The potential for MBAs to occur has been noted for a few decades [78, 112, 130, 150], and the hunt for their signature in nature followed [24, 38, 115, 141, 184–186, 209, 210]. There has been discussion about what causes more than one basin of attraction, including space, stochasticity, and predation. Below we examine two mechanisms that may reveal priority effects: strong interference competition [175], and *intraguild predation*, an indirect interaction which has elements of both competition and predation [165]. We first investigate interference competition using a three-species Lotka–Volterra model, and then with a model of resource competition.

8.2 Lotka–Volterra Competition and MBA

You have already seen in this book the case of two-species Lotka–Volterra competition where an attractor–repeller, or saddle, arises when interspecific competition is greater than intraspecific competition. When this is the case, each species can suppress the other, if it gets a head start. This is a priority effect.

Having a larger negative effect on your competitor than on yourself may not be too unusual. Examples that come immediately to mind are cases where species compete preemptively for space (territories, or substrate surface) or for a resource with a unidirectional flow (drifting prey or particles in a stream, or light coming down on a forest canopy), then one species may gain the upper hand by acquiring a large proportion of resources first, that is, preempting resources.

In human economic systems, businesses can have direct negative effects on each other through questionable business practices. For instance, a larger company can temporarily flood a local market with below-cost merchandise, and drive out smaller competitors, prior to raising prices again. Note that it requires the raised prices for its long term equilibrium, but uses temporary below-market prices to eliminate competitors. In contrast, economies of scale can provide a very different kind of mechanism by which larger businesses can outcompete smaller businesses at equilibrium.

Here we explore how MBA can arise in a simple competitive system. We use a three-species Lotka–Volterra model to illustrate how strong interference competition may reveal priority effects. We use a slightly different representation of the Lotka–Volterra competition model. Note that the sign of each α must be negative, because we are adding them — this is similar to the notation in Chapter 7, but differs from previous treatment of Lotka–Volterra competition (Chapters 3, 5).

$$
\begin{aligned}
\frac{\mathrm{d}N_1}{\mathrm{d}t} &= r_1 N_1 \left(1 + \alpha_{11} N_1 + \alpha_{12} N_2 + \alpha_{13} N_3\right) \\
\frac{\mathrm{d}N_2}{\mathrm{d}t} &= r_2 N_2 \left(1 + \alpha_{21} N_1 + \alpha_{22} N_2 + \alpha_{23} N_3\right) \\
\frac{\mathrm{d}N_3}{\mathrm{d}t} &= r_3 N_3 \left(1 + \alpha_{31} N_1 + \alpha_{32} N_2 + \alpha_{33} N_3\right)
\end{aligned}
$$

Note two aspects of the above equations. First note that within the parentheses, the N_i are all arranged in the same order — N_1, N_2, N_3. This reflects their relative positions in a food web matrix, and reveals the row-column relevance of the subscripts of the αs. Second, note that $\alpha_{ii} = 1/K_i$, and in some sense K_i

results from a particular α_{ii}.[3] We can represent these equations as

$$\frac{dN_i}{dt} = N_i\left(r_i + \sum_{j=1}^{3} r_i\alpha_{ij}N_j\right) \tag{8.1}$$

in a fashion similar to what we saw in Chapter 7.

Another aspect of three species Lotka–Volterra models that we will state, without demonstration, is that *the equilibria can depend on* r_i. This contrasts with what we learned about the two-species model, which depended solely on the α_{ij}. Indeed, our example uses parameter values in which variation in r contributes to the equilibria.

Consider a particular set of simulations (Fig. 8.2).

- Species differ in their intrinsic rates of increase ($r_1 = 0.6$, $r_2 = 1$, $r_3 = 2$).
- Species with higher r also have slightly greater negative impacts on themselves ($a_{11} = 0.001$, $a_{22} = 0.00101$, $a_{33} = 0.00102$); this constitutes a tradeoff between maximum growth rate and carrying capacity.
- All species have twice as big a negative effect on each other as they do on themselves ($a_{ij,i\neq j} = 0.002$); this creates unstable equilibria, which happen to be attractor–repellers or saddles.
- Initial abundances are random numbers drawn from a normal distribution with a mean of 200, and a standard deviation of 10; in addition we also start them at precisely equal abundances ($N_i = 200$).
- Species 1 has the highest carrying capacity (smallest α_{ii}), and would therefore often considered the best competitor; in the absence of priority effects, we would otherwise think that it could always displace the others.

Now ponder the results (Fig. 8.2). Does each species win in at least one simulation? Which species wins most often? Does winning depend on having the greatest initial abundance? Make a 3×3 table for all species pair combinations (column = highest initial abundance, row=winner), and see if there are any combinations that never happen.

Note that when they start at equal abundances (Fig. 8.2), the species with the intermediate carrying capacity and intermediate r displaces the other two species. However, note also that (i) with a little stochasticity in initial conditions (slight variation in N_i at $t = 0$), this simple model generates unpredictable outcomes, and (ii) initial abundance does not determine everything.

It is critical to realize that this is occurring in part because species have larger negative competitive effects on others than they have on themselves. In this case the effects are direct, because the model is Lotka–Volterra competition. The effects may also be indirect, when species compete for more than one limiting resource. MacArthur [120] showed, for instance, that when generalists and specialists compete so that not all resources are available to all species, alternative stable states occur [115]. After we work through the code for the Lotka–Volterra example we just discussed, we take up MBA in the context of resource competition.

[3] We might also note is that it differs from notation typically used in textbooks for two-species models.

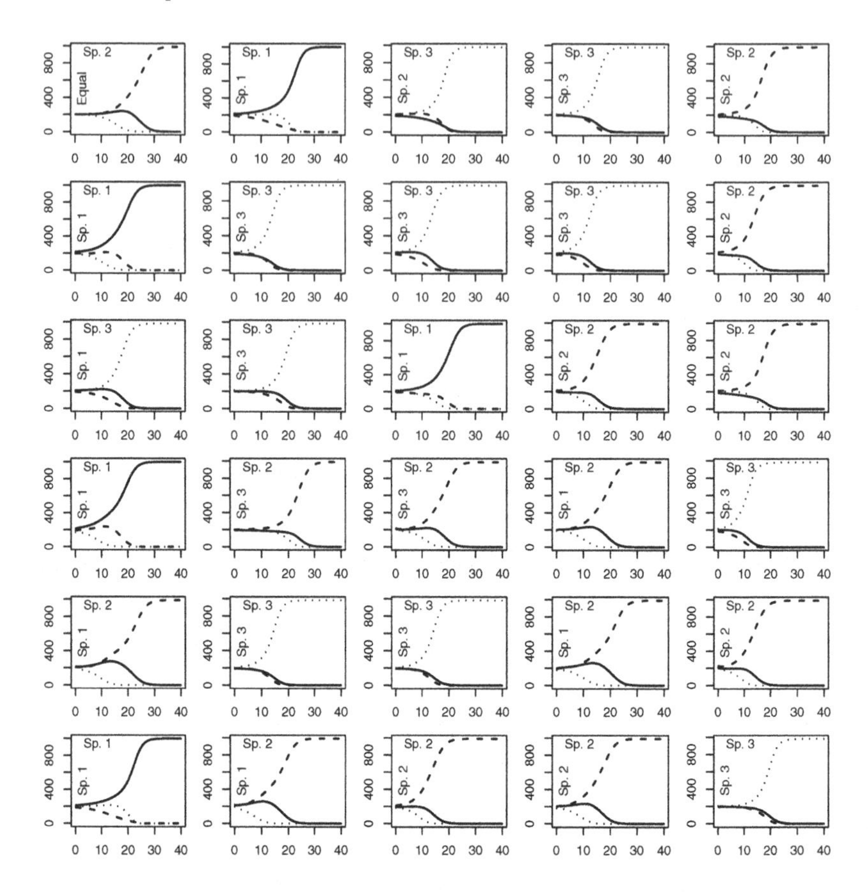

Fig. 8.2: Interaction between strong interference competition and initial abundances with three competing species (solid line - sp. 1; dashed line - sp. 2; dotted line - sp. 3). The species with the highest initial abundance is indicated in vertical orientation at the beginning of each trajectory ("Equal" indicates all species started at $N = 200$). The eventual winner is indicated in horizontal orientation at the top of each graph. See text for more details.

8.2.1 Working through Lotka–Volterra MBA

Here we create a function for multi-species Lotka–Volterra competition, taking advantage of matrix operations. Note that we can represent the three species as we would one, $\dot{\boldsymbol{N}} = \boldsymbol{rN}(1 - \boldsymbol{\alpha N})$, because the $\boldsymbol{\alpha N}$ actually becomes a matrix operation, `a %*% N`.[4]

```
> lvcompg <- function(t, n, parms) {
+     r <- parms[[1]]; a <- parms[[2]]
+     dns.dt <- r * n * (1 - (a %*% n))
```

[4] Recall that %*% is matrix multiplication in R because by default, R multiplies matrices and vectors element-wise.

```
+     return(list(c(dns.dt)))
+ }
```

We are going to use one set of parameters, but let initial abundances vary stochastically around the unstable equilibrium point, and examine the results.

Next we decide on the values of the parameters. We will create intrinsic rates of increase, *r*s, and intraspecific competition coefficients, α_{ii}, that correspond roughly to an $r - K$ tradeoff, that is, between maximum relative growth rate (r) and carrying capacity ($1/\alpha_{ii}$). Species 1 has the lowest maximum relative growth rate and the weakest intraspecific density dependence.

Following these ecological guidelines, we create a vector of *r*s, and a matrix of αs. We then put them together in a *list*,[5] and show ourselves the result.

```
> r <- c(r1 = 0.6, r2 = 1, r3 = 2)
> a <- matrix(c(a11 = 0.001, a12 = 0.002, a13 = 0.002, a21 = 0.002,
+     a22 = 0.00101, a23 = 0.002, a31 = 0.002, a32 = 0.002,
+     a33 = 0.00102), nrow = 3, ncol = 3)
> parms <- list(r, a); parms

[[1]]
 r1  r2  r3
0.6 1.0 2.0

[[2]]
      [,1]    [,2]    [,3]
[1,] 0.001 0.00200 0.00200
[2,] 0.002 0.00101 0.00200
[3,] 0.002 0.00200 0.00102
```

Next we get ready to simulate the populations 24 times. We set the time, t, and the *mean* and *standard deviation* of the initial population sizes. We then create a matrix of initial population sizes, with one set of three species' n_0 for each simulation. This will create a 3×24 matrix, where we have one row for each species, and each column is one of the initial sets of population sizes.

```
> t = seq(0, 40, by = 0.1); ni <- 200; std = 10
> N0 <- sapply(1:30, function(i) rnorm(3, mean = ni, sd = std))
```

Now let's replace the first set of initial abundances to see what would happen if they start out at precisely the same initial abundances. We can use that as a benchmark.[6]

```
> N0[, 1] <- ni
```

When we actually do the simulation, we get ready by first creating a graphics device (and adjust the margins of each graph). Next we tell R to create a graph layout to look like a 6×4 matrix of little graphs. Finally, we run the simulation, calling one column of our initial population sizes at a time, integrate with the

[5] A list is a specific type of R object.

[6] R's recycling rule tells it to use the single value of `ni` for all three values in the first column of `N0`.

ODE solver, and plot the result, 24 times. As we plot the result, we also record which species has the greatest initial abundance.

```
> par(mar = c(2, 2, 1, 1))
> layout(matrix(1:30, ncol = 5))
> for (i in 1:30) {
+     lvout <- ode(N0[, i], t, lvcompg, parms)
+     matplot(t, lvout[, 2:4], type = "l", lwd = 1.5, col = 1)
+     if (all(N0[, i] == 200)) {
+         text(3, 500, "Equal", srt = 90)
+     }
+     else {
+         text(3, 500, paste("Sp.", which.max(N0[, i])), srt = 90)
+     }
+     lastN <- lvout[nrow(lvout), 2:4]
+     text(3, max(lastN), paste("Sp.", which.max(lastN)), adj = c(0,
+         1))
+ }
```

8.3 Resource Competition and MBA

Above, we explored how simple Lotka–Volterra competition could result in unstable equilibria, causing saddles, and multiple basins of attraction. Here we take a look an example of how resource competition can do the same thing. Recall that *resource competition is an indirect interaction*, where species interact through shared resources. This particular example results in a type of MBA scenario, *hysteresis*, where gradual changes in the external environment result in abrupt and sometimes catastrophic changes in the biological system (Fig. 8.4).

Scheffer and colleagues [184] provide evidence that anthropogenically enriched (eutrophic) lakes can shift away from dominance by submerged macrophytes[7] rooted in substrate, into systems completely dominated by floating plants such as duckweed (*Lemna* spp.) and water fern (*Azolla* spp.). Submerged macrophytes can extract nutrients out of both sediments and the water column. At low, typically unpolluted, nutrient levels, submerged plants can draw down water nitrogen levels to a very low level, below levels tolerated by duckweed and water fern. At high nutrient levels, floating plants are no longer limited by water column nitrogen levels, and can create deep shade that kills submerged vegetation. Aside from killing these wonderful submerged macrophytes, the loss of this structure typically alters the rest of the lake food web.

Uncertainties arise in this scenario at intermediate levels of eutrophication (Fig. 8.3). As stated above, submerged plants dominate at low nitrogen supply rates, and floating plants dominate at high nitrogen supply rates. However, at intermediate supply rates (~ 1.5–$2.5\,\mathrm{mg\,L^{-1}}$), the outcome depends on priority effects.

[7] "Macrophyte" is a term often used for aquatic rooted vascular plants.

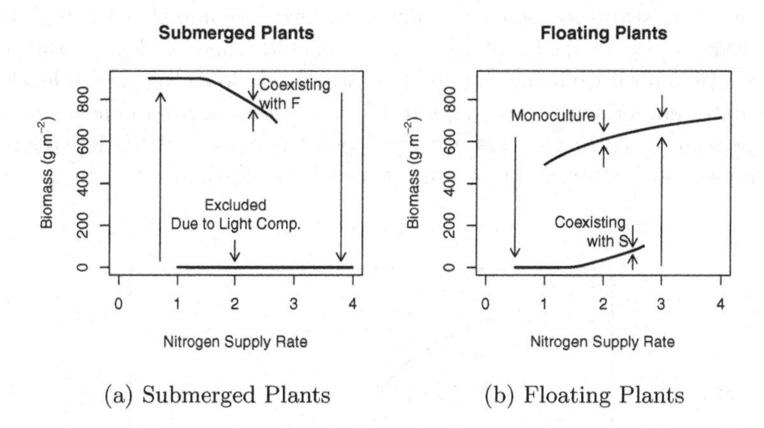

(a) Submerged Plants (b) Floating Plants

Fig. 8.3: The outcome of competition depends on both nutrient loading and priority effects. The lines provide equilibrial biomasses that depend on nitrogen supply rate for (a) submerged plants, and (b) floating plants. Thus both lines in (a) refer to possible equilibria for submerged plants, and both lines in (b) refer to possible equilibria for floating plants. (S = submerged plants, F = floating plants).

When submerged plants initially dominate a lake prior to eutrophication, they continue exclude floating plants as nitrogen supply rates increase, up until about 1.5 mg L^{-1} (Fig. 8.3a). If supply rates go higher (~ 1.5–2.5 mg L^{-1}), submerged plants continue to dominate, because their high abundance can draw nitrogen levels down quite low. Above this level, however, floating plants reach sufficiently high abundance to shade out the submerged plants which then become entirely excluded. Thus, at 2.5 mg L^{-1} we see a catastrophic shift in the community.

Once the system is eutrophic, and dominated by floating plants, reducing the nitrogen supply rate does not return the system to its original state in a way that we might expect (Fig. 8.3a). Once the floating plants have created sufficient shade, they can suppress submerged plants even at low nitrogen levels. If changes in watershed management bring nitrogen supply rates back down, submerged plants cannot establish until supply rate falls to 1 mg L^{-1}. This is well below the level at which submerged plants could dominate, if they were abundant to begin with.

This is an excellent example of a system prone to *hysteresis* (Fig. 8.4). Gradual changes in an external driver (e.g., nitrogen runoff) cause catastrophic shifts in the internal system, because at intermediate levels of the driver, the internal system is influenced by priority effects. Hysteresis is often defined as a context-dependent response to an environmental driver. As a different example, imagine that the state variable in Figure 8.4 is annual precipitation, and the driver is average annual temperature. Imagine that over many years, the regional average temperature increases. At first, increased temperature has little effect on precipitation. Once precipitation reaches a particular threshold, precipitation drops dramatically, and additional increased temperature has little effect. When the temperature starts to come back down, however, we find that the high levels of precipitation do not return at the same threshold where we lost it. Rather, it does not return until we bring temperature way back down to original levels. At intermediate temperatures (grey region, Fig. 8.4), *precipitation depends on what the temperature used to be.* This history dependence, or context dependence is the hallmark of hysteresis. It is important to understand that, in principle,

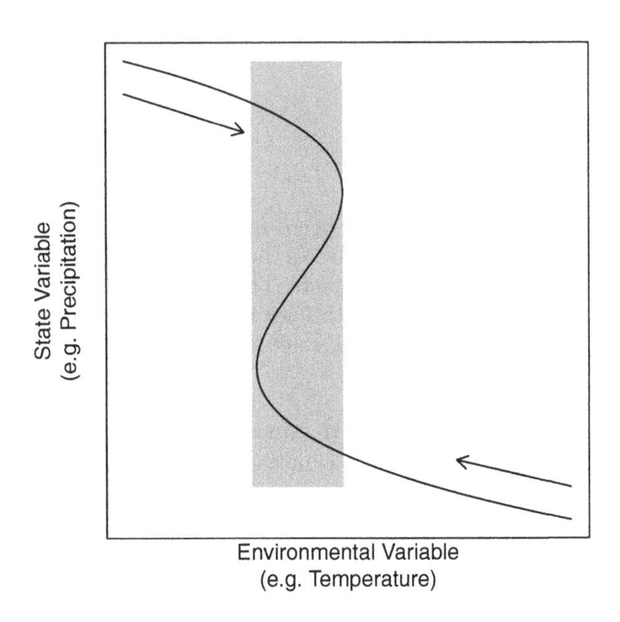

Fig. 8.4: Hysteresis. When the system changes from left to right, the threshold value of the predictor of catastrophic change is greater than when the system moves from right to left. At intermediate levels of the driver, the value of the response variable depends on the history of the system. Each value, either high or low, represents an alternative, stable, basin of attraction. The arrows in this figure represent the direction of change in the environmental driver.

this is *not* the result of a time lag. It is *not the case* that this pattern is due to a lagged or slow response by the state variable. Rather, these alternative basins (Fig. 8.4, solid lines in the grey area) represent permanent stable states from which the response variable cannot ever emerge, without some external force, such as very large changes in the environmental driver. Time lags may be important in other, different, circumstances, but are not, in principle, related to hysteresis.

8.3.1 Working through resource competition

Scheffer and colleagues represented submerged and floating plant interactions in the following manner, where F and S are floating and submerged plants, respectively [184].

$$\frac{\mathrm{d}F}{\mathrm{d}t} = r_f F \frac{n}{n+h_f} \frac{1}{1+a_f F} - l_f F \tag{8.2}$$

$$\frac{\mathrm{d}S}{\mathrm{d}t} = r_s S \frac{n}{n+h_s} \frac{1}{1+a_s S + bF + W} - l_s S \tag{8.3}$$

$$\tag{8.4}$$

As usual, r represents the maximum per capita rate of increase for F and S respectively. Thus rF is exponential growth of floating plants. This exponential growth is modified by terms for nitrogen (n) limited growth, and light limited growth. This modification results in type II functional responses. Here we discuss these modifications.

Nitrogen limitation The first factor to modify exponential growth above is $n/(n+h_x)$, water column nitrogen limitation. It varies from 0–1 and is a function of water column nitrogen concentration, n. The half saturation constant h controls the shape of the relation: when $h = 0$ there is no nitrogen limitation, that is, the term equals 1.0 even at very low n. If $h > 0$, then the growth rate is a Michaelis-Menten type saturating function where the fraction approaches zero as $n \rightarrow 0$, but increases toward 1.0 (no limitation) at high nutrient levels. For submerged plants, $h_s = 0$ because they are never limited by water column nitrogen levels because they derive most of the nitrogen from the substrate.

Nitrogen concentration Nitrogen concentration, n, is a simple saturating function that declines with increasing plant biomass which achieves a maximum N in the absence of any plants.

$$n = \frac{N}{1+q_s S + q_f F} \tag{8.5}$$

The nutrient concentration, n, depends not only on the maximum N nutrient concentration, but also on the effect of submerged and floating plants which take up nitrogen out of the water at rates $1/(1+q_S S)$ and $1/\left(1+q_f F\right)$ respectively; $1/q$ is the half-saturation constant.

Light limitation The second factor to modify exponential growth is light limitation, $1/(1+aF)$; $1/a_f$ and $1/a_s$ are *half-saturation constants* — they determine the plant densities at which the growth rates are half of the maxima; b represents the shade cast by a single floating plant, and W represents the light intercepted by the water column.

Loss The second terms in the above expressions $l_f F$ and $l_s S$ are simply density independent loss rates due to respiration or mortality.

Scheffer and colleagues very nicely provide units for their parameters and state variables (Table 8.1).

Table 8.1: Parameter and variable units and base values. Plant mass (g) is dry weight.

Parameter/Variable	Value	Units
F, S	(varies)	$\mathrm{g\,m^{-2}}$
N, n	(varies)	$\mathrm{mg\,L^{-1}}$
a_s, a_f	0.01	$(\mathrm{g\,m^{-2}})^{-1}$
b	0.02	$(\mathrm{g\,m^{-2}})^{-1}$
q_s, q_f	0.075, 0.005	$(\mathrm{g\,m^{-2}})^{-1}$
h_s, h_f	0.0, 0.2	$\mathrm{mg\,L^{-1}}$
l_s, l_f	0.05	$\mathrm{g\,g^{-1}\,day^{-1}}$
r_s, r_f	0.5	$\mathrm{g\,g^{-1}\,day^{-1}}$

Consider the meanings of the parameters (Table 8.1). What is a, and why is $1/a_s = 1/a_f$? Recall that they are half-saturation constants of light limited growth; their identical values indicate that both plants become self-shading at the same biomasses. What is q? It is the per capita rate at which plants pull nitrogen out of the water. Why is $q_s > q_f$? This indicates that a gram of submerged plants can pull more nitrogen out of the water than a gram of floating plant. Last, why is $h_s = 0$? Because submerged plants grow independently of the nitrogen content in the water column.

Code for Scheffer *et al.*

Now we are set to model this in R, using a built-in function for the ODEs, `scheffer`. First, let's set the parameters (Table 8.1), time, initial abundances, and see what we have.

```
> p <- c(N = 1, as = 0.01, af = 0.01, b = 0.02, qs = 0.075,
+     qf = 0.005, hs = 0, hf = 0.2, ls = 0.05, lf = 0.05, rs = 0.5,
+     rf = 0.5, W = 0)
> t <- 1:200
> Initial <- c(F = 10, S = 10)
```

We then run the solver, and plot the result.

```
> S.out1 <- ode(Initial, t, scheffer, p)
> matplot(t, S.out1[, -1], type = "l")
> legend("right", c("F", "S"), lty = 1:2, bty = "n")
```

From this run, at these nutrient levels, we observe the competitive dominance of the submerged vegetation (Fig. 8.5a). Let's increase nitrogen and see what happens.

```
> p["N"] <- 4
> S.out2 <- ode(Initial, t, scheffer, p)
> matplot(t, S.out2[, -1], type = "l")
```

Ah-ha! At high nutrient levels, floating vegetation triumphs (Fig. 8.5b). So where are the cool multiple basins of attraction? We investigate that next.

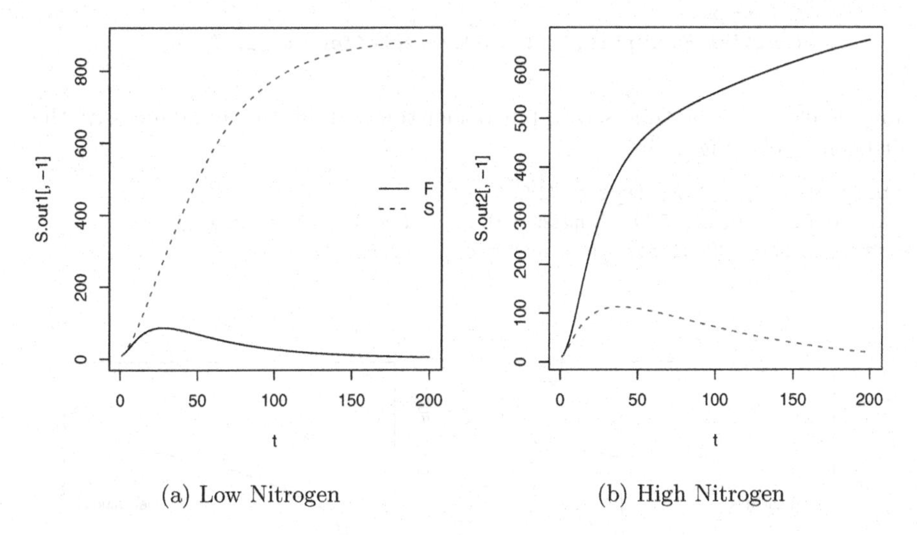

(a) Low Nitrogen

(b) High Nitrogen

Fig. 8.5: The outcome of competition depends on the nutrient loading.

Let's mimic nature by letting the effect of *Homo sapiens* increase gradually with increasing over-exploitation of the environment. We will vary N, increasing it slowly, and hang on to only the final, asymptotic abundances, at the final time point.

```
> N.s <- seq(0.5, 4, by = 0.1); t <- 1:1000
> S.s <- t(sapply(N.s, function(x) {
+     p["N"] <- x
+     ode(Initial, t, scheffer, p)[length(t), 2:3]
+ }))
```

Now we plot, not the time series, but rather the asymptotic abundances *vs.* the nitrogen levels (Fig. 8.6a).

```
> matplot(N.s, S.s, type = "l")
> legend("topright", c("F", "S"), lty = 1:2, bty = "n")
> arrows(0.5, 500, 2, 500, length = 0.1, lwd = 3, col = "grey")
> text(0.5, 500, "Increasing N", adj = c(0, -0.5))
```

Now we can see this catastrophic shift at around 2.7 mg N L^{-1} (Fig. 8.6a). As nitrogen increases, we first see a gradual shift in community composition, but then, wham!, all of a sudden, a small additional increase at ≈ 2.7 causes dominance by floating plants, and the loss of our submerged plants.

Now let's try to fix the situation by reducing nitrogen levels. We might implement this by asking upstream farmers to use no-till practices, for instance [211]. This is equivalent to starting at high floating plant abundances, low submerged plant abundances, and then see what happens at different nitrogen levels.

```
> Initial.Eutrophic <- c(F = 600, S = 10)
> S.s.E <- t(sapply(N.s, function(x) {
```

```
+     p["N"] <- x
+     ode(Initial.Eutrophic, c(1, 1000), scheffer, p)[2, 2:3]
+ }))
```

Now we plot, not the time series, but rather the asymptotic abundances *vs.* the nitrogen levels (Fig. 8.6b).

```
> matplot(N.s, S.s.E, type = "l")
> arrows(4, 500, 2, 500, length = 0.1, lwd = 3, col = "grey")
> text(4, 500, "Declining N", adj = c(1, -0.5))
```

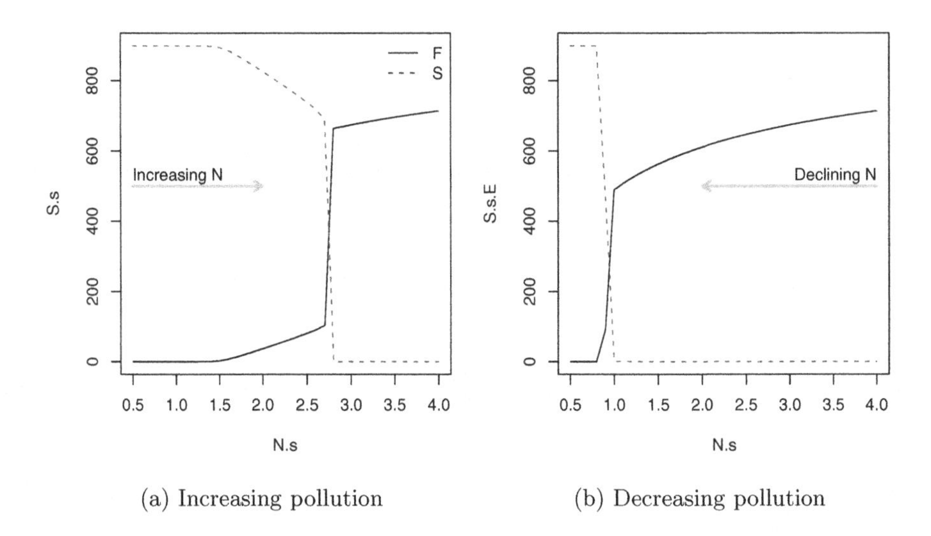

(a) Increasing pollution (b) Decreasing pollution

Fig. 8.6: The outcome of competition depends on the history of nutrient loading.

Wait a second! If we start at high floating plant biomass (Fig. 8.6b), the catastrophic shift takes place at a much lower nitrogen level. This is telling us that from around N =0.9–2.7, the system has two stable basins of attraction, or alternative stable states. It might be dominated either by floating plants or by submerged plants. This is often described as *hysteresis*, where there is more than one value of the response for each value of the predictor variable.

Now let's represent these changes for the two state variables in the aquatic plant model. First we represent the floating plants. Here we plot the low abundance state for the floating plants, adjusting the figure margins to accommodate all abundances, and then add in the high abundance data (Fig. 8.3b).

```
> plot(N.s[1:23], S.s[1:23, 1], type = "l", lwd = 2, xlim = c(0,
+   4), ylim = c(0, 900), main = "Floating Plants", ylab =
+   expression("Biomass (g m"^-2 * ")"), xlab = "Nitrogen Supply Rate")
> lines(N.s[-(1:5)], S.s.E[-(1:5), 1], lwd = 2)
```

Here we reinforce the concepts of multiple basins and hysteresis, by showing where the attractors are. I will use arrows to indicate these basins. At either high nitrogen or very low nitrogen, there is a single, globally stable attractor. At low nutrients, only submerged plants exist regardless of starting conditions. At high nutrients, only floating plants persist. Let's put in those arrows (Fig. 8.3b).

```
> arrows(3, 10, 3, 620, length = 0.1); arrows(3, 820, 3, 720, length = 0.1)
> arrows(0.5, 620, 0.5, 50, length = 0.1)
```

Next we want arrows to indicate the alternative basins of attraction at intermediate nitrogen supply rates. Floating plants might be at kept at low abundance at intermediate nitrogen supply rates if submerged plants are abundant (Fig. 8.6b). Let's indicate that with a pair of arrows.

```
> arrows(2.5, -10, 2.5, 60, length = 0.1)
> arrows(2.5, 200, 2.5, 100, length = 0.1)
> text(2.5, 100, "Coexisting\nwith S", adj = c(1.1, 0))
```

Alternatively, if submerged plants were at low abundance, floating plants would get the upper hand by lowering light levels, which would exclude submerged plants altogether (Fig. 8.3b). Let's put those arrows in (Fig. 8.3b).

```
> arrows(2, 480, 2, 580, length = 0.1); arrows(2, 750, 2, 650, length = 0.1)
> text(2, 700, "Monoculture", adj = c(1.1, 0))
```

Now let's repeat the exercise with the submerged plants. First we plot the high abundance state, and then add the low abundance state (Fig. 8.3a).

```
> plot(N.s[1:23], S.s[1:23, 2], type = "l", lwd = 2, xlim = c(0,
+   4), ylim = c(0, 900), main = "Submerged Plants",
+   ylab = expression("Biomass (g m"^-2 * ")"), xlab = "Nitrogen Supply Rate")
> lines(N.s[-(1:5)], S.s.E[-(1:5), 2], lwd = 2)
```

Now we highlight the global attractors that occur at very low or very high nitrogen supply rates (Fig. 8.3a).

```
> arrows(0.7, 30, 0.7, 830, length = 0.1)
> arrows(3.8, 830, 3.8, 30, length = 0.1)
```

Next we highlight the local, alternative stable equilibria that occur at intermediate nitrogen supply rates; either the submerged plants are dominating due to nitrogen competition, and achieving high abundance,

```
> arrows(2.3, 650, 2.3, 750, length = 0.1)
> arrows(2.3, 900, 2.3, 800, length = 0.1)
> text(2.4, 900, "Coexisting\nwith F", adj = c(0, 1))
```

or they are excluded entirely, due to light competition (Fig. 8.3a).

```
> arrows(2, 130, 2, 30, length = 0.1)
> text(2, 140, "Excluded\nDue to Light Comp.", adj = c(0.5,
+     -0.3))
```

Once again, we see what underlies these alternative states, or basins (Fig. 8.3). One population gains a numerical advantage that results in an inordinately large negative effect on the loser, and this competitive effect comes at little cost to the dominant species. At intermediate nitrogen supply rates, the submerged vegetation can reduce ambient nitrogen levels in the water column to undetectable levels because it gets most of its nitrogen from sediments. On the other hand, if floating plants can ever achieve high densities (perhaps due to a temporary nutrient pulse), then the shade they cast at intermediate supply rates prevents lush growth of the submerged plants. As a consequence, the submerged plants can never grow enough to draw nitrogen levels down to reduce the abundance of the floating plants.

8.4 Intraguild Predation

Intraguild predation (IGP) differs from omnivory only in degree (Chapter 7). In omnivory, a predator shares a resource with one or more of its prey (Fig. 8.4). Thus the top predator feeds relatively high on the food chain, getting most of its energy or resources by eating its competitor ($a > 0.5$ in Fig. 8.4). An extension of this is the case of *intraguild predation*, in which a species preys upon one or more of its competitors (Fig. 8.4). Intraguild predation is thus refers to the case in which the top predator gets most of its energy or resources from the more basal resource, eating lower on the food chain ($a < 0.5$ in Fig. 8.4). The distinction is not qualitative, but rather quantitative. If both consumer species prey upon each other, then we could make the argument that the name we ascribe to it depends entirely upon one's perspective. In such a case, however, we generally refer to the relations as intraguild predation.

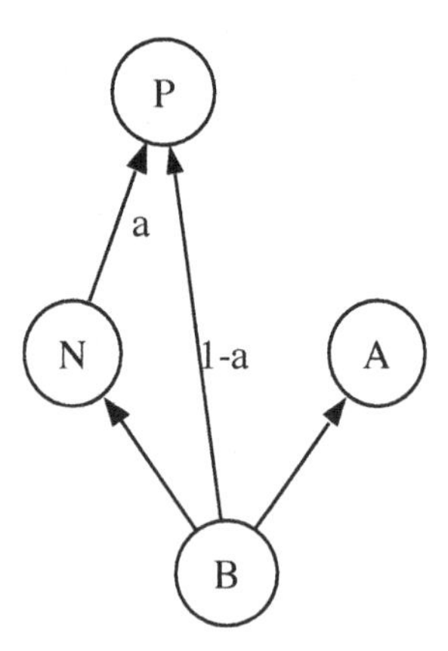

Fig. 8.7: We typically use "omnivory" when $a > 0.5$, and "intraguild predation" when $a < 0.5$. If we remove A from this model, then the species represent those of Holt and Polis [79].

8.4.1 The simplest Lotka–Volterra model of IGP

We can extend our good ol' Lotka–Volterra competition model to describe intraguild predation. All we do is add a term onto each competitor. For the competitor that gets eaten (the "IG-prey"), we subtract mass action predation, with a constant attack rate, a. For the top predator (the "IG-predator"), we add this same term, plus an conversion efficiency, $b \ll 1$.

$$\frac{\mathrm{d}N_1}{\mathrm{d}t} = r_1 N_1 (1 - \alpha_{11} N_1 - \alpha_{12} N_2) + baN_1 N_2 \tag{8.6}$$

$$\frac{\mathrm{d}N_2}{\mathrm{d}t} = r_2 N_2 (1 - \alpha_{21} N_1 - \alpha_{22} N_2) - aN_1 N_2 \tag{8.7}$$

Here a is attack rate of the IG-predator, N_1, on the IG-prey, N_2; b is the conversion efficiency of the prey into predator growth. Recall that this is the classic type I predator functional response of Lotka–Volterra predation.

Let's work through a little logic.

- In the absence of the other species, each species will achieve its usual carrying capacity, $1/\alpha_{ii}$.
- If we could have stable coexistence without IG-predation, then adding predation will increase the risk of extinction for the prey, and increase the abundance (if only temporarily) for the predator.
- If the poorer competitor is able to feed on the better competitor, this has the potential to even the scales.
- If the poor competitor is also the prey, then — forget about it — the chances of persistence by the IG-prey are slim indeed.

Now let's move on to a model of intraguild predation with resource competition.

8.4.2 Lotka–Volterra model of IGP with resource competition

Here we introduce a simple IGP model where the competition between consumers is explicit resource competition [79], rather than direct competition as in the Lotka–Volterra model above. The resource for which they compete is a logistic population.

$$\frac{\mathrm{d}P}{\mathrm{d}t} = \beta_{PB}\alpha_{BP}PB + \beta_{PN}\alpha_{NP}PN - m_P P \tag{8.8}$$

$$\frac{\mathrm{d}N}{\mathrm{d}t} = \beta_{NB}\alpha_{BN}BN - m_N N - \alpha_{NP}PN \tag{8.9}$$

$$\frac{\mathrm{d}B}{\mathrm{d}t} = rB(1 - \alpha_{BB}B) - \alpha_{BN}BN - \alpha_{BP}PB \tag{8.10}$$

Recall that the units for attack rate, α, are number of prey (killed) per individual of prey per individual of predator; the units for conversion efficiency, β, are number of predators (born) per number of prey (killed, and presumably eaten and assimilated). The consumers in this model have a type I functional response (mass action). The basal resource species exhibits logistic population growth.

Holt and Polis show analytically that five equilibria are present [79].

1. All species have zero density.
2. Only the resource, B, is present, at $B = K$.
3. Only the resource, B, and IG-prey, N, are present.
4. Only the resource, B, and IG-predator, P, are present.
5. All species present.

We will explore how initial conditions influence the outcomes of this simple IGP model. We will focus on the last three equilibria, with two or three species present.

Are there lessons we can apply from the previous competition models? In those cases, we tended to get MBAs when the negative effects each species on its competitors was greater than its negative effects on itself. How can we apply that to predation?

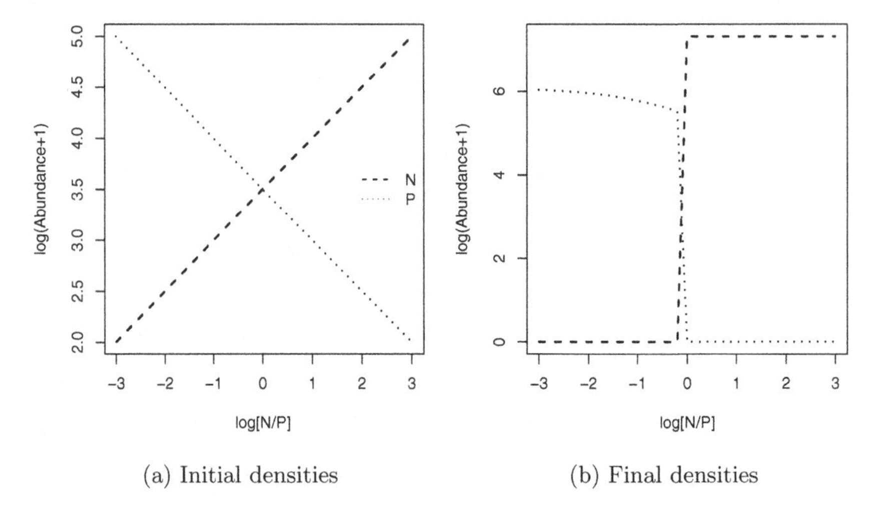

(a) Initial densities (b) Final densities

Fig. 8.8: Initial abundance (a) determines whether the IG-predator, P, or the IG-prey, N, win (b). Parameters are those set with the vector `params1` (see below).

Let us think about net effects of the IG-predator and IG-prey, both competition and consumption. Recall that the IG-prey must be the superior competitor — this means that, given similar attack rates on the basal resource B ($\alpha_{BN} = \alpha_{BP}$), the IG-prey must have a greater conversion efficiency ($\beta_{NB} > \beta_{PB}$).[8] In the absence of IG-predation, the superior competitor would always exclude the inferior competitor. However, if we add predation to suppress the superior competitor, then the IG-predator could win. If the relationship is such that each species has a larger net effect on each other than on themselves, we see that initial abundance (Fig. 8.8a) determines the winners (Fig. 8.8b). This allows either species to win, but not coexist.

[8] Think of conversion efficiency as the effect of the prey i on the predator j, β_{ij}.

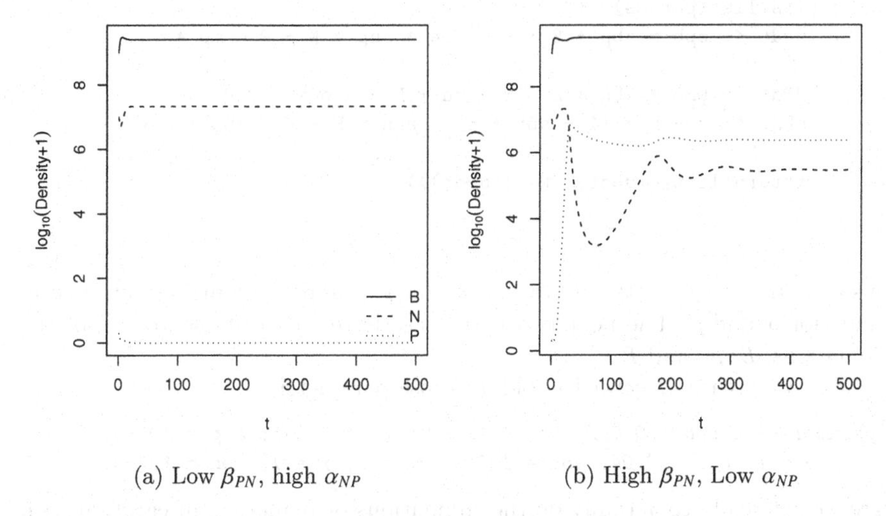

(a) Low β_{PN}, high α_{NP} (b) High β_{PN}, Low α_{NP}

Fig. 8.9: Conversion efficiencies and attack rates control coexistence. (a) With low conversion effciency, and high attack rates, species do not coexist. (b) By reducing the attack rate of the predator on the prey ($\alpha_{NP} = 10^{-4} \rightarrow 10^{-7}$), and increasing the direct benefit of prey to the predator ($\beta_{PN} = 10^{-5} \rightarrow 0.5$), we get coexistence. Parameters are otherwise the same as in Fig. 8.11 (see `params2` below).

How might we get coexistence between IG-predator and IG-prey? We have already provided an example where the IG-prey is the better resource competitor (Fig. 8.8). To reduce the negative effect of the IG-predator on the IG-prey, we can reduce attack rate. However, when we do that, the predator cannot increase when it is rare (Fig. 8.9a). If we further allow the predator to benefit substantially from each prey item (increasing conversion efficiency), then we see that the IG-predator can increase when rare, but eliminate the prey. Indeed, these are the essential components suggested by Holt and Polis: species coexist when their negative effects on each other are weaker (of smaller magnitude) than their negative effects on themselves. In a consumer–resource, predator-prey context, this can translate to reduced attack rates, and greater efficiency of resource use.

8.4.3 Working through an example of intraguild predation

To play with IBP in R, we start by examining an R function for the above Lotka–Volterra intraguild predation model, `igp`.

```
> igp

function (t, y, params)
{
    B <- y[1]
    N <- y[2]
    P <- y[3]
```

```
    with(as.list(params), {
        dPdt <- bpb * abp * B * P + bpn * anp * N * P - mp *
            P
        dNdt <- bnb * abn * B * N - mn * N - anp * N * P
        dBdt <- r * B * (1 - abb * B) - abn * B * N - abp * B *
            P
        return(list(c(dBdt, dNdt, dPdt)))
    })
}
```

This code uses three-letter abbreviations (α_{NP} = anp). The first letter, a or b, stands for α and β. The next two lower case letters correspond to one of the populations, B, N, and P.

Next, we create a vector to hold all those parameters.

```
> params1 <- c(bpb = 0.032, abp = 10^-8, bpn = 10^-5, anp = 10^-4,
+     mp = 1, bnb = 0.04, abn = 10^-8, mn = 1, r = 1, abb = 10^-9.5)
```

Here we get ready to actually do the simulations or numerical integration with ode. We set the time, and then we set four different sets (rows) of initial population sizes, label them, and look at them.

```
> t = seq(0, 60, by = 0.1)
> N.init <- cbind(B = rep(10^9, 4), N = 10^c(2, 5, 3, 4), P = 10^c(5,
+     2, 3, 4))
```

Now we integrate the population dynamics and look at the results. Here we first set up a graphics device with a layout of four figures and fiddle with the margins. We then use a for-loop to integrate and plot four times. Then we add a legend.

```
> quartz(, 4, 4)
> layout(matrix(1:4, nr = 2))
> par(mar = c(4, 4, 1, 1))
> for (i in 1:4) {
+     igp.out <- ode(N.init[i, 1:3], t, igp, params1)
+     matplot(t, log10(igp.out[, 2:4] + 1), type = "l", lwd = 2,
+         ylab = "log(Abundance)")
+ }
```

Clearly, initial abundances affect which species can coexist (Fig. 8.10). If either consumer begins with a big advantage, it excludes the other consumer. In addition, if they both start at low abundances, the IG-prey, N, excludes the predator; if they start at moderate abundances, the IG-predator, P, wins.

Now we need to get more thorough and systematic. The above code and its results show us the dynamics (through time) of particular scenarios. This is good, because we need to see how the populations change through time, just to see if unexpected things happen, because sometimes unexpected dynamics happen. A complementary way to analyze this model is to vary initial conditions more systematically and more thoroughly, and then simply examine the end points, rather than each entire trajectory over time. It is a tradeoff — if we

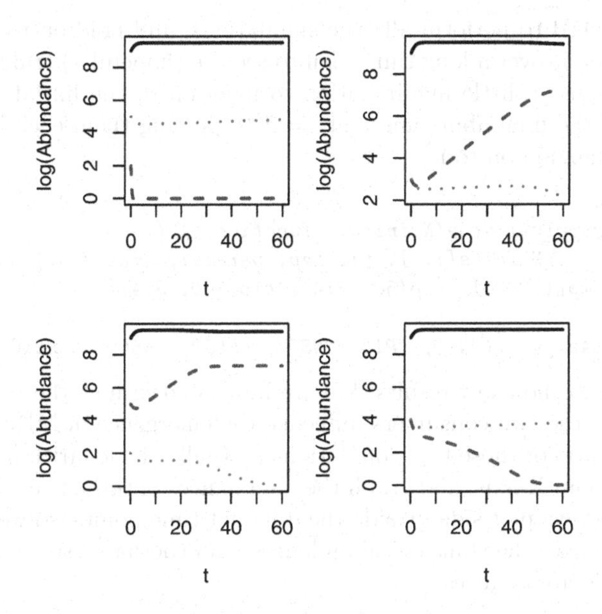

Fig. 8.10: Dynamics of Lotka Volterra intraguild predation, with differing initial abundances. See code for parameter values. Solid line - basal resource, dashed line - IG-prey, dotted line - IG-predator.

want to look at a lot of different initial conditions, we can't also look at the dynamics.

In the next sections, we examine the effects of *relative* abundance of the two consumers, and then of their *absolute* abundances.

8.4.4 Effects of relative abundance

First we will vary the relative abundances of the IG-prey and IG-predator, N and P. We create a slightly more complete set of initial abundances, with B constant, and N increases as P decreases.

```
> logNP <- seq(2, 5, by = 0.1)
> N.inits <- cbind(B = rep(10^9, length(logNP)), N = 10^logNP,
+     P = 10^rev(logNP))
```

We see (scatterplot matrix not shown) that we do have negative covariation in the starting abundances in the two consumer species, the IG-prey and IG-predator.

Next, we need to perform all[9] the simulations, and hold on to all the endpoints.[10] We do it over a long time span to see the (hopefully) truly asymptotic outcomes. We use a little manipulation to hang on to the initial abundances, at $t = 50$ and the final abundances at $t = 500$, putting them each in their own column and hanging on to it.

```
> t1 <- 1:500
> MBAs <- t(sapply(1:nrow(N.inits), function(i) {
+     tmp <- ode(N.inits[i, ], t1, igp, params1, hmax = 0.1)
+     cbind(tmp[1, 3:4], tmp[50, 3:4], tmp[500, 3:4])
+ }))
> colnames(MBAs) <- c("N1", "P1", "N50", "P50", "N500", "P500")
```

Now we need to show our results. We are interested in how the relative initial abundances of the two consumers influence the emergence of MBA. Therefore, let's put the ratio of those two populations (actually the logarithm of the ratio, $log[N/P]$[11]) on an X-axis, and graph the abundances of those two species on the Y-axis. Finally, we plot side by side the different time points, so we can see the initial abundances, the transient abundances, and (perhaps) something close to the asymptotic abundances.

```
> matplot(log10(N.inits[, "N"]/N.inits[, "P"]), log10(MBAs[,
+ 1:2] + 1), type = "l", col = 1, lty = 2:3, lwd = 2, ylab = "log(Abundance+1)",
+ xlab = "log[N/P]")
> legend("right", c("N", "P"), lty = 2:3, col = 1, bty = "n")
```

```
> matplot(log10(N.inits[, "N"]/N.inits[, "P"]), log10(MBAs[,
+ 3:4] + 1), type = "l", col = 1, lty = 2:3, lwd = 2, ylab = "log(Abundance+1)",
+ xlab = "log[N/P]")
```

```
> matplot(log10(N.inits[, "N"]/N.inits[, "P"]), log10(MBAs[,
+ 5:6] + 1), type = "l", col = 1, lty = 2:3, lwd = 2, ylab = "log(Abundance+1)",
+ xlab = "log[N/P]")
```

It is still amazing to me that different initial abundances can have such a dramatic effect (Fig. 8.11). It is also interesting that they take so long to play out. It all really just makes you wonder about the world we live in.

8.4.5 Effects of absolute abundance

Now let's hold relative abundance constant and equal, and vary absolute abundance. Recall that in our first explorations, we found different outcomes, de-

[9] Recall that `sapply` and related functions "apply" a function (in this case a simulation) to each element of the first argument (in this case each row number of the initial abundance matrix).

[10] We transpose the output matrix (`t()`) merely to keep the populations in columns. We also use the `hmax` argument in `ode` to make sure the ODE solver doesn't try to take steps that are too big.

[11] logarithms of ratios frequently have much nicer properties than the ratios themselves. Compare `hist(log(runif(100)/runif(100)))` vs. `hist(runif(100)/runif(100))`.

pending on different total abundances. Now instead of varying N and P in opposite order, we have them covary positively.

```
> logAbs <- seq(2, 7, by = 0.2)
> N.abs.inits <- cbind(B = rep(10^9, length(logAbs)), N = 10^logAbs,
+     P = 10^logAbs)
```

Now we simulate[12] the model, using the same basic approach as above, setting the time, and hanging on to three different time points.

```
> t1 <- 1:500
> MBA.abs <- t(sapply(1:nrow(N.abs.inits), function(i) {
+     tmp <- ode(N.abs.inits[i, ], t1, igp, params1, hmax = 0.1)
+     cbind(tmp[1, 3:4], tmp[50, 3:4], tmp[500, 3:4])
+ }))
> colnames(MBAs) <- c("N1", "P1", "N50", "P50", "N500", "P500")
```

We plot it as above, except that now we simply use log_{10}-abundances on the x-axis, rather than the ratio of the differing abundances.

```
> layout(matrix(1:3, nr = 1))
> matplot(log10(N.abs.inits[, "N"]), log10(MBA.abs[, 1:2] +
+ 1), type = "l", main = "Initial Abundances (t=1)", col = 2:3,
+ lty=2:3,lwd=2,ylab="log(Abundance+1)",xlab=expression(log[10]("N")))
> legend("right", c("N", "P"), lty = 2:3, col = 2:3, bty = "n")
> matplot(log10(N.abs.inits[, "N"]), log10(MBA.abs[, 3:4] +
+ 1), type = "l", main = "At time = 50)", col = 2:3, lty = 2:3,
+ lwd = 2, ylab = "log(Abundance+1)", xlab = expression(log[10]("N")))
> matplot(log10(N.abs.inits[, "N"]), log10(MBA.abs[, 5:6] +
+ 1), type = "l", main = "At time = 500", col = 2:3, lty = 2:3,
+ lwd = 2, ylab = "log(Abundance+1)", xlab = expression(log[10]("N")))
```

8.4.6 Explanation

Now, ...we have to explain it! Let's begin with what we think we know from Lotka–Volterra competition — each species has a bigger effect on the others than on itself. How do we apply that here. First let's look at the per capita direct effects, the parameters for each interaction.

$$\begin{array}{cccc} & B & N & P \\ B & r - r\alpha_{BB}B & -\alpha_{BN} & -\alpha_{BP} \\ N & \beta_{NB}\alpha_{BN} & 0 & -\alpha_{NP} \\ P & \beta_{PB}\alpha_{BP} & \beta_{PN}\alpha_{NP} & 0 \end{array} \tag{8.11}$$

Then we calculate the values for these and ask if competitors have larger effects on each other than they do on themselves.

[12] Unfortunately 'simulate' may mean 'integrate,' as it does here, or any other kind of made up scenario.

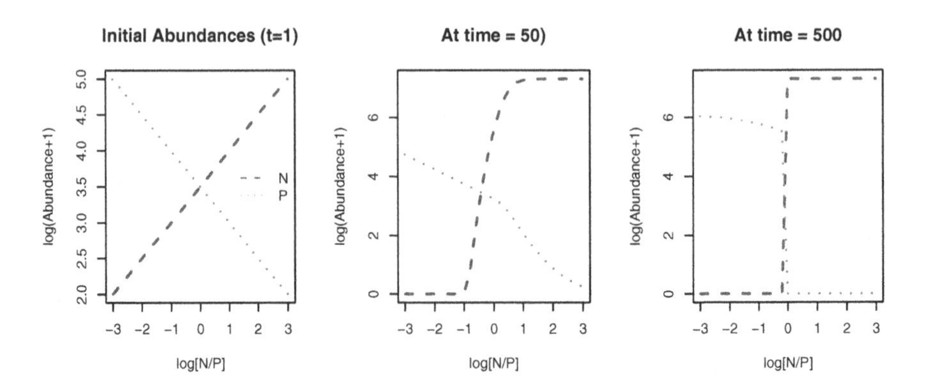

Fig. 8.11: Initial, transient, and near-asymptotic abundances of the intraguild prey, N, and predator, P, of Lotka–Volterra intraguild predation, with differing initial abundances.

```
> params1

      bpb       abp       bpn       anp        mp       bnb       abn
3.200e-02 1.000e-08 1.000e-05 1.000e-04 1.000e+00 4.000e-02 1.000e-08
       mn         r       abb
1.000e+00 1.000e+00 3.162e-10

> with(as.list(params1), {
+     rbind(B = c(r - r * abb * 10^9, -abn, -abp), N = c(bnb *
+         abn, 0, -anp), P = c(bpb * abp, bpn * anp, 0))
+ })

       [,1]   [,2]   [,3]
B 6.838e-01 -1e-08 -1e-08
N 4.000e-10  0e+00 -1e-04
P 3.200e-10  1e-09  0e+00

> with(as.list(params1), {
+     rbind(B = c(r - r * abb * 10^9, -abn, -abp), N = c(bnb,
+         0, -anp), P = c(bpb, bpn, 0))
+ })

    [,1]   [,2]   [,3]
B 0.6838 -1e-08 -1e-08
N 0.0400  0e+00 -1e-04
P 0.0320  1e-05  0e+00
```

So, from this we are reminded that the per capita direct effects on B, the basal resource, by both consumers are the same. N, the IG-prey, however, benefits more per capita, and so can attain a higher population size, and therefore could persist, and also exclude P. Thus it has a larger indirect negative effect on P than on itself. P, on the other hand, could have a huge direct negative effect on N. To achieve this effect, however, P has to have a sufficiently large population size. That is exactly why we get the results we do. If N starts out as relatively abundant, it reduces B and probably excludes P. If, on the other hand, P is abundant, they can have a large direct negative effect on N, and exclude N.

Holt and Polis suggest that coexistence is more likely when (i) the IG-prey is the better competitor (as we have above) and (ii) *the IG-predator benefits substantially from feeding on the IG-prey*, that is, when the conversion efficiency of prey into predators, β_{PN}, is relatively large.

Let's try increasing β_{PN} to test this idea. Let's focus on the ASS where the predator is excluded, when both species start out at low abundances (Fig. 8.10, upper right panel). We can focus on *the invasion criterion* by starting N and B at high abundance and then test whether the predator can invade from low abundance. We can really ramp up β_{PN} to be 100 times β_{PB}. This might make sense if N nutrient value is greater than B. In real food chains this seems plausible, because the C:N and C:P ratios of body tissue tend to decline as one moves up the food chain [193]; this makes animals more nutritious, in some ways, than plants.

```
> params2 <- params1
> params2["anp"] <- 10^-7
```

Now we numerically integrate the model, and plot it.

```
> t <- 1:500
> N.init.1 <- c(B = 10^9, N = 10^7, P = 1)
> trial1 <- ode(N.init.1, t, igp, params2)
> matplot(t, log10(trial1[, -1] + 1), type = "l", col = 1,
+     ylab = quote(log[10] * ("Density+1")))
> legend("bottomright", c("B", "N", "P"), lty = 1:3, bty = "n")
```

Whoa, dude! Fig. 8.9a reveals very different results from those in Fig. 8.11, but makes sense, right? We make the predator benefit a lot more from each prey item, then the predator doesn't need to be a good competitor, and can persist even if the IG prey reduces the basal resource level. Our next step is to rein the predator back in. One way to do this is to reduce the *attack rate*, so that the predator has a smaller per capita direct effect on the prey. It still benefits from eating prey, but has a harder time catching them. Let's change α_{NP} from 10^-4 to 10^-7 and see what happens.

```
> params2["bpn"] <- 0.5
> trial2 <- ode(N.init.1, t, igp, params2)
> matplot(t, log10(trial2[, -1] + 1), type = "l", col = 1,
+     ylab = quote(log[10] * ("Density+1")))
```

Now that we have allowed the predator to benefit more from individual prey (Fig. 8.9b), but also made it less likely to attack and kill prey, we get coexistence regardless of the predator starting at low abundance. Additional exploration would be nice, but we have made headway. In particular, it turns out that this model of intraguild predation yields some very interesting cases, and the outcomes depend heavily on the productivity of the system (i.e., the carrying capacity of B). Nonetheless, we have explored the conditions that help facilitate coexistence of the consumers — the IG-prey is the superior exploitative competitor, and the IG-predator benefits substantively from the prey.

8.5 Summary

A few points are worth remembering:

- Alternative stable equibria, alternative stable states, and multiple basins of attraction are defined generally as mulitple attractors in a system defined by a particular suite of species in a environment in which external fluxes are constant.
- Hysteresis is typically an example of alternative stable states that are revealed through gradual changes in an external driver, such as temperature, or nutrient supply rate.
- Alternative stable equilibria will have low invasibility by definition; this lack of invasibility might come about through large direct negative effects (high attack rate, or aggression), or through a preempted resource (e.g. light interception, an exclusive substitutable resource, or allelopathy). There could also be life history variation, where long lived adults prevent colonization by less competitive juveniles, or juveniles vulnerable to predation [165].
- Alternative stable equilibria seem to be more common when species have relatively larger negative effects on each other and weaker negative effects on themselves.

Problems

8.1. General
Compare and contrast the terms "alternative stable states" and "multiple basins of attraction." Define each and explain how the terms are similar and how they differ.

8.2. Lotka–Volterra competition
(a) Explain what we learn from Figure 8.2 regarding how growth rate, initial abundance and intraspecific density dependence (or carrying capacity) influence outcomes. Specifically, which of these best predicted the final outcome of competition? Which was worst? Explain.
(b) Explain in non-mathematical terms why strong interference allows for priority effects.
(c) Create a simulation to more rigorously test the conclusions you drew in part (a) above.

8.3. Resource competition
(a) Explain hysteresis.
(b) Alter the equation for submerged plants to represent the hypothetical situation in which submerged plants get most or all of their resources from the water column. Explain your rationale, and experiment with some simulations. What

would you predict regarding (i) coexistence and (ii) hysteresis? How could you test your predictions?

8.4. Intraguild Predation
(a) Use Figure 8.8 to explain how initial abundances influence outcomes. Are there initial conditions that seem to result in all species coexisting? Are there things we should do to check this?
(b) Explain how high attack rates and low conversion efficiencies by the top predator create alternative stable states.

9

Competition, Colonization, and Temporal Niche Partitioning

In this chapter, we will explore and compare models in which transient dynamics at one spatial or temporal scale result in long-term coexistence at another. All these models assume that species coexist because there exists at least a brief window in time during which each species has an advantage.

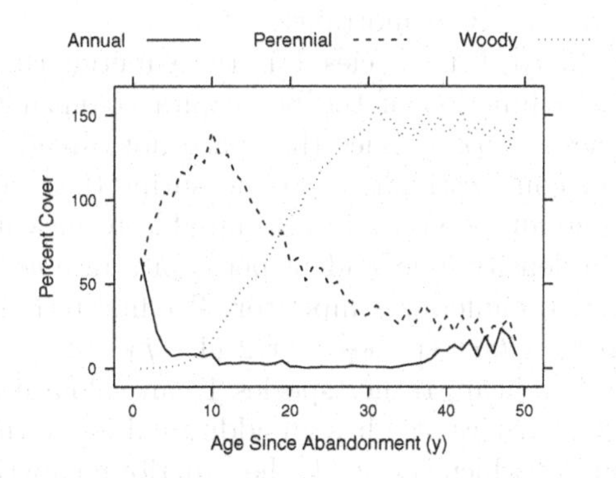

Fig. 9.1: Successional trajectory of annual and perennial herbaceous and woody plants in the Buell-Small Succession Study (http://www.ecostudies.org/bss/). These are mean percent cover from 480 plots across 10 fields, sampled ever 1–2 years.

We begin with a simple model of the *competition–colonization tradeoff*, explore *habitat destruction and the extinction debt*, and then examine a model that adds an important subtlety: *finite rates of competitive exclusion*. We finish up with an introduction to the *storage effect*

9.1 Competition–colonization Tradeoff

Models of coexistence via this tradeoff have been around for awhile [4, 73, 80, 108, 111, 187, 189]. In this tradeoff, species coexist because all individuals die, and therefore all species have to have some ability to colonize new space. Therefore, this means that species have two ways of being successful. Successful species may be very good at colonizing open sites, or they may be very good at displacing other species from a given site. These two extremes setup the basic tradeoff *surface*, wherein species coexist when they make this tradeoff in a manner in which none of them have too superior a combination of both traits.

Here we provide the two-species model of this phenomenon [73]. We explored the basis of this model back in our chapter on metapopulation dynamics. Similarly, here we focus on the case where the state variable is the *proportion of available sites*, rather than N. In addition, we extend this to two species using [73]. Here we represent the proportion of sites occupied by each of two species,

$$\frac{\mathrm{d}p_1}{\mathrm{d}t} = c_1 p_1 (1 - p_1) - m_1 p_1 \tag{9.1}$$

$$\frac{\mathrm{d}p_2}{\mathrm{d}t} = c_2 p_2 (1 - p_1 - p_2) - m_2 p_2 - c_1 p_1 p_2 \tag{9.2}$$

where p_i is the proportion of available sites occupied by species i, and c_i and m_i are the per capita colonizing and mortality rates of species i. Note that m is some combination of inherent senescense plus a background disturbance rate; we will refer to these simply as mortality.

As represented in eq. 9.1, species 1 is the superior competitor. The rate of increase in p_1 is a function of the per capita colonizing ability times the abundance of species 1 ($c_1 p_1$) times the space not already occupied by that species ($1 - p_1$). One could estimate c_1 by measuring the rate at which an open site is colonized, assuming one would also be able to measure p_1. The rate of decrease is merely a density-independent per capita rate m_1.

Eq. 9.2 represents the inferior competitor. The first term includes only space that is unoccupied by *either* species 1 or 2 ($1 - p_1 - p_2$). Note that species 1 does not have this handicap; rather species 1 can colonize a site occupied by species 2. The second species also has an additional loss term, $c_1 p_1 p_2$ (eq. 9.2). This term is the rate at which species 1, the superior competitor, colonizes sites occupied by species 2, *and immediately displaces it.* Note that species 1 is not influenced at all by species 2.

Competition–colonization tradeoff model

Here we implement in R a function of ODEs for eqs. 9.1, 9.2.

```
> compcol <- function(t, y, params) {
+     p1 <- y[1]
+     p2 <- y[2]
+     with(as.list(params), {
+         dp1.dt <- c1 * p1 * (1 - p1) - m1 * p1
+         dp2.dt <- c2 * p2 * (1 - p1 - p2) - m2 * p2 - c1 *
+             p1 * p2
+         return(list(c(dp1.dt, dp2.dt)))
+     })
+ }
```

At equilibrium, species 1 has the same equilibrium as in the Levins single species metapopulation model.

$$p_1^* = 1 - \frac{m_1}{c_1} \tag{9.3}$$

The abundance of species 1 increases with its colonizing ability, and decreases with its mortality rate.

Species 2 is influenced by species 1 — how do we know? We see the species 1 appears in the equation for species 2. Let's solve for p_2^* now.

$$\begin{aligned}
0 &= c_2 p_2 (1 - p_1 - p_2) - m_2 p_2 - c_1 p_1 p_2 \\
0 &= p_2 (c_2 - c_2 p_1 - c_2 p_2 - m_2 - c_1 p_1) \\
0 &= c_2 - c_2 p_1 - c_2 p_2 - m_2 - c_1 p_1 \\
c_2 p_2 &= c_2 - c_2 p_1 - m_2 - c_1 p_1 \\
p_2^* &= 1 - p_1^* - \frac{m_2}{c_2} - \frac{c_1}{c_2} p_1^* && (9.4) \\
& && (9.5)
\end{aligned}$$

In addition to the trivial equilibrium ($p_2^* = 0$), we see that the nontrivial equilibrium depends on the equilibrium of species 1. This equilibrium makes intuitive sense, in that species 2 cannot occupy sites already occupied by species 1, and like species one is limited by its own mortality and colonization rates ($-m_2/c_2$). It is also reduced by a bit due to those occasions when both species colonize the same site ($(c_1/c_2)p_1$), but only species 1 wins.

Substituting p_1^* into that equilibrium, we have the following.

$$p_2^* = 1 - \left(1 - \frac{m_1}{c_1}\right) - \frac{m_2}{c_2} - \frac{c_1}{c_2}\left(1 - \frac{m_1}{c_1}\right) \tag{9.6}$$

$$p_2^* = \frac{m_1}{c_1} - \frac{m_2 - m_1 + c_1}{c_2} \tag{9.7}$$

What parallels can you immediately draw between the equilibrium for the two species? The form of the equilibrium is quite similar, but with two additions for species 2. The numerator of the correction term includes its own mortality, just like species 1, but its mortality is adjusted downward (reduced) by the mortality of species 1. Thus the greater the mortality rate of species 1, greater is the opportunity for species 2. This term is also adjusted upward by the colonizing ability of species 1; the greater species 1's colonizing ability, the more frequently it lands on and excludes (immediately) species 2, causing a drop in species 2's equilibrium abundance. In order to focus on the competition–colonization tradeoff, it is common to assume mortality is the same for both species. Tradeoffs with regard to martality may also be quite important, especially if high mortality is correlated with high colonization rate, and negatively correlated with competitive ability.

If we assume $m_1 = m_2$, perhaps to focus entirely on the competition–colonization tradeoff, we can simplify eq. 9.7 further to examine when species 2 can invade (i.e. $p_2^* > 0$). Eq. 9.7 can simplify to

$$\frac{m}{c_1} > \frac{c_1}{c_2} \tag{9.8}$$

How can we interpret these? Species 2 can invade if the space not occupied by species 1 (m/c_1) is greater the species 1's ability to colonize an open patch faster than species 2 (c_1/c_2). An alternative representation ($mc_2 > c_1^2$) shows that species two can persist if mortality is high (but cannot exceed c_1), or if species 2's colonization rate is high. That seems fairly restrictive, on the face of it. However, if we assume that species 2 is the better colonizer, then this simply specifies how much better it has to be; it also indicates that increasing disturbance (and hence mortality) will enhance the abundance of the species which can recolonize those disturbances.

Thus, this model predicts that these species can coexist, even though one is a superior competitor. Further, it predicts that species 2 will become relatively more abundant as mortality increases.

Estimating colonization and mortality rates

Just exactly what corresponds to a "site" is not always defined, although site-based models such as these have often been used in a qualitative manner to describe the dynamics plant communities [83,152,153,202]. Indeed, such models could describe systems such as a single field, where a "site" is a small spot of ground, sufficient for the establishment of an individual [202]. Alternatively, a "site" could be an entire field in a large region of mixed successional stages. For the time being, let us continue to focus on a single field as a collection of small plots (e.g., 0.1×0.1 m), each of which might hold 1–3 potentially reproductive individuals. There may be several ways to estimate model parameters for these sorts of models [152,202], and here we try to estimate c and m directly. Assume that we clear all plants from 100 0.1×0.1 m plots. In the first year, we find that annuals (e.g., ragweed) showed up in 95 plots, and perennials (e.g. goldenrod) showed up in 60 plots. The following year, we find that 70 of the plots have perennials and 90 plots have annuals. Most plots contain both annuals and perennials.

If we make sufficient assumptions, we could easily estimate c and m. Further, in making these assumptions and making estimates, we provide an example that allows one to think critically about the ways in which the assumptions are close to reality or are misleading.

Let us assume that

1. perennials are virtually everywhere in this field ($p \approx 1$) and annual seeds are abundant throughout the soil ($p \approx 1$, then $cp \approx c$,
2. these small plots do not contribute much to the propagule pool, and that they receive all of their propagules from the ubiquitous *rain of propagules* from the surrounding vegetation.
3. in the first two years, these two species do not interact strongly.

Clearly these assummptions are wrong, but perhaps not too wrong, and will allow us to calculate some back-of-the-envelope estimates of parameters. With these assumption, we can estimate c and m using the *propagule rain* metapopulation model, $\dot{p} = c(1-p) - mp$. The discrete time version of this is simply the difference equation[1]

$$p_{t+1} = p_t + c_d(1 - p_t) - m_d p_t \tag{9.9}$$

where c_d and m_d are the discrete time versions of the colonization and mortality constants.

With only two parameters, we need only a small amount of data to estimate these constants — remember the old rule from algebra? We need two equations to estimate two unknowns. Based on our data above, we have

$$0.5 = 0 + c_d\,(1 - 0) - m_d\,(0)$$
$$0.7 = 0.5 + c_d\,(1 - 0.5) - m_d\,(0.5)$$

and very quickly we find that $c_d = 0.5$ and, given that, we find that $m_d = 0.1$.

We have another species as well, and we can estimate c_d and m_d for species 2 as well.

$$0.95 = 0 + c_d\,(1 - 0) - m_d\,(0)$$
$$0.90 = 0.95 + c_d\,(1 - 0.95) - m_d\,(0.95)$$

and very quickly we find that $c_d = 0.95$ and, given that, we find that $m_d \approx 0.1$.

What do the dynamics look like if we assume no interaction? Both species rise to achieve their equilibria (Fig. 9.2a). We have fit the data assuming that the species did not interact, and clearly this is not true, but we have begun the process of thinking critically about what we mean.

Let us assume that the species start interacting in year 2 — what would the dynamics look like? The difference equation for species 2 then becomes

$$p_{2,t+1} = p_{2,t} + c_{2,d}(1 - p_{2,t}) - m_{1,d}p_{2,t} - c_{1,d1}p_{1,t}p_{2,t} \tag{9.10}$$

where we include the subscript for each species. The dynamics differ radically when we include competitive exclusion (Fig. 9.2b). The species reach very different equilibria.

Let us extrapolate these dynamics (Fig. 9.2) to secondary succession in general (e.g., Fig. 9.1). Our model results seem qualitatively consistent with what we know about successional trajectories. With a lot of open sites, perhaps early in secondary or primary succession, good colonizers get in quickly and fill the site. They are subsequently replaced by the competitively superior species. This is a classic view of *succession*, with pioneer and climax species. Note that our example focused on small plots within a single field, but we could apply it to a much larger scale. This could approximate a *landscape mosaic* composed of patches of different sucessional ages, in which species of different dispersal and competitive abilities persist in the landscape, because they occupy patches of different ages.

[1] In general, we can create a difference equation from any differential equation, $\dot{N} = F(N)$, where $N_{t+1} = N_t + F(N)$ but where parameter estimates will differ somewhat.

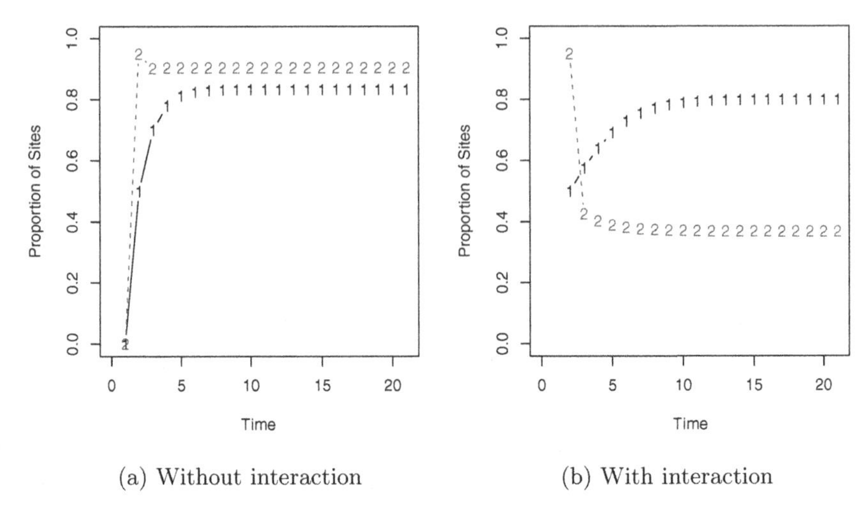

(a) Without interaction (b) With interaction

Fig. 9.2: Dynamics of perennials (sp. 1) and annuals (sp. 2) without and with interaction. The interaction assumes that species 1 excludes species 2 instantaneously whenever they come into contact ($c_{d,1} = 0.5$, $c_{d,2} = 0.95$, $m_{d,1} = m_{d,2} = 0.1$)

Estimating and using c_d and m_d

Let us code the data we derive above, and project over 20 years. First without interaction.

```
> cd1 <- 0.5; cd2 <- 0.95; md1 <- 0.1; md2 <- 0.1
```

We create a big enough matrix, and perform the projection.

```
> t <- 20
> ps <- matrix(0, nrow = t + 1, ncol = 2)
> for (i in 1:t) ps[i + 1, ] <- {
+     p1 <- ps[i, 1] + cd1 * (1 - ps[i, 1]) - md1 * ps[i, 1]
+     p2 <- ps[i, 2] + cd2 * (1 - ps[i, 2]) - md2 * ps[i, 2]
+     c(p1, p2)  }
> matplot(0:t + 1, ps, type = "b", ylab = "Proportion of Sites",
+     xlab = "Time", xlim = c(0, t + 1), ylim = c(0, 1))
```

Now assume they interact from year 2 onward.

```
> ps2 <- matrix(0, nrow = t + 1, ncol = 2)
> ps2[1, ] <- ps[2, ]
> for (i in 1:t) ps2[i + 1, ] <- {
+     p1 <- ps2[i, 1] + cd1 * ps2[i, 1] * (1 - ps2[i, 1]) -
+         md1 * ps2[i, 1]
+     p2 <- ps2[i, 2] + cd2 * ps2[i, 2] * (1 - ps2[i, 2]) -
+         md2 * ps2[i, 2] - cd1 * ps2[i, 2]
+     c(p1, p2)   }
> matplot(1:t + 1, ps2[-(t + 1), ], type = "b", ylab = "Proportion of
+     Sites", xlab = "Time", xlim = c(0, t + 1), ylim = c(0, 1))
```

Habitat destruction

Nee and May [146] later showed that, given these tradeoffs, an interesting phenomenon arose. If species coexist via this competition–colonization tradeoff, destruction of habitat increases the abundance of the *inferior competitor*. How does it do this? First let's derive this analytically, and then consider it from an intuitive point of view.

We can alter the above equations to include habitat destruction, D.

$$\frac{dp_1}{dt} = c_1 p_1 (1 - D - p_1) - m_1 p_1$$
$$\frac{dp_2}{dt} = c_2 p_2 (1 - D - p_1 - p_2) - m_2 p_2 - c_1 p_1 p_2$$

We can then solve for the equilibria.

$$p_1^* = 1 - D - \frac{m_1}{c_1} \tag{9.11}$$
$$p_2^* = \frac{m_1}{c_1} - \frac{m_2}{c_2} - \frac{c_1}{c_2}\left(1 - D - \frac{m_1}{c_1}\right) \tag{9.12}$$
$$\tag{9.13}$$

What does this mean to us? First note that habitat destruction has a simple and direct negative effect on the abundance of species 1. For species 2, we see the first two terms are unaltered by habitat destruction. The third and last term represents the proportion of colonization events won by the superior competitor, species 1, and thus depends on the abundance of species 1. Because habitat destruction has a direct negative effect on species 1, this term shows that habitat destruction *can increase the abundance of inferior competitors by negatively affecting the superior competitor.*

We must also discuss what this does *not* mean. Typically, an ecologist might imagine that *disturbed* habitat favors the better colonizer, by making more sites available for good colonizers, and perhaps creating microhabitat conditions that favor rapid growth (e.g., pulse of high resources). This is different than habitat destruction, which removes entirely the habitat in question. Imagine a parking lot is replacing a grassland, or suburban sprawl is replacing forest; the habitat is shrinking — rather than merely being disturbed — and this has a negative impact on the better competitor.

Multispecies competition–colonization tradeoff and habitat destruction

The work of David Tilman and his colleagues in grassland plant communities at the Cedar Creek Natural History Area (a NSF-LTER site) initially tested predictions from the R^* model, and its two-resource version, the resource ratio model [200,201,218,222]. They found that soil nitrogen was really the only resource for which the dominant plant species (prairie grasses) competed strongly. If this is true, then the single resource R^* model of competition predicted that the best competitor would eliminate the other species, resulting in monocultures of the best competitor. The best competitors were little bluestem and big bluestem (*Schizachyrium scoparium* and *Andropogon gerardii*), widespread dominants of mixed and tall grass prairies, and R^* predicted that they should exclude everything else. However, they observed that the control plots, although dominated by big and little bluestem, were also *the most diverse.* While big and little bluestem did dominate prairies, the high diversity of these communities directly contradicts the R^* model of competition.[2] Another pattern they observed was that weaker competitors colonized and became abundant in abandoned fields sooner than the the better competitors. It turned out that variation in dispersal abilities might be the key to understanding both of these qualitative patterns [65, 162, 174].

In an effort to understand how bluestem-dominated prairies could maintain high diversity in spite of single resource competition, Tilman generalized the [73] equations to include n species [202].

$$\frac{dp_i}{dt} = c_i p_i \left(1 - \sum_{j=1}^{i} p_j\right) - m_i p_i - \left(\sum_{j=1}^{i-1} c_j p_j p_i\right) \tag{9.14}$$

where the last term describes the negative effect on species i of all species of superior competitive ability.

Multispecies competition–colonization model

Here we create an R function for eq. 9.14.

```
> compcolM <- function(t, y, params) {
+     S <- params[["S"]]
+     D <- params[["D"]]
+     with(params, list(dpi.dt <- sapply(1:S, function(i) {
+         params[["ci"]][i] * y[i] * (1 - D - sum(y[1:i])) -
+             params[["m"]][i] * y[i] - sum(params[["ci"]][0:(i -
+             1)] * y[0:(i - 1)] * y[i])
+     })))
+ }
```

This code seems strange, that is, unlike previous systems of ODEs in which we wrote each separate equation out. The above code merely implements the strict hierarchy of eq. 6.12, and is inspired by a similar approach by Roughgarden [181]. It also allows us to specify, on the fly, the number of species we want. We also sneak in a parameterization for habitat destruction, D, and we will address that later.

One goal was to explain the successional patterns of grasses in his study area, the low nutrient grassland/savanna of Minnesota sand plains. Following early succession, the common perennial prairie grass species seemed to form an abundance hierarchy based on competitive ability: the best competitors were the most abundant, and the worst competitors were least abundant. A caricature

[2] For different views see [40, 204].

of a generic species abundance distribution is the geometric distribution, where each species, with rank i makes up a constant, declining, fraction of the total density of all individuals (see Chapter 10 for more detail).[3] Specifically, the proportional abundance of each species i can be calculated as a function of proportional abundance of the most abundant species, d, and the species rank, i.

$$p_i = d(1-d)^{i-1} \tag{9.15}$$

Thus if the most abundant species makes up 20% of the assemblage ($d = 0.20$), the second most abundant species makes up 20% of the remaining 80%, or $0.2(1-0.2)^1 = 0.16 = 16\%$ of the community. Tilman *et al.* [202] showed that if all species experience the same loss rate, then species abundances will conform to a geometric distribution when colonization rates conform to this rule

$$c_i = \frac{m}{(1-d)^{2i-1}} \tag{9.16}$$

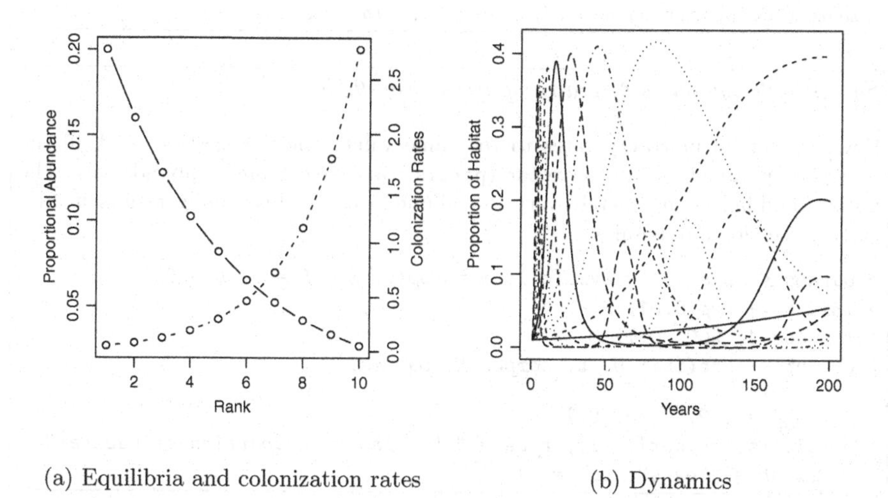

(a) Equilibria and colonization rates (b) Dynamics

Fig. 9.3: (a) Rank–abundance distribution and the colonization rates that create them ($m = 0.04$). (b) Successional dynamics with the competition–colonization tradeoff, from low initial abundances. Here, equilibrium abundance of the best competitor is 20% ($d = 0.2$), mortality is 4% ($m = 0.04$), and colonization rates are determined by eq. 9.16, resulting, at equilibrium, in a geometric species rank–abundance distribution.

[3] The most abundant species has rank equal to 1.

Calculating rank–abundance distributions and colonization rates (Fig. 9.3a)

Here we select 10 species, with the most abundant species equaling 20% of the biomass in the community, and specify a common mortality or disturbance rate m. We then create *expressions* for eqs. 9.15 and 9.16.

```
> S <- 10
> ranks <- 1:S
> d <- 0.2
> m = 0.04
> geo.p <- expression(d * (1 - d)^(ranks - 1))
> ci <- expression(m/(1 - d)^(2 * ranks - 1))
```

Next we create a plot with two y-axes.

```
> par(mar = c(5, 4, 1, 4), mgp = c(2, 0.75, 0))
> plot(ranks, eval(geo.p), type = "b", ylab = "Proportional Abundance",
+     xlab = "Rank", xlim = c(1, S))
> par(new = TRUE)
> plot(ranks, eval(ci), type = "b", axes = FALSE, ann = FALSE,
+     lty = 2)
> axis(4)
> mtext("Colonization Rates", side = 4, line = 2)
```

Sucessional dynamics of prairie grasses (Fig. 9.3b)

Here, we set all mortality rates to the same value, one per species, pool all the necessary parameters into a vector (**params**), and select initial abundances. The initial abundances are merely very low abundances — this merely results in fun, early successional dynamics.

```
> params <- list(ci = eval(ci), m = rep(m, S), S = S, D = 0)
> init.N <- rep(0.01, S)
> t = seq(1, 200, 0.1)
> cc.out <- ode(init.N, t, compcolM, params)

> par(mgp = c(2, 0.75, 0))
> matplot(t, cc.out[, -1], type = "l", ylab = "Proportion of Habitat",
+     xlab = "Years", col = 1)
```

Tilman and colleagues [203] startled folks when they showed that a common scenario of habitat destruction led to a perfectly counterintuitive result: that habitat destruction led to the very slow but deterministic *loss of the best competitor*. It was unsettling not only that the dominant species would be lost first (a result demonstrated by Nee and May [146]), but also that the loss of dominant species will take a long time. This implied that we would not realize the true cost of habitat destruction until long after the damage was done. These two conclusions posed a problem for conservationists.

- Species that were thought safe from extirpation at local scales — the best competitors — could actually be the ones most likely to become extinct, and further,

- That the process of extinction may take a long time to play out, and so the data demonstrating this loss might require a long time to collect (Fig. 9.4).

These two predictions constitute the original conception for *extinction debt.* However, the term has become more broadly used to described the latter phenomenon, that extinction due to habitat destruction may take a long time, and current patterns may be a function of past land use [75, 114].

We see (Fig. 9.4) the predictable, if counterintuitive, result: the most abundant species, the competitive dominant, becomes extinct over a long period of time, and the next best competitor replaces it as the most abundant species.

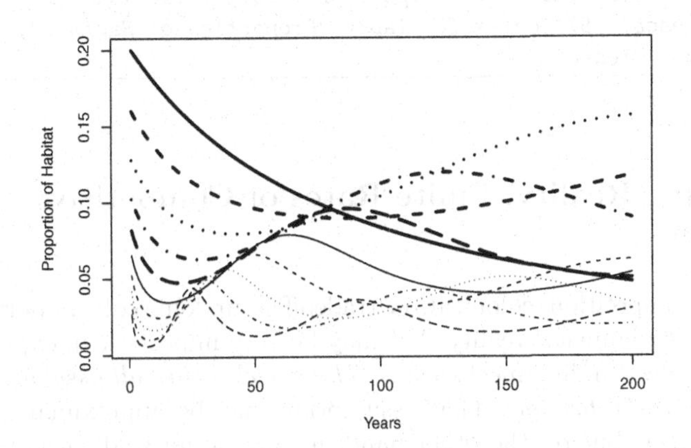

Fig. 9.4: Extinction debt. Destruction of 25% of the habitat causes the loss of the competitive dominant (wide solid line). Parameters the same as in Fig. 9.3b, but initial abundances are equilibrium abundances in the absence of habitat destruction. The second best competitor (wide dashed line) will eventually become the most common species.

Competition is a local phenomenon, and the better competitor can typically hold onto a given site; however, individuals of *all* species eventually die. Therefore, for two species to actually coexist in a landscape, even the best competitor must colonize some new space at some point. If habitat destruction reduces habitat availability too far, the worst colonizer (i.e., the best competitor) will be unable to disperse effectively to new habitat.

Extinction debt (Fig. 9.4)

We use the same functions and parameters as above. We add habitat destruction for a quarter of the available habitat, which is greater that the equilibrium for the dominant species, and will result in the slow loss of the dominant species. We also start the species off at their equilibrium abundances, determined by the geometric distribution.

```
> params["D"] <- 0.25
> init.N <- eval(geo.p)
> cchd.out <- ode(init.N, t, compcolM, params)
> matplot(t, cchd.out[, -1], type = "l", lty = 1:5, lwd = rep(c(3,
+     1), each = 5), col = 1, ylab = "Proportion of Habitat",
+     xlab = "Years")
```

9.2 Adding Reality: Finite Rates of Competitive Exclusion

While the competition–colonization tradeoff is undoubtedly important, it ignores some fundamental reality that may be very important in explaining patterns and understanding mechanisms. *The models above all assume competitive exclusion is instantaneous.* That assumption may be approximately or qualitatively correct, but on the other hand, it may be misleading. Given the implications of *extinction debt* for conservation, it is important to explore this further. Indeed, Pacala and Rees did so [152], and came to very different conclusions than did Tilman *et al.* [202]. This section explores the work of Pacala and Rees [152].

If we look at species in the real world, a couple of observations arise, with respect to tradeoff of species of different successional status. *First*, species characterized by high dispersal ability are also often characterized by high maximum growth rates, related to high metabolic and respiration rates, and allocation to reproductive tissue. These we refer to as *r-selected* species [119,122]. *Second*, we observe that when deaths of individuals free up resources, individuals with high maximum growth rates can take advantage of those high levels of resources to grow quickly and reproduce. *Third*, we observe that the arrival of a propagule of a superior competitor in the vicinity of a poor competitor does not result in the instantaneous draw down of resource levels and exclusion of the poor competitor. Rather, the poor competitor may continue to grow and reproduce even in the presence of the superior competitor *prior to the reduction of resources to equilibrium levels.* It is only over time, and in the absence of disturbance, that better resource competitors will tend to displace individuals with good colonizing ability and high maximum growth rates.

Pacala and Rees [152] wanted to examine the impact of finite rates of competitive exclusion on the competition–colonization tradeoff. Implicit in this is the role of maximum growth rate as a trait facilitating coexistance in the landscape. High growth rate can allow a species to reproduce prior to resource reduction and competitive exclusion. This creates an ephemaral niche, and Pacala and Rees referred to this as the *successional niche.* Species which can take good advantage of the successional niche are thus those with the ability to disperse to, and reproduce in, sites where resources have not yet been depleted by superior competitors. To facilitate their investigation, Pacala and Rees added *finite rates of succession* to a simple two species competition–colonization model.

Possible community states

They envisioned succession on an open site proceeding via three different pathways. They identified five possible states of the successional community (Fig. 9.5).

1. *Free* — Open, unoccupied space.
2. *Early* — Occupied by only the early sucessional species.
3. *Susceptible* — Occupied by only the late successional species and susceptible to invasion because resource levels have not yet been driven low enough to exclude early successional species.
4. *Mixed* — Occupied by both species, and in *transition* to competitive exclsuion.
5. *Resistant* — Occupied by only the late successional species and resistant to invasion because resource levels have been driven low enough to exclude early successional species.

Pathways

Given the five states, succession can then proceed along any of three pathways (Fig 9.5):

1. Free → Early → Mixed → Resistant,
2. Free → Susceptible → Mixed → Resistant,
3. Free → Susceptible → Resistant.

In this context, Pacala and Rees reasoned that the competition–colonization tradeoff focuses on mutually exclusive states, and assumes there are only Free, Early, and Resistant states. In contrast, if species coexist exclusively via the competition–maximum growth rate tradeoff, then we would observe only Free, Mixed, and Resistant states. They showed that these two mechanisms are not mutually exclusive and that the roles of finite rates of competitive exclusion and the successional niche in maintaining diversity had been underestimated.

Note that the interpretation of this model is thus a little different than other models that we have encountered. It is modeling the dynamics of *different community states.* It makes assumptions about species traits, but tracks the frequency of different community states. Technically speaking, any metapopulation model is doing this, but in the contexts we have seen, the states of different patches of the environment were considered to be completely correlated with the abundance of each species modeled. Here we have five different state variables, or possible states, and only two species.

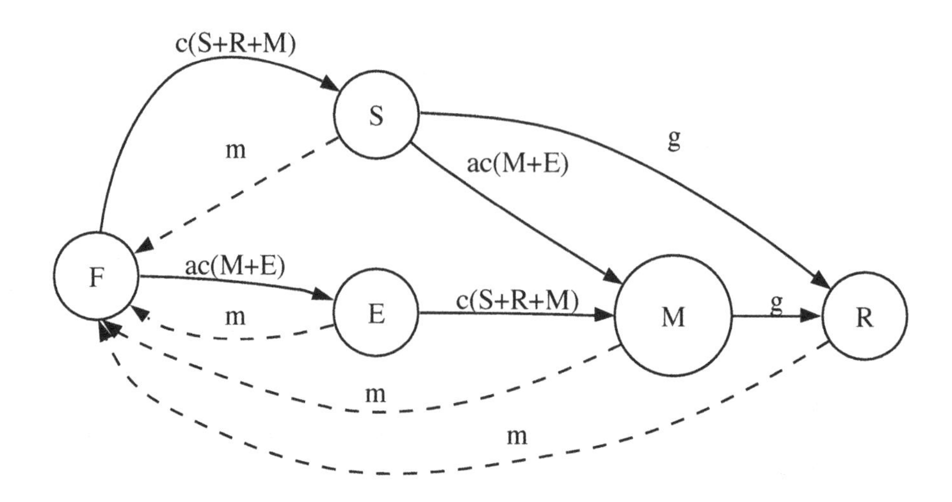

Fig. 9.5: The state variables in the [152] model. Dashed lines indicate mortality; the larger size of the Mixed state merely reminds us that it contains two species instead of one. Each pathway is labelled with the per capita rate from one state to the other. For instance the rate at which Mixed sites are converted to Resistant sites is $g(M)$, and the rate at which Free sites are converted to Early sites is $ac(M+E)$

The traits of the two species that create these four states are embedded in this model with four parameters, c, α, m, γ. There is a base colonization rate, c, relative colonization rate of the poor competitor, α, mortality (or disturbance rate), m, and the rate of competitive exclusion, γ. We could think of γ as the rate at which the better competitor can grow and deplete resources within a small patch. We model the community states as follows, using F to indicate a fifth state of Free (unoccupied) space.

$$\frac{dS}{dt} = [c(S+R+M)]F - [\alpha c(M+E)]S - \gamma S - mS \tag{9.17}$$

$$\frac{dE}{dt} = [\alpha c(M+E)]F - [c(S+R+M)]E - mE \tag{9.18}$$

$$\frac{dM}{dt} = [\alpha c(M+E)]S + [c(S+R+M)]E - \gamma M - mM \tag{9.19}$$

$$\frac{dR}{dt} = \gamma(S+M) - mR \tag{9.20}$$

$$F = 1 - S - E - M - R \tag{9.21}$$

Successional niche model

In addition to representing the original successiona niche model, we can also slip in a parameters for habitat destruction, D. As with the above model, habitat destruction D is simply a value between 0–1 that accounts for the proportion of the habitat destroyed. Pacala and Rees [152] didn't do that, but we can add it here. We also have to ensure that F cannot be negative.

```
> succniche <- function(t, y, params) {
+     S <- y[1]
+     E <- y[2]
+     M <- y[3]
+     R <- y[4]
+     F <- max(c(0, 1 - params["D"] - S - E - M - R))
+     with(as.list(params), {
+         dS = c * (S + R + M) * F - a * c * (M + E) * S -
+             g * S - m * S
+         dR = g * (S + M) - m * R
+         dM = a * c * (M + E) * S + c * (S + R + M) * E -
+             g * M - m * M
+         dE = a * c * (M + E) * F - c * (S + R + M) * E -
+             m * E
+         return(list(c(dS, dE, dM, dR)))
+     })
+ }
```

Now we can examine the dynamics of this model. When we make the rate of competitive exclusion very high, the model approximates the simple competition–colonization tradeoff (Fig. 9.5)[4] The susceptible and mixed states are not apparent, and the better competitor slowly replaces the good colonizer.

[4] When $\gamma = 5$, this means that in one year, exclusion will be 99.3% complete, because it is a pure negative exponential process, where $X_t = X_0 e^{\gamma t}$. Similarly, recall our calculation of doubling time, $t = log(X)/r$, where X is the relative size of the population (e.g., 2, if the population doubles); here $r < 0$ and time is < 1.

Dynamics of the successional niche model with a high rate of competitive exclusion (Fig. 9.6a)

For no particular reason, we pretend that the poor competitor is 7× as fast at colonizing ($\alpha = 7$), and that the rate of competitive exclusion is very high, $\gamma = 5$. We assume mortality is low, and there is no habitat destruction.

```
> params.suc <- c(a = 7, c = 0.2, g = 5, m = 0.04, D = 0)
```

Next we let time be 50 y, and initial abundances reflect a competitive advantage to the early successional species, and run the model.

```
> t = seq(0, 100, 0.1)
> init.suc <- c(S = 0.01, E = 0.03, M = 0, R = 0)
> ccg.out <- data.frame(ode(init.suc, t, succniche, params.suc))
```

Last we plot our projections.

```
> matplot(t, ccg.out[, -1], type = "l", ylab = "Relative Frequency",
+     xlab = "Time", ylim = c(0, 1), col = 1)
> legend("topright", colnames(ccg.out)[5:2], lty = 4:1, bty = "n")
```

Now let's slow down competitive exclusion, that is, we will slow the transition from M to R. We set γ to be a small number, so that only 10% of the mixed plots become resistant plots over one year ($\gamma = 0.1$). When we slow down competitive exclusion, we see that we get persistence of both species (Fig. 9.6b). First, we get greater frequency of the mixed state, where we always see a large fraction of the mixed habitat occupied by both species (state M), relative to the case with rapid competitive exclusion (Fig. 9.6a). In addition, we see a higher proportion of habitat in early succession phase (state E).

Dynamics of the successional niche model with a low rate of competitive exclusion (Fig. 9.6b)

Here we slow down the rate of competitive exclusion, and 90% of the mixed sites stay mixed after a one year interval.

```
> params.suc["g"] <- 0.1
> exp(-0.1)

[1] 0.9048

> ccg.out <- data.frame(ode(init.suc, t, succniche, params.suc))
```

We plot our projections.

```
> matplot(t, ccg.out[, -1], type = "l", ylab = "Relative Frequency",
+     xlab = "Time", ylim = c(0, 1), col = 1)
```

Now let's imagine that mortality rate increases from 1% (Fig. 9.6a) to 10% (Fig. 9.6c). ($m = 0.04$ *vs.* $m = 0.105$). With this moderate disturbance rate, the system takes longer to approach an equilibrium, and also has a higher frequency of sites with just the early sucessional species (Fig. 9.6c). This reflects the mechanisms underlying the *intermediate disturbance hypothesis* [37]. At very

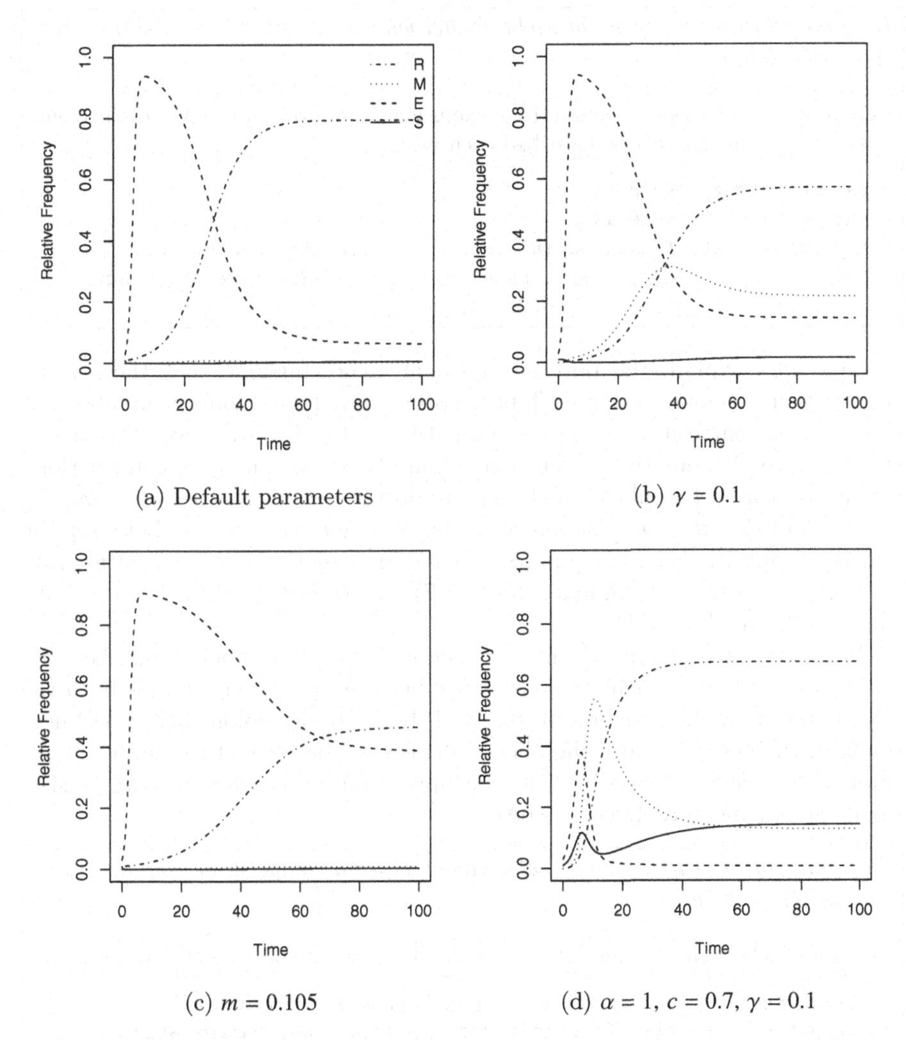

(a) Default parameters

(b) $\gamma = 0.1$

(c) $m = 0.105$

(d) $\alpha = 1$, $c = 0.7$, $\gamma = 0.1$

Fig. 9.6: Unless otherwise noted in the figure, $\alpha = 7$, $c = 0.2$, $\gamma = 5$, $m = 0.04$. (a) Competition–colonization (high rate of competitive exclusion), (b) Competition–colonization and the successional niche (slower competitive exclusion), (c) Competition–colonization with intermediate rather than low disturbance, and (d) Successional niche (equal colonizing ability).

low disturbance rates, the best competitor prevails. At moderate disturbance rates, however, we have far more habitat that supports both *pioneer* species (good colonizers) as well as *climax* species (superior competitor). Note that slowing down competitive exclusion (Fig. 9.6b) or increasing the disturbance rate (Fig. 9.6c) result in similar frequencies of both species, but via different mechanisms.

Dynamics of the successional niche model with an intermediate disturbance rate (Fig. 9.6c)

Here we have a low rate of competitive exclusion (as above), but higher disturbance rates (10% of the habitat is disturbed each year).

```
> params.suc["g"] <- 5
> params.suc["m"] <- 0.105
> ccg.out <- data.frame(ode(init.suc, t, succniche, params.suc))
> matplot(t, ccg.out[, -1], type = "l", ylab = "Relative Frequency",
+     xlab = "Time", ylim = c(0, 1), col = 1)
```

Now let's explore the pure successional niche. Imagine that there is no competition–colonization tradeoff: both species have high colonization rates, but the superior competitor retains its competitive edge. However, we also assume that the rate of competitive exclusion is finite ($\gamma \ll \infty$). The pure competition–colonization model would predict that we do not get coexistence. *In contrast, we will find that the successional niche model allows coexistence.* Let's set the relative competitive ability equal ($\alpha = 1$) and increase the base colonization rate to the higher of the two original rates ($c = 7$). We also let the rate of competitive exclusion be small ($\gamma < 1$).

What do the dynamics of the pure successional niche model look like (Fig. 9.6d)? We see that we achieve coexistence because the system retains both the mixed state and the resistant state. With both species colonizing everywhere (high c), the successional niche allows coexistence because of the finite rate of competitive exclusion. Species 1 now occupies a pure *successional niche* persisting in the mixed state M (Fig. 9.6d).

Dynamics of the successional niche model with no competition–colonization tradeoff (Fig. 9.6d)

Here we equal and high colonization rates, and a slow rate of competitive exclusion.

```
> params.suc <- c(a = 1, c = 0.7, g = 0.1, m = 0.04, D = 0)
> ccg.out <- data.frame(ode(init.suc, t, succniche, params.suc))
> matplot(t, ccg.out[, -1], type = "l", ylab = "Relative Frequency",
+     xlab = "Time", ylim = c(0, 1), col = 1)
```

Let's consider *the successional niche* analytically, by effectively eliminating colonization limitation for either species ($c >> 1$, $\alpha = 1$). If there is no colonization limitation, then propagules of both species are everywhere, all the time. This means that states F, E, and S no longer exist, because you never have free space or one species without the other. Therefore M is one of the remaining states. The only monoculture that exists is R, because in R, both propagules may arrive, but the early successional species 1 cannot establish. Therefore the two states in the pure successional niche model are M and R.

How does M now behave? M can only increase when R dies back. M will decrease through competitive exclusion. We might also imagine that M would decrease through its own mortality; we have stipulated, however, that colonization is not limiting. Therefore, both species are always present, if only as propagules. The rates of change for M and R, therefore are,

$$\frac{\mathrm{d}M}{\mathrm{d}t} = mR - \gamma M$$
$$R = 1 - M$$

The equilibria for the two states can be found by first setting $\dot{M} = 0$ and substituting $1 - M$ in for R, as

$$0 = m(1 - M) - \gamma M$$
$$M* = \frac{m}{\gamma + m}$$

making $R* = \gamma/(\gamma + m)$.

Up until now, we have focused on the frequencies of the four *states*, rather than the frequencies of the two types of species (early successional and competitive dominant). If we are interested in the relative abundances of the two types of species, we merely have to make assumptions about how abundant they are in each different state. If we assume that they are equally abundant in each state, then the abundance of the early successional species is $E + M$ and the abundance of the competitive dominant is $S + M + R$.

Let us finally investigate extinction debt with this model. We should first verify that we get extinction debt with a "pure" competition–colonization tradeoff (large γ, large α). We should then reduce γ and α to make the successional niche the primmary mechanism, and see what happens to the pattern of extinction.

We can compare patterns of extinction under these two scenarios (Fig. 9.7, without and with the successional niche). In doing so, we find *a complete reversal of our conclusions*: when the primary mechanism of coexistence is the successional niche, we find that the competitive dominant persists, rather than the early successional species. (Recall that when colonization rates are equal, the rate of extinction *must* be slow in order to achieve coexistance even without habitat destruction).

These opposing predictions highlight the important of getting the mechanism right. They also illustrate the power of simplistic models to inform understanding.

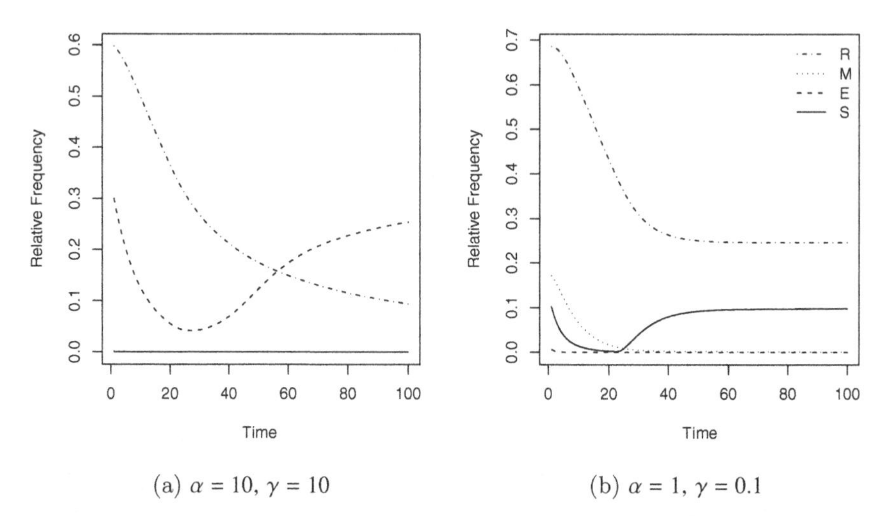

(a) $\alpha = 10$, $\gamma = 10$ (b) $\alpha = 1$, $\gamma = 0.1$

Fig. 9.7: Extinction dynamics beginning equilibrium abundances. (a) Relying on the competition–colonization tradeoff results in a loss of the competitive dominant, (b) Relying on the successional niche tradeoff results in a persistance of the competitive dominant.

Dynamics of the extinction debt with the successional niche model (Fig. 9.7)

Here we approximate the competitive-colonization model with unequal colonization rates, and a very high rate of competitive exclusion. We then find, through brute force, the equlibria for our parameter set by integrating a long time and keeping the last observations as the equilibria.

```
> params.suc1 <- c(a = 10, c = 0.1, g = 10, m = 0.04, D = 0)
> Xstar1 <- ode(init.suc, 1:500, succniche, params.suc1)[500, -1]
```

We then create habitat destruction, and plot the result.

```
> params.suc1D <- c(a = 10, c = 0.1, g = 10, m = 0.04,
+     D = as.numeric(Xstar1["R"]))
> t = 1:100
> ccg.out1 <- data.frame(ode(Xstar1, t, succniche, params.suc1D))
> matplot(t, ccg.out1[, -1], type = "l", col = 1,
+     ylab = "Relative Frequency", xlab = "Time")
```

Next, we include the successional niche, by making colonization rates equal and high, and γ small. We then find our equilibria, without habitat destruction.

```
> params.suc2 <- c(a = 1, c = 1, g = 0.1, m = 0.04, D = 0)
> Xstar2 <- ode(init.suc, 1:500, succniche, params.suc2)[500,
+     -1]
```

We then create habitat destruction, and plot the result.

```
> params.suc2D <- c(a = 1, c = 0.7, g = 0.1, m = 0.04,
+     D = as.numeric(Xstar1["R"]))
> ccg.out2 <- data.frame(ode(Xstar2, t, succniche, params.suc2D))
> matplot(t, ccg.out2[, -1], type = "l", ylab = "Relative Frequency",
+     xlab = "Time", col = 1)
> legend("topright", colnames(ccg.out2[5:2]), lty = 4:1, bty = "n")
```

9.3 Storage effect

What if all of this is wrong? What if none of these tradeoffs underlie coexistance? Jim Clark and his colleagues [31] examined dispersal traits of co-occurring deciduous forest trees (fecundity, dispersal), and successional status, and found no evidence that early successional species had higher dispersal capacity. This suggests a lack of support for competition–colonization tradeoffs. Rather, they found evidence that *asynchronous success in reproduction* of these *long-lived organisms* allowed them to coexist. Together, these traits constitute a mechanism referred to as the *storage effect* [28, 217].

In the storage effect, competing species can *store* energy for reproduction until favored conditions arise [28, 217]. Assumptions include:

Variable environment Each species encounters both favorable and unfavorable periods for reproduction.

Buffered population growth Each species stores energy in a resistant stage (e.g., long lived adults, seeds, spores, eggs) between favorable periods.

Environment–competition covariation The same conditions that favor reproduction for a particular species also increase competition intensity for that species. If, for instance, winter rains favor a particular desert annual, that desert annual will experience the greatest intraspecific competition following a wet winter precisely because of its large population size.

One prediction of the storage effect is that small population sizes will be more variable (have higher CV, coefficient of variation) than large populations of competing species [89].

In one sense, the storage effect constitutes a temporal niche [28]. That is, the theory simply stipulates that different species succeed at different times. For rare species to coexist with common species, their relative success needs to be somewhat greater than the relative success of common species. In the absence of environmental variability, species would not coexist.

Here we provide one set of equations describing the dynamics of the storage effect for each species i in the community [26].

$$N_{i,t+1} = (1 - d)N_{i,t} + R_{i,t}N_{i,t} \qquad (9.22)$$

$$R_{i,t} = e^{E_{i,t} - C_{i,t}} \qquad (9.23)$$

$$E_{i,t} = F(X_{i,t}) \qquad (9.24)$$

$$C_{i,t} = \sum_{i=1}^{S} \alpha_i e^{E_{i,t}} N_{i,t} \qquad (9.25)$$

$$(9.26)$$

Here, E is an unspecified function of the environment, $X_{i,t}$, so that E specifically differs among species and across time. We can think of $exp(E_{i,t})$ (the first growth factor in $R_{i,t}$) as the maximum per capita reproductive rate for species i, at time t, in the absence of competition. It is determined by the environment at time t.

The other growth factor, $exp(-C_{i,t})$, is the effect of competition. It allows us to intensify competition at large population sizes (e.g., during favorable conditions), and lessen competition at low population sizes (e.g., during poor conditions). These two factors allow us to represent independently the positive and negative effects of the environment on an organism's capacity to grow (E_i), and also to represent how competition intensity covaries with population density (C_i). There are many other representations of the storage effect [29, 217], but this is a simple and convenient one [26].

The storage effect is a special case of *lottery* models [143]. Originally developed for reef fishes [183], lottery models are considered general models for other systems with important spatial structure such as forest tree assemblages. Conditions associated with lottery systems [30] include:

1. Juveniles (seedlings, larvae) establish territories in suitable locations and hold this territory for the remainder of their lives. Individuals in non-suitable sites do not survive to reproduce.
2. Space is limiting; there are always far more juveniles than available sites.
3. Juveniles are highly dispersed such that their relative abundances and their spatial distributions are independent of the distribution of parents.

These conditions facilitate coexistence because the same amount of open space has a greater benefit for rare species than for common species. While the invulnerable nature of successful establishment slows competitive exclusion, permanent nonequilibrium coexistence does not occur unless there is a storage effect, such as with overlapping generations, where a reproductive stage (resting eggs, seeds, long lived adults) buffers the population during unfavorable environmental conditions, and negative environment–competition covariation.

9.3.1 Building a simulation of the storage effect

Here we simulate one rare and one common species, wherein the rare species persists only via the storage effect. We conclude with a simple function, `chesson`, that simplifies performing more elaborate simulations.

Fluctuating environment

First we create a variable environment. Environments are frequently noisy and also temporally autocorrelated — we refer to a special type of scale-independent autocorrelation as *red noise* or $1/f$ *noise* ("one over 'f' noise") [67]. Here we use simply *white noise*, which is not temporally autocorrelated, but rather, completely random at the time scale we are examining.

```
> years <- 100
> t <- 1:years
> variability = 4
> env <- rnorm(years, m = 0, sd = variability)
> plot(t, env, type = "l")
```

Differential responses to the environment

A key part of the storage effect is that species have differential reproduction in response to a fluctuating environment (Fig. 9.8). However, species can differ for all sorts of reasons. Therefore, we will let our two species have different *average* fitness. We do this specifically because in the absence of the stochastic temporal niche of the storage effect, the species with the higher fitness would eventually replace the rare species. We want to show that the storage effect allows coexistance in spite of this difference. We will let these fitnesses be

```
> w.rare <- 0.5
> w.comm <- 1
```

but as we will see in the simulation, competitive exclusion does not happen — the species coexist. For the example we are building (Fig. 9.8), we will pretend that

- our rare species grows best when the environment (maybe rainfall) is above average,
- our common species grows best when the environment is below average, and,
- both grow under average conditions; their niches overlap, and we will call the overlap rho, ρ.

As merely a starting point, we let overlap, ρ, be equal to the standard deviation of our environmental variabiity.

```
> rho <- sd(env)
```

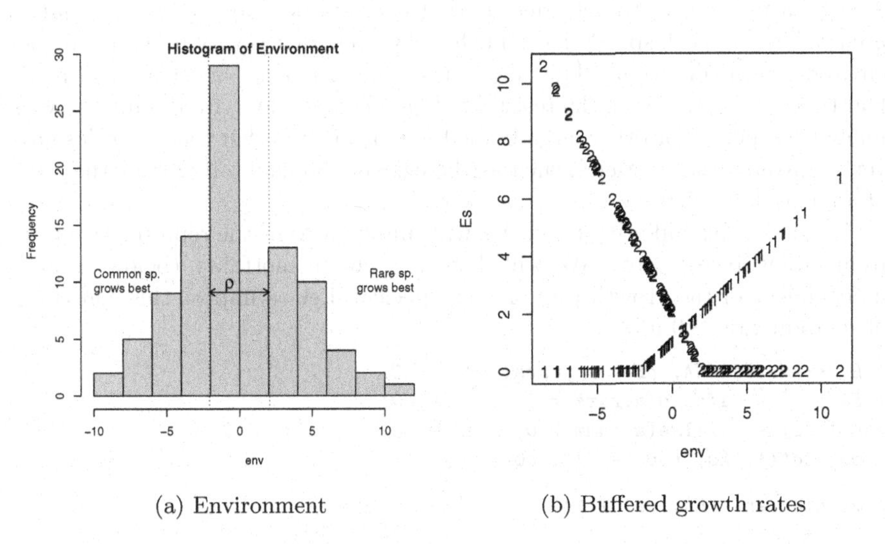

(a) Environment (b) Buffered growth rates

Fig. 9.8: Environmental variability, niche overlap (ρ), and the resulting buffered population growth rates.

Code for a pretty histogram (Fig. 9.8)

Here we simply create a pretty histogram.

```
> hist.env <- hist(env, col = "lightgray",
+                  main = "Histogram of Environment")
> abline(v = c(c(-rho, rho)/2), lty = 3)
> arrows(x0 = -rho/2, y0 = mean(hist.env[["counts"]]), x1 = rho/2,
+     y1 = mean(hist.env[["counts"]]), code = 3, length = 0.1)
> text(0, mean(hist.env[["counts"]]), quote(italic(rho)), adj = c(1.5,
+     0), cex = 1.5)
> text(min(hist.env[["breaks"]]), mean(hist.env[["counts"]]),
+     "Common sp.\ngrows best", adj = c(0, 0))
> text(max(hist.env[["breaks"]]), mean(hist.env[["counts"]]),
+     "Rare sp.\ngrows best", adj = c(1, 0))
```

To quantitfy reproduction as a function of the environment, we will simply let each species growth rate be, in part, the product of its fitness and the environment, with the sign appropriate for each species.

```
> a.rare <- (env + rho/2) * w.rare
> a.comm <- -(env - rho/2) * w.comm
```

This will allow the rare species to have highest reproduction when the environment variable is above average, and the common species to have high reproduction when the environmental variable is below average. It also allows them to share a zone of overlap, ρ, when they can both reproduce (Fig. 9.8a).

Buffered population growth

A key feature of the storage effect is that each species has *buffered population growth*. That is, each species has a life history stage that is very resistant to poor environmental conditions. This allows the population to persist even in really bad times. In some cases, the resistant stage may be a long-lived adult, as with many tree species, or other large-bodied organisms. In other cases, species have very resistant resting stages, such as the eggs of zooplankton [20], or the seeds of annual plants [54].

To model this buffering effect, we will simply prevent the reproductive rates from falling below zero. (We will, however, create mortality (below) that is independent of the growth rate of each species). Let us impose this constraint of reproduction ≥ 0 now.

```
> Es <- matrix(NA, nrow = years, ncol = 2)
> Es[, 1] <- ifelse(a.rare > 0, a.rare, 0)
> Es[, 2] <- ifelse(a.comm > 0, a.comm, 0)
> matplot(t, Es, type = "l", col = 1)

> matplot(env, Es, col = 1)
```

As we said, however, organisms will die. Let us create a variable for community-wide mortality, δ, as if a disturbance kills a constant fraction of the community.

```
> d <- 0.1
```

Covariance between competition and environment

We also want to assume that species compete for shared, limiting resources. Individuals have negative effects on each other. As a result, the more individuals of all species there are (increasing N_{total}), the more negative the total effect is. To account for this, we will stipulate a per capita negative effect α of any individual on any other. Therefore, in good times (high N_{total}), the effect of competition increases. In contrast, when times are bad, and N is small, competition is low. This is what Chesson and colleagues mean by *covariation between competition and the environment.*

Eq. (9.22) provides a reasonable way to represent the competitve effect $C_{i,t}$. Here we simplify further, and assume that the per capita effect of competition on growth is constant through time. However, to emphasize that point that one species has higher average fitness, we let the rare species experience greater per capita effects of competition. For our example, let us set $\alpha_{rare} = 0.0002, \alpha_{comm} = 0.0001$.

```
> alpha <- c(2 * 1e-05, 1e-05)
```

Thus, these α are the species-specific effects of all individuals on the rare and common species.

Simulating dynamics

Finally, we simulate these dynamics. We should create matrices to hold stuff as we simulate each year, for N, C, and R. Unlike E, these are simplest to collect as we simulate N, year by year.

```
> Ns <- matrix(NA, nrow = years + 1, ncol = 2)
> Cs <- matrix(NA, nrow = years, ncol = 2)
> Rs <- matrix(NA, nrow = years, ncol = 2)
```

Next we initialize our populations at t_0.

```
> Ns[1, ] <- c(1000, 1e+05)
```

Finally, we run the for-loop

```
> for (i in 1:years) Ns[i + 1, ] <- {
+     juveniles <- sum(exp(Es[i, ]) * Ns[i, ])
+     Cs[i, ] <- alpha * juveniles
+     Rs[i, ] <- exp(Es[i, ] - Cs[i, ])
+     (1 - d) * Ns[i, ] + Rs[i, ] * Ns[i, ]
+ }
```

and plot the populations.

```
> matplot(c(0, t), Ns, type = "b", log = "y")
```

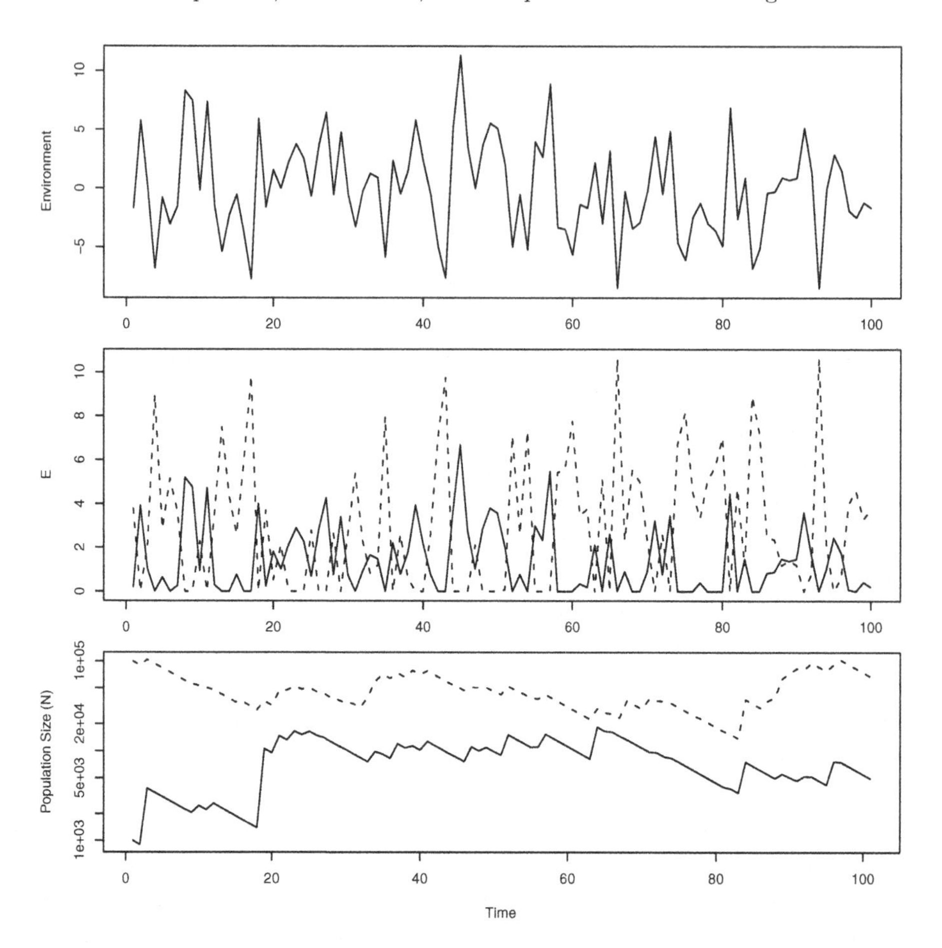

Fig. 9.9: A simulation of coexistence via the storage effect. E (middle panel) is maximum environment-mediated reproduction, in the absence of competition. See text and Fig. 9.8 for more information about species responses to the environment and average fitness.

Examining characteristics of the storage effect

Let us go back and examine a few of the characteristics that we should observe, if the storage effect is operating. First, note that above we showed differential responses to the environment, incomplete niche overlap, and buffered growth (Fig. 9.8).

Next, we will try to examine the environment-competition covariation. This is not trivial, and papers are written about how to estimate this. For now, recall that in Chapter 3, we began with an examination of negative density-dependence. Here we quantify the magnitude of this negative effect, as our "effect of competition." Let us invent a new value, ν, to measure how much the observed growth rate is affected by large population sizes,

$$\nu_{i,t} = \log\left(\frac{R_{max}}{R_{i,t}}\right) \tag{9.27}$$

where $R_{i,t}$ is the observed annual population growth rate, N_{t+1}/N_t, for species i, and R_{max} is the maximum of these.

To measure the covariation, we will find first $N_{total,t}$, $R_{i,t}$, and $R_{i,max}$.

```
> Nt1 <- rowSums(Ns)[1:years]
> R.obs <- Ns[-1, ]/Ns[-(years + 1), ]
> Rmax <- apply(R.obs, 2, max)
```

Now we calculate ν_i, and estimate the covariance.

```
> nu <- log(t(Rmax/t(R.obs)))
> colnames(nu) <- c("nu.rare", "nu.comm")
> var(Nt1, nu)

     nu.rare nu.comm
[1,]   534.1   745.2
```

This illustrates that both populations exhibit positive covariation between the quality of the environment (defined operationally as N_{total}) and the intensity of competition.

Last, recall that we stated above that the CV (coefficent of variation) should be greater for rare species than for common species. If we check that for our populations (eliminating the first half of the time series),

```
> apply(Ns[round(years/2):years, ], 2, function(x) sd(x)/mean(x) *
+     100)

[1] 44.62 55.07
```

we see that, indeed, the rare species has a higher CV. Examination of the time series (Fig. 9.9) confirms this.

To facilitate playing more games, the function `chesson` provides an easy wrapper for storage effect simulations (Fig. 9.10). Please try `?chesson` at the Rprompt.

Here we run the `chesson` model, and calculate the overlap, ρ, for each simulation.

```
> outA <- chesson(years = 500, specialization = 1, spread = 0.1)
> outB <- chesson(years = 500, specialization = 5, spread = 0.67)
> outA$overlap

[1] 0.9172

> outB$overlap

[1] 0.1234
```

By specifying greater specialization and greater spread between the environmental optima of the species pair in the second model, we have reduced niche overlap (Fig. 9.10a). Overlap in this model is the area under both species fitness-independent response curves. Note that large differences in overall fitness can alter effective overlap described by the density-independent reproductive rate, $E_{i,t}$ (Fig. 9.10b).

```
> matplot(outB[["env"]], outB[["Es"]], pch = 1:2, xlim = c(-0.6,
+   0.6), ylab = "Density-independent Reproduction", xlab = "Environment")
> matplot(outA[["env"]], outA[["Es"]], pch = c(19, 17), add = TRUE)
```

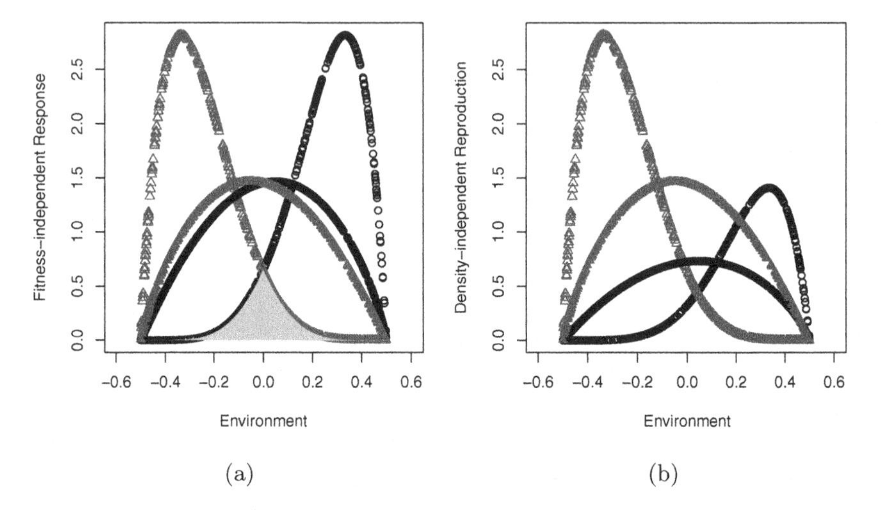

Fig. 9.10: Species responses to the environment, using the **chesson** model. Relative to the pair of species represented by solid circles, the pair of species with open symbols shows greater difference between optimal environments (greater **spread**), and narrower niches (greater **specialization**). (a) The underlying Beta probability density distributions; the grey *area* under the curves of the more differentiated species is ρ, the degree of niche overlap. (b) Density-independent reproduction (the parameter $E_{i,t}$ from eq. 9.22). (grey/red triangles - common species, black circles - rare species; open symbols - highly differentiated species, solid symbols - similar species). See text and help page (?**chesson**) for more details.

9.4 Summary

This chapter focused far more than previous chapters on the *biological importance of temporal dynamics.*

- In the framework of this chapter, the dynamics of succession result from the processes of mortality or disturbance, dispersal, and competitive exclusion. This framework can be applied over a broad range of spatial and temporal scales.
- Coexistence is possible via tradeoffs between competition, dispersal, and growth rate; the level of disturbance can influence the relative abundances of co-occuring species.
- The consequences of habitat destruction depend critically on the mechanisms underlying coexistence.

- The storage effect is an example of temporal niche differentiation. It depends on differential responses to the environment, buffered population growth, and covariation between competition intensity and population size.

Problems

Competition, colonization, and the successional niche

9.1. Basic interpretation
(a) Explain each of the paramters c, α, γ and m. Explain what each does in the model.

9.2. Two models in one?
(a) Given the model of Pacala and Rees, explain which parameters you would manipulate and to what values you would set them to make it a pure competition–colonization model.
(b) Given the model of Pacala and Rees, explain which parameters you would manipulate and to what values you would set them to make it a pure successional niche model.

9.3. For each "pure" model, explain how transient dynamics at the local scale result in a steady state at the large scale.

9.4. How would you evaluate the relative importance of these two mechanisms in maintaining biodiversity through successional trajectories and at equilibrium?

Storage effect

9.5. Develop a two-species example of the storage effect, in which you manipulate both (i) fitness differences, and (ii) environmental variation. Show how these interact to determine relative abundances of the two species.

10

Community Composition and Diversity

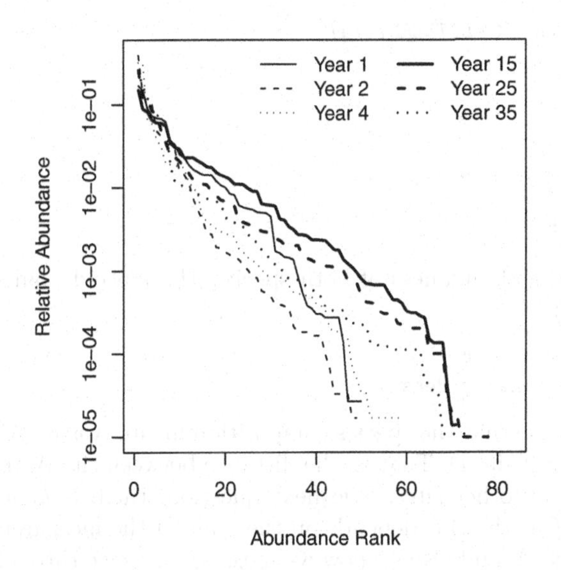

Fig. 10.1: Empirical rank–abundance distributions of successional plant communities (old-fields) within the temperate deciduous forest biome of North America. "Year" indicates the time since abandonment from agriculture. Data from the Buell-Small succession study (http://www.ecostudies.org/bss/)

It seems easy, or at least tractable, to compare the abundance of a single species in two samples. In this chapter, we introduce concepts that ecologists use to compare entire communities in two samples. We focus on two quantities: *species composition*, and *diversity*. We also discuss several issues related to this, including species–abundance distributions, ecological neutral theory, diversity partitioning, and species–area relations. Several packages in R include functions for dealing specifically with these topics. Please see the "Environmetrics" link within the "Task Views" link at any CRAN website for downloading Rpackages. Perhaps the most comprehensive (including both diversity and composition) is the **vegan** package, but many others include important features as well.

10.1 Species Composition

Species composition is merely the set of species in a site or a sample. Typically this includes some measure of abundance at each site, but it may also simply be a list of species at each site, where "abundance" is either presence or absence. Imagine we have four sites (A–D) from which we collect density data on two species, *Salix whompii* and *Fraxinus virga.* We can enter hypothetical data of the following abundances.

```
> dens <- data.frame(Salwho = c(1, 1, 2, 3), Fravir = c(21,
+     8, 13, 5))
> row.names(dens) <- LETTERS[1:4]
> dens
```

```
  Salwho Fravir
A      1     21
B      1      8
C      2     13
D      3      5
```

Next, we plot the abundances of both species; the plotted points then are the sites (Fig. 10.2).

```
> plot(dens, type = "n")
> text(dens, row.names(dens))
```

In Fig. 10.2, we see that the species composition in site A is most different from the composition of site D. That is, the distance between site A and D is greater than between any other sites. The next question, then, is *how* far apart are any two sites? Clearly, this depends on the scale of the measurement (e.g., the values on the axes), and also on how we measure distance through multivariate space.

10.1.1 Measures of abundance

Above we pretended that the abundances were absolute densities (i.e., 1 = one stem per sample). We could of course represent all the abundances differently. For instance, we could calculate *relative density*, where each species in a sample is represented by the *proportion* of the sample comprised of that species. For site A, we divide each species by the sum of all species.

```
> dens[1, ]/sum(dens[1, ])
```

```
   Salwho Fravir
A 0.04545 0.9545
```

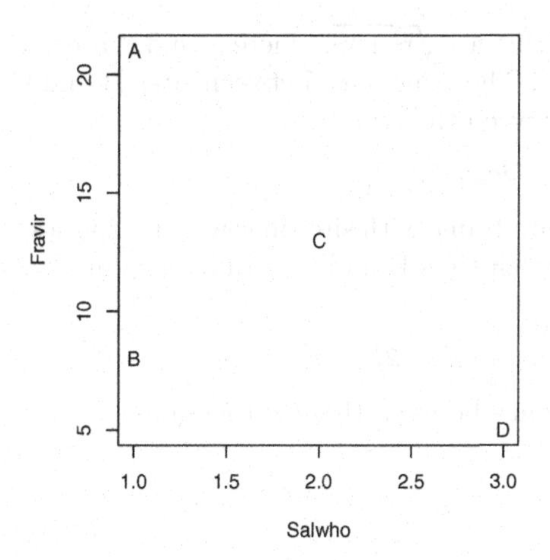

Fig. 10.2: Hypothetical species composition for four sites (A–D).

We see that *Salix* makes up about 5% of the sample for Site A, and *Fraxinus* makes up about 95% of the sample. Once we calculate relative densities for each species at each site, this eliminates differences in total density at each site because all sites then total to 1.

We could also calculate relative measures for any type of data, such as biomass or percent cover.

In most instances, *relative density* refers to the density of a species *relative* to the other species in a sample (above), but it can also be density in a sample relative to other samples. We would thus make each *species* total equal 1, and then its abundance at each site reflects the proportion of a species total abundance comprised by that site. For instance, we can make all *Salix* densities relative to each other.

```
> dens[, 1]/sum(dens[, 1])
```

```
[1] 0.1429 0.1429 0.2857 0.4286
```

Here we see that sites A and B both have about 14% of all *Salix* stems, and site D has 43%.

Whether our measures of abundance are absolute or relative, we would like to know how different samples (or sites) are from each other. Perhaps the simplest way to describe the difference among the sites is to calculate the *distances* between each pair of sites.

10.1.2 Distance

There are many ways to calculate a *distance* between a pair of sites. One of the simplest, which we learned in primary school, is *Euclidean distance.* With two species, we have two dimensional space, which is Fig. 10.2. The Euclidean distance between two sites is merely the length of the vector connecting those sites. We calculate this as $\sqrt{x^2+y^2}$, where x and y are the (x,y) distances between a pair of sites. The x distance between sites B and C is the difference in *Salix* abundance between the two sites,

```
> x <- dens[2, 1] - dens[3, 1]
```

where **dens** is the data frame with sites in rows, and species in different columns. The y distance between sites B and C is difference in *Fraxinus* abundance between the two sites.

```
> y <- dens[2, 2] - dens[3, 2]
```

The Euclidean distance between these is therefore

```
> sqrt(x^2 + y^2)
```

```
[1] 5.099
```

Distance is as simple as that. We calculate all pairwise Euclidean distances between sites A–D based on 2 species using built-in functions in R.

```
> (alldists <- dist(dens))
```

```
       A      B      C
B 13.000
C  8.062  5.099
D 16.125  3.606  8.062
```

We can generalize this to include any number of species, but it becomes increasingly harder to visualize. We can add a third species, *Mandragora officinarum*, and recalculate pairwise distances between all sites, but now with three species.

```
> dens[["Manoff"]] <- c(11, 3, 7, 5)
> (spp3 <- dist(dens))
```

```
       A      B      C
B 15.264
C  9.000  6.481
D 17.205  4.123  8.307
```

We can plot species abundances as we did above, and **pairs(dens)** would give us all the pairwise plots given three species. However, what we really want for species is a 3-D plot. Here we load another package[1] and create a 3-D scatterplot.

```
> pairs(dens)# not shown
> library(scatterplot3d)
> sc1 <- scatterplot3d(dens,  type='h', pch="",
+        xlim=c(0,5), ylim=c(0, 25), zlim=c(0,15))
> text(sc1$xyz.convert(dens), labels=rownames(dens))
```

In three dimensions, Euclidean distances are calculated the same basic way, but we add a third species, and the calculation becomes $\sqrt{x^2+y^2+z^2}$. Note that we take the square root (as opposed to the cube root) because we originally *squared* each distance. We can generalize this for two sites for R species as

[1] You can install this package from any R CRAN mirror.

$$D_E = \sqrt{\sum_{i=1}^{R} (x_{ai} - x_{bi})^2} \tag{10.1}$$

Of course, it is difficult (impossible?) to visualize arrangements of sites with more than three axes (i.e., > 3 species), but we can always *calculate* the distances between pairs of sites, regardless of how many species we have.

There are many ways, in addition to Euclidean distances, to calculate distance. Among the most commonly used in ecology is *Bray–Curtis* distance, which goes by other names, including *Søorenson* distance.

$$D_{BC} = \sum_{i=1}^{R} \frac{|x_{ai} - x_{bi}|}{x_{ai} + x_{bi}} \tag{10.2}$$

where R is the number of species in all samples. Bray–Curtis distance is merely the total difference in species abundances between two sites, divided by the total abundances at each site. Bray–Curtis distance (and a couple others) tends to result in more intuitively pleasing distances in which both common and rare species have relatively similar weights, whereas Euclidean distance depends more strongly on the most abundant species. This happens because Euclidean distances are based on squared differences, whereas Bray–Curtis uses absolute differences. Squaring always amplifies the importance of larger values. Fig. 10.3 compares graphs based on Euclidean and Bray–Curtis distances of the same raw data.

Displaying multidimensional distances

A simple way to display distances for three or more species is to create a plot in two dimensions that attempts to arrange all sites so that they are *approximately* the correct distances apart. In general this is impossible to achieve precisely, but distances can be approximately correct. One technique that tries to create an optimal (albiet approximate) arrangement is *non-metric multidimensional scaling*. Here we add a fourth species (*Aconitum lycoctonum*) to our data set before plotting the distances.

```
> dens$Acolyc <- c(16, 0, 9, 4)
```

The non-metric multidimensional scaling function is in the **vegan** package. It calculates distances for us using the original data. Here we display Euclidean distances among sites (Fig. 10.3a).

```
> library(vegan)
> mdsE <- metaMDS(dens, distance = "euc", autotransform = FALSE,
+       trace = 0)
> plot(mdsE, display = "sites", type = "text")
```

Here we display Bray–Curtis distances among sites (Fig. 10.3b).

```
> mdsB <- metaMDS(dens, distance = "bray", autotransform = FALSE,
+       trace = 0)
> plot(mdsB, display = "sites", type = "text")
```

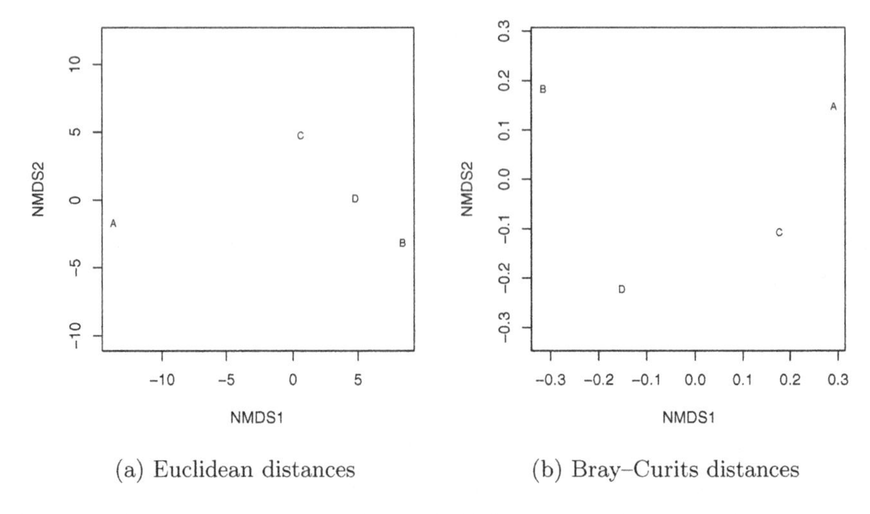

(a) Euclidean distances (b) Bray–Curits distances

Fig. 10.3: Nonmetric multidimensional (NMDS) plots showing approximate distances between sites. These two figures display the same raw data, but Euclidean distances tend to emphasize differences due to the more abundant species, whereas Bray-Curtis does not. Because NMDS provides iterative optimizations, it will find slightly different arrangements each time you run it.

10.1.3 Similarity

Sometimes, we would like to know how *similar* two communities are. Here we describe two measures of similarity, *percent similarity*, and *Sørensen similarity* [124].

Percent similarity may be the simplest of these; it is simply the sum of the minimum percentages for each species in the community. Here we convert each species to its relative abundance; that is, its proportional abundance at each site. To do this, we treat each site (row) separately, and then divide the abundance of each species by the sum of the abundances at each site.

```
> (dens.RA <- t(apply(dens, 1, function(sp.abun) sp.abun/sum(sp.abun))))

   Salwho Fravir Manoff Acolyc
A 0.02041 0.4286 0.2245 0.3265
B 0.08333 0.6667 0.2500 0.0000
C 0.06452 0.4194 0.2258 0.2903
D 0.17647 0.2941 0.2941 0.2353
```

Next, to compare two sites, we find the minimum relative abundance for each species. Comparing sites A and B, we have,

```
> (mins <- apply(dens.RA[1:2, ], 2, min))

 Salwho  Fravir  Manoff  Acolyc
0.02041 0.42857 0.22449 0.00000
```

Finally, we sum these, and multiply by 100 to get percentages.

```
> sum(mins) * 100

[1] 67.35
```

The second measure of similarity we investigate is Sørensen's similarity,

$$S_s = \frac{2C}{A + B} \tag{10.3}$$

where C is the number of species two sites have in common, and A and B are the number of species at each site. This is equivalent to dividing the shared species by the average richness.

To calculate this for sites A and B, we could find the species which have non-zero abundances both sites.

```
> (shared <- apply(dens[1:2, ], 2, function(abuns) all(abuns !=
+     0)))

Salwho Fravir Manoff Acolyc
  TRUE   TRUE   TRUE  FALSE
```

Next we find the richness of each.

```
> (Rs <- apply(dens[1:2, ], 1, function(x) sum(x >
+     0)))

A B
4 3
```

Finally, we divide the shared species by the summed richnesses and multiply by 2.

```
> 2 * sum(shared)/sum(Rs)

[1] 0.8571
```

Sørensen's index has also been used in the development of more sophisticated measures of similarity between sites [144, 164].

10.2 Diversity

To my mind, there is no more urgent or interesting goal in ecology and evolutionary biology than understanding the determinants of biodiversity. Biodiversity is many things to many people, but we typically think of it as a measure of the variety of biological life present, perhaps taking into account the relative abundances. For ecologists, we most often think of *species diversity* as some quantitative measure of the variety or number of different species. This has direct analogues to the genetic diversity within a population [45, 213], and the connections between species and genetic diversity include both shared patterns and shared mechanisms. Here we confine ourselves entirely to a discussion of species diversity, the variety of different species present. Consider this example.

Table 10.1: Four hypothetical stream invertebrate communities. Data are total numbers of individuals collected in ten samples (sums across samples). Diversity indices (Shannon-Wiener, Simpson's) explained below.

Species	Stream 1	Stream 2	Stream 3	Stream 4
Isoperla	20	50	20	0
Ceratopsyche	20	75	20	0
Ephemerella	20	75	20	0
Chironomus	20	0	140	200
Number of species (R)	4	3	4	1
Shannon-Wiener H	1.39	1.08	0.94	0
Simpson's S_D	0.75	0.66	0.48	0

We have four stream insect communities (Table 10.1). Which has the highest "biodiversity"?

We note that one stream has only one species — clearly that can't be the most "diverse" (still undefined). Two streams have four species — are they the most diverse? Stream 3 has more bugs in total (200 *vs.* 80), but stream 1 has a more equal distribution among the four species.

10.2.1 Measurements of variety

So, how shall we define "diversity" and measure this variety? There are many mathematical expressions that try to summarize biodiversity [76, 92]. The inquisitive reader is referred to [124] for a practical and comprehensive text on measures of species diversity. Without defining it precisely (my pay scale precludes such a noble task), let us say that diversity indices attempt to quantify

- the probability of encountering different species at random, or,
- the uncertainty or multiplicity of possible community states (i.e., the entropy of community composition), or,
- the variance in species composition, relative to a multivariate centroid.

For instance, a simple count of species (Table 10.1) shows that we have 4, 3, 4, and 1 species collected from streams 1–4. The larger the number of species, the less certain we could be about the identity of an individual drawn blindly and at random from the community.

To generalize this further, imagine that we have a species pool[2] of R species, and we have a sample comprised of only one species. In a sample with only one species, then we know that the possible *states* that sample can take is limited to one of only R different possible states — the abundance of one species is 100% and all others are zero. On the other hand, if we have two species then the community could take on $R(R-1)$ different states — the first species could be any one of R species, and the second species could be any one of the other species, and all others are zero. Thus increasing diversity means increasing

[2] A *species pool* is the entire, usually hypothetical, set of species from which a sample is drawn; it may be all of the species known to occur in a region.

the possible states that the community could take, and thus increasing our uncertainty about community structure [92]. This increasing lack of information about the system is a form of *entropy*, and increasing diversity (i.e., increasing multiplicity of possible states) is increasing entropy. The jargon and topics of statistical mechanics, such as entropy, appear (in 2009) to be an increasingly important part of community ecology [71, 171].

Below we list three commonly used diversity indices: species richness, Shannon-Wiener index, and Simpson's diversity index.

- Species richness, R, the count of the number of species in a sample or area; the most widely used measure of biodiversity [82].
- Shannon-Wiener diversity.[3]

$$H' = -\sum_{i=1}^{R} p_i \ln(p_i) \tag{10.4}$$

  where R is the number of species in the community, and p_i is the relative abundance of species i.
- Simpson's diversity. This index is (i) the probability that two individuals drawn from a community at random will be different species [147], (ii) the initial slope of the species-individuals curve [98] (e.g., Fig. 10.6), and (iii) the expected variance of species composition (Fig. 10.5) [97, 194].

$$S_D = 1 - \sum_{i=1}^{R} p_i^2 \tag{10.5}$$

  The summation $\sum_{i=1}^{R} p_i^2$ is the probability that two individuals drawn at random are the same species, and it is known as Simpson's "dominance." Lande found that this Simpon's index can be more precisely estimated, or estimated accurately with smaller sample sizes, than either richness or Shannon-Wiener [97].

These three indices are actually directly related to each other — they comprise estimates of *entropy*, the amount of disorder or the multiplicity of possible states of a system, that are directly related via a single constant [92]. However, an important consequence of their differences is that richness depends most heavily on rare species, Simpson's depends most heavily on common species, and Shannon-Wiener stands somewhere between the two (Fig. 10.4).

Relations between number of species, relative abundances, and diversity

This section relies heavily on code and merely generates Fig. 10.4.

Here we display diversities for communities with different numbers and relative abundances of species (Fig. 10.4). We first define functions for the diversity indices.

[3] Robert May stated that this index is connected by merely an "ectoplasmic thread" to information theory [135], but there seems be a bit more connection than that.

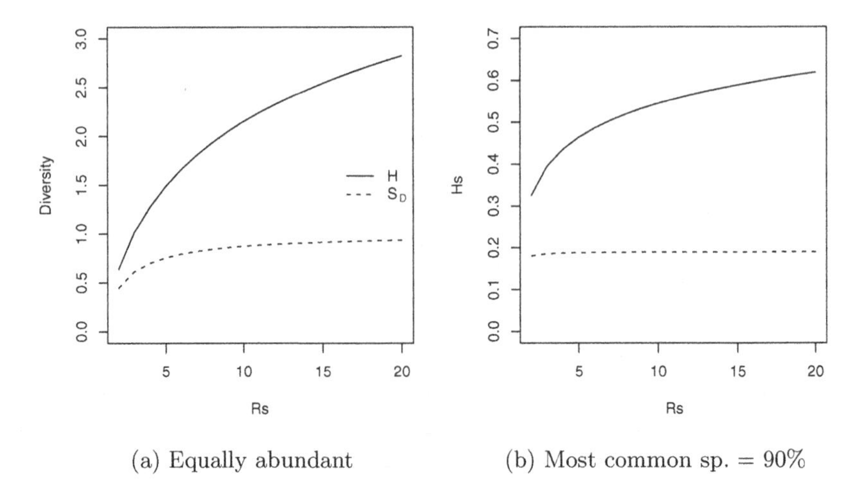

(a) Equally abundant (b) Most common sp. = 90%

Fig. 10.4: Relations between richness, Shannon-Weiner, and Simpson's diversities (note difference in y-axis scale between the figures). Communities of 2–20 species are composed of either equally abundant species (a) or with the most common species equal to 90% of the community (b).

```
> H <- function(x) {
+     x <- x[x > 0]
+     p <- x/sum(x)
+     -sum(p * log(p))
+ }
> Sd <- function(x) {
+     p <- x/sum(x)
+     1 - sum(p^2)
+ }
```

Next we create a *list* of communities with from 1 to 20 *equally* abundant species, and calculate H and S_D for each community.

```
> Rs <- 2:20
> ComsEq <- sapply(Rs, function(R) (1:R)/R)
> Hs <- sapply(ComsEq, H)
> Sds <- sapply(ComsEq, Sd)
> plot(Rs, Hs, type = "l", ylab = "Diversity", ylim = c(0,
+     3))
> lines(Rs, Sds, lty = 2)
> legend("right", c(expression(italic("H")), expression(italic("S"["D"]))),
+     lty = 1:2, bty = "n")
```

Now we create a *list* of communities with from 2 to 25 species, where one species always comprises 90% of the community, and the remainder are equally abundant rare species. We then calculate H and S_D for each community.

```
> Coms90 <- sapply(Rs, function(R) {
+     p <- numeric(R)
```

```
+     p[1] <- 0.9
+     p[2:R] <- 0.1/(R - 1)
+     p
+ })
> Hs <- sapply(Coms90, H)
> Sds <- sapply(Coms90, Sd)
> plot(Rs, Hs, type = "l", ylim = c(0, 0.7))
> lines(Rs, Sds, lty = 2)
```

Simpson's diversity, as a variance of composition

This section relies heavily on code.

Here we show how we would calculate the variance of species composition. First we create a pretend community of six individuals (rows) and 3 species (columns). Somewhat oddly, we identify the degree to which each individual is comprised of each species; in this case, individuals can be only one species.[4] Here we let two individuals be each species.

```
> s1 <- matrix(c(1, 1, 0, 0, 0, 0, 0, 0, 1, 1, 0,
+     0, 0, 0, 0, 0, 1, 1), nr = 6)
> colnames(s1) <- c("Sp.A", "Sp.B", "Sp.C")
> s1

     Sp.A Sp.B Sp.C
[1,]    1    0    0
[2,]    1    0    0
[3,]    0    1    0
[4,]    0    1    0
[5,]    0    0    1
[6,]    0    0    1
```

We can plot these individuals in community space, if we like (Fig. 10.5a).

```
> library(scatterplot3d)
> s13d <- scatterplot3d(jitter(s1, 0.3), type = "h",
+     angle = 60, pch = c(1, 1, 2, 2, 3, 3), xlim = c(-0.2,
+         1.4), ylim = c(-0.2, 1.4), zlim = c(-0.2,
+         1.4))
> s13d$points3d(x = 1/3, y = 1/3, z = 1/3, type = "h",
+     pch = 19, cex = 2)
```

Next we can calculate a *centroid*, or multivariate mean — it is merely the vector of species means.

```
> (centroid1 <- colMeans(s1))

  Sp.A   Sp.B   Sp.C 
0.3333 0.3333 0.3333 
```

[4] Imagine the case where the columns are traits, or genes. In that case, individuals could be characterized by affiliation with multiple columns, whether traits or genes.

Given this centroid, we begin to calculate a variance by (i) subtracting each species vector (0s, 1s) from its mean, (ii) squaring each of these deviates, and (3) summing to get the sum of squares.

```
> (SS <- sum(sapply(1:3, function(j) (s1[, j] -
+      centroid1[j]))^2))
```

```
[1] 4
```

We then divide this sum by the number of individuals that were in the community (N)

```
> SS/6
```

```
[1] 0.6667
```

We find that the calculation given above for Simpson's diversity returns exactly the same number. We would calculate the relative abundances, square them, add them, and subtract that value from 1.

```
> p <- c(2, 2, 2)/6
> 1 - sum(p^2)
```

```
[1] 0.6667
```

In addition to being the variance of species composition, this number is also the probability that two individuals drawn at random are different species. As we mentioned above, there are other motivations than these to derive this and other measures of species diversity, based on entropy and information theory [92] — and they are all typically correlated with each other.

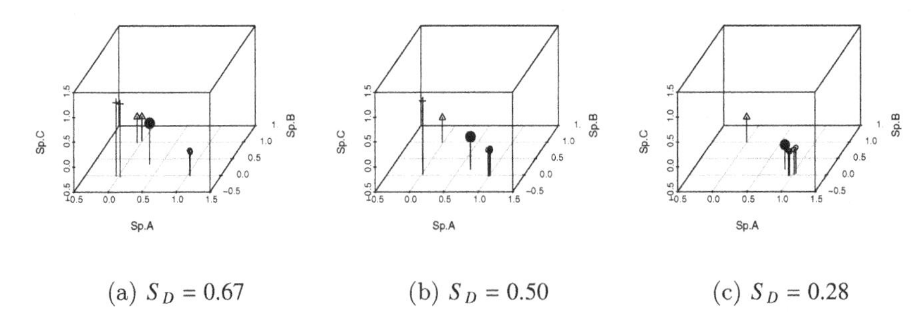

(a) $S_D = 0.67$ (b) $S_D = 0.50$ (c) $S_D = 0.28$

Fig. 10.5: Plotting three examples of species composition. The centroid of each composition is a solid black dot. The third example (on right) has zero abundances of species C. Simpson's diversity is the variance of these points around the centroid. Individual points are not plotted at precisely 0 or 1 — they are plotted with a bit of jitter or noise so that they do not overlap entirely.

10.2.2 Rarefaction and total species richness

Rarefaction is the process of generating the relationship between the number of species *vs.* the number of individuals in one or more samples. It is typically determined by randomly resampling individuals [59], but could also be determined by resampling samples. Rarefaction allows direct comparison of the richness of two samples, corrected for numbers of individuals. This is particularly important because *R* depends heavily on the number of individuals in a sample. Thus rarefaction finds the general relation between the number(s) of species *vs.* number of individuals (Fig. 10.6), and is limited to less than or equal to the number of species you actually observed. A related curve is the *species-accumulation* curves, but this is simply a useful but haphazard accumulation of new species (a cumulative sum) as the investigator samples new individuals.

Another issue that ecologists face is trying to estimate the *true* number of species in an area, given the samples collected. This number of species would be larger than the number of species you observed, and is often referred to as *total species richness* or the *asymptotic richness.* Samples almost always find only a subset of the species present in an area or region, but we might prefer to know how many species are really there, in the general area we sampled. There are many ways to do this, and while some are better than others, none is perfect. These methods estimate minimum numbers of species, and assume that the unsampled areas are homogeneous and similar to the sampled areas.

Before using these methods seriously, the inquisitive reader should consult [59, 124] and references at http://viceroy.eeb.uconn.edu/EstimateS. Below, we briefly explore an example in R.

An example of rarefaction and total species richness

Let us "sample" a seasonal tropical rainforest on Barro Colorado Island (BCI) http://ctfs.si.edu/datasets/bci/). Our goal will be to provide baseline data for later comparison to other such studies.

We will use some of the data from a 50 ha plot that is included in the `vegan` package [36, 151]. We will pretend that we sampled every tree over 10 cm dbh,[5] in each of 10 plots scattered throughout the 50 ha area. What could we say about the forest within which the plots were located? We have to consider the scale of the sampling. Both the *experimental unit* and the *grain* are the 1 ha plots. Imagine that the plots were scattered throughout the 50 ha plot, so that the extent of the sampling was a full 50 ha.[6] First, let's pretend we have sampled 10 1 ha plots by extracting the samples out of the larger dataset.

```
> library(vegan)
> data(BCI)
> bci <- BCI[seq(5, 50, by = 5), ]
```

[5] "dbh" is diameter at 1.37 m above the ground.

[6] See John Wiens' foundational paper on spatial scale in ecology [220] describing the meaning of grain, extent, and other spatial issues.

Next, for each species, I sum all the samples into one, upon which I will base rarefaction and total richness estimation.

Next we combine all the plots into one sample (a single summed count for each species present), select numbers of individuals for which I want rarefied samples (multiples of 500), and then perform rarefaction for each of those numbers.

```
> N <- colSums(bci)
> subs3 <- c(seq(500, 4500, by = 500), sum(N))
> rar3 <- rarefy(N, sample = subs3, se = T, MARG = 2)
```

Next we want to graph it, with a few bells and whistles. We set up the graph of the 10 plot, individual-based rarefaction, and leave room to graph total richness estimators as well (Fig. 10.6).

```
> plot(subs3, rar3[1, ], ylab = "Species Richness",
+      axes = FALSE, xlab = "No. of Individuals",
+      type = "n", ylim = c(0, 250), xlim = c(500,
+          7000))
> axis(1, at = 1:5 * 1000)
> axis(2)
> box()
> text(2500, 200, "Individual-based rarefaction (10 plots)")
```

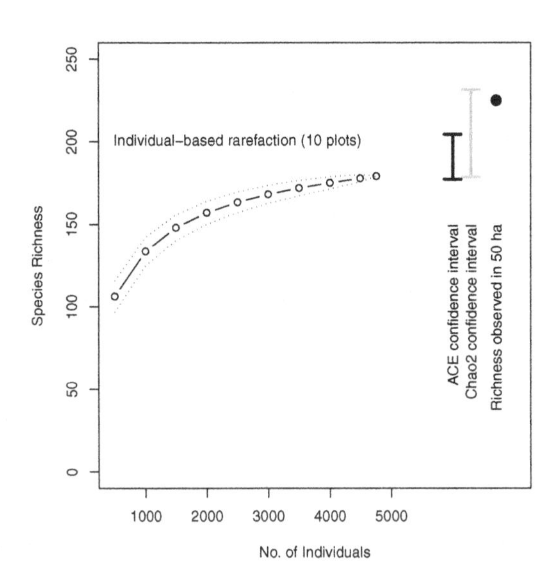

Fig. 10.6: Baseline tree species richness estimation based on ten 1 ha plots, using individual-based rarefaction. We also include two different total richness estimators, ACE and Chao 2, and the observed total tree richness in the 50 ha plot for comparison.

Here we plot the expected values and also ± 2 SE.

```
> lines(subs3, rar3[1, ], type = "b")
> lines(subs3, rar3[1, ] + 2 * rar3[2, ], lty = 3)
> lines(subs3, rar3[1, ] - 2 * rar3[2, ], lty = 3)
```

Next we hope to estimate the minimum total number of species (asymptotic richness) we might observe in the area around (and in) our 10 plots, if we can assume that the surrounding forest is homogeneous (it would probably be best to extrapolate only to the 50 ha plot). First, we use an *abundance-based coverage estimator*, *ACE*, that appears to give reasonable estimates [124]. We plot intervals, the expected values ± 2 SE (Fig. 10.6).

```
> ace <- estimateR(N)
> segments(6000, ace["S.ACE"] - 2 * ace["se.ACE"],
+     6000, ace["S.ACE"] + 2 * ace["se.ACE"], lwd = 3)
> text(6000, 150, "ACE estimate", srt = 90, adj = c(1,
+     0.5))
```

Next we use a frequency-based estimator, Chao 2, where the data only need to be presence/absence, but for which we also need multiple sample plots.

```
> chaoF <- specpool(bci)
> segments(6300, chaoF[1, "chao"] - 2 * chaoF[1,
+     "chao.se"], 6300, chaoF[1, "chao"] + 2 * chaoF[1,
+     "chao.se"], lwd = 3, col = "grey")
> text(6300, 150, "Chao2 estimate", srt = 90, adj = c(1,
+     0.5))
```

Last we add the observed number of tree species (over 10 cm dbh) found in the entire 50 ha plot.

```
> points(6700, dim(BCI)[2], pch = 19, cex = 1.5)
> text(6700, 150, "Richness observed in 50 ha",
+     srt = 90, adj = c(1, 0.5))
```

This shows us that the total richness estimators did not overestimate the total number of species within the extent of this relatively homogenous sample area (Fig. 10.6).

If we wanted to, we could then use any of these three estimators to compare the richness in this area to the richness of another area.

10.3 Distributions

In addition to plotting species in multidimensional space (Fig. 10.3), or estimating a measure of diversity or richness, we can also examine the *distributions* of species abundances.

Like any other vector of numbers, we can make a histogram of species abundances. As an example, here we make a histogram of tree densities, where each species has its own density (Fig. 10.7a). This shows us what is patently true for nearly all ecological communities — *most species are rare.*

10.3.1 Log-normal distribution

Given general empirical patterns, that most species are rare, Frank Preston [168,170] proposed that we describe communities using the logarithms of species abundances (Fig. 10.7).[7] This often reveals that a community can be described approximately with the normal distribution applied to the log-transformed data, or the *log-normal ditribution*. We can also display this as a *rank–abundance distribution* (Fig. 10.7c). To do this, we assign the most abundant species as rank = 1, and the least abundant has rank = R, in a sample of R species, and plot log-abundance *vs.* rank.

May [129] described several ways in which common processes may drive log-normal distributions, and cause them to be common in data sets. Most commonly cited is to note that log-normal distributions arise when each observation (i.e., each random variable) results from the product of independent factors. That is, if each species' density is determined by largely independent factors which act multiplicatively on each species, the resulting densities would be log-normally distributed.

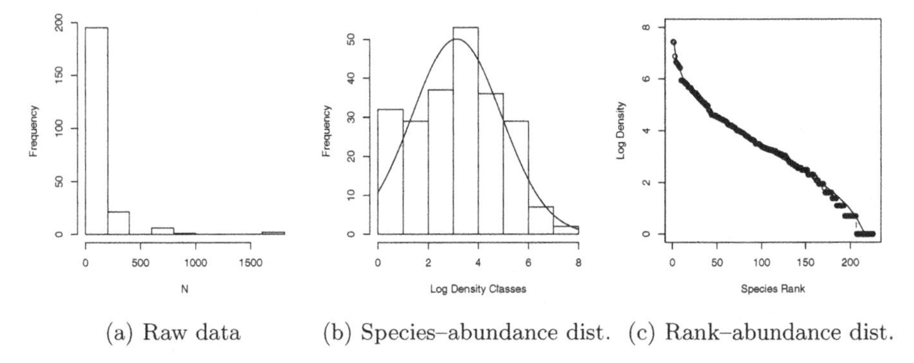

(a) Raw data (b) Species–abundance dist. (c) Rank–abundance dist.

Fig. 10.7: Three related types of distributions of tree species densities from Barro Colorado Island [36]. (a) Histogram of raw data, (b) histogram of log-transformed data; typically referred to as the "species–abundance distribution," accompanied here with the normal probability density function, (c) the "rank–abundance distribution," as typically presented with the log-transformed data, with the complement of the cumulative probability density function (1-pdf) [129]. Normal distributions were applied using the mean and standard deviation from the log-transformed data, times the total number of species.

[7] Preston used base 2 logs to make his histogram bins, and his practice remains standard; we use the natural log.

Log-normal abundance distributions (Fig. 10.7)

We can plot tree species densities from Barro Colorado Island [36], which is aailable online, or in the **vegan** package. First we make the data available to us, then make a simple histogram.

```
> data(BCI)
> N <- sort(colSums(BCI), decr = TRUE)
> hist(N, main = NULL)
```

Next we make a *species–abundance distribution*, which is merely a histogram of the log-abundances (classically, base 2 logs, but we use base e). In addition, we add the normal probability density function, getting the mean and standard deviation from the data, and plotting the expected number of species by multiplying the densities by the total number of species.

```
> hist(log(N), xlab = "Log Density Classes", main = NULL)
> m.spp <- mean(log(N))
> sd.spp <- sd(log(N))
> R <- length(N)
> curve(dnorm(x, m.spp, sd.spp) * R, 0, 8, add = T)
```

Next we create the *rank–abundance distribution*, which is just a plot of log-abundances *vs.* ranks.

```
> plot(log(N), type = "b", ylim = c(0, 8), main = NULL,
+      xlab = "Species Rank", ylab = "Log Density")
> ranks.lognormal <- R * (1 - pnorm(log(N), m.spp,
+      sd.spp))
> lines(ranks.lognormal, log(N))
```

We can think of the rank of species i as the total number of species that are more abundant than species i. This is essentially the opposite (or complement) of the integral of the species–abundance distribution [129]. That means that if we can describe the species abundance distribution with the normal density function, then 1-cumulative probability function is the rank.

10.3.2 Other distributions

Well over a dozen other types of abundance distributions exist to describe abundance patterns, other than the log-normal [124]. They can all be represented as *rank–abundance distributions*.

The *geometric distribution*[8] (or pre-emption niche distribution) reflects a simple idea, where each species pre-empts a constant fraction of the remaining niche space [129, 145]. For instance, if the first species uses 20% of the niche space, the second species uses 20% of the remaining 80%, etc. The frequency of the ith most abundant species is

[8] This probability mass function, $P_i = d(1-d)^{i-1}$, is the probability distribution of the number of attempts, i, needed for one success, if the independent probability of success on one trial is d.

$$N_i = \frac{N_T}{C} d\,(1-d)^{i-1} \tag{10.6}$$

where d is the abundance of the most common species, and C is just a constant to make $\sum N_i = N_T$, where $C = 1 - (1-d)^{S_T}$. Thus this describes the geometric rank–abundance distribution.

The *log-series distribution* [55] describes the expected frequency of species with n individuals,

$$F\,(S_n) = \frac{\alpha x^n}{n} \tag{10.7}$$

where α is a constant that represents diversity (greater α means greater diversity); the α for a diverse rainforest might be 30–100. The constant x is a fitted, and it is always true that $0.9 < x < 1.0$ and x increases toward 0.99 as $N/S \rightarrow 20$ [124]. x can be estimated from $S/N = [(1-x)/x] \cdot [-\ln(1-x)]$. Note that this is not described as a rank–abundance distribution, but species abundances can nonetheless be plotted in that manner [129].

The log-series rank–abundance distribution is a bit of a pain, relying on the standard exponential integral [129], $E_1(s) = \int_s^{\infty} exp(-t)/t\,dt$. Given a range of N, we calculate ranks as

$$F\,(N) = \alpha \int_s^{\infty} exp(-t)/t\,dt \tag{10.8}$$

where we can let $t = 1$ and $s = N \log\,(1 + \alpha/N_T)$.

The log-series distribution has the interesting property that the total number of species in a sample of N individuals would be $S_T = \alpha \log(1 + N/\alpha)$. The parameter α is sometimes used as a measure of diversity. If your data are log-series distributed, then α is approximately the number of species for which you expect 1 individual, because $x \approx 1$. Two very general theories predict a log-series distribution, including neutral theory, and maximum entropy. Oddly, these two theories both predict a log-series distribution, but make opposite assumptions about niches and individuals (see next section).

MacArthur's *broken stick distribution* is a classic distribution that results in a very even distribution of species abundances [118]. The number of individuals of each species i is

$$N_i = \frac{N_T}{S_T} \sum_{n=i}^{S_T} \frac{1}{n} \tag{10.9}$$

where N_T and S_T are the total number of individuals and species in the sample, respectively. MacArthur described this as resulting from the simultaneous breakage of a stick at random points along the stick. The resulting size fragments are the N_i above. MacArthur's broken stick model is thus both a *stochastic* and a *deterministic* model. It has a simulation (stick breakage) that is the direct analogue of the deterministic analytical expression.

Other similarly tactile stick-breaking distributions create a host of different rank–abundance patterns [124, 206]. In particular, the stick can be broken *sequentially*, first at one random point, then at a random point along one of two newly broken fragments, then at an additional point along any one of the *three* broken fragments, *etc.*, with $S_T - 1$ breaks creating S_T species. The critical difference between the models then becomes *how each subsequent fragment is selected.* If the probability of selecting each fragment is related directly to its size, then this becomes identical to MacArthur's broken stick model. On the other hand, if each subsequent piece is selected randomly, regardless of its size, then this results in something very similar to the log-normal distribution [196, 205]. Other variations on fragment selection generate other patterns [206].

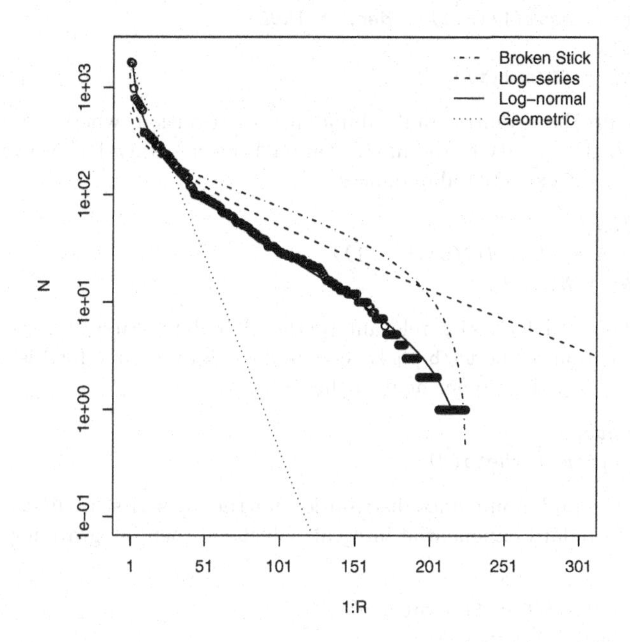

Fig. 10.8: A few common rank–abundance distributions, along with the BCI data [36]. The log-normal curve is fit to the data, and the broken stick distribution is always determined by the number of species. Here we let the geometric distribution be determined by the abundance of the most common species. The log-series was plotted so that it matched qualitatively the most abundant species.

Generating other rank–abundance distributions

We can illustrate the above rank–abundance distributions as they might relate to the BCI tree data (see previous code). We start with MacArthur's broken stick model. We use cumulative summation backwards to add all $1/n_i$, and then re-sort it by rank (cf. eq. 10.9).

```
> N <- sort(colSums(BCI), decr = TRUE)
> f1 <- sort(cumsum(1/(R:1)), decr = TRUE)
> Nt <- sum(N)
> NMac <- Nt * f1/sum(f1)
```

Next, we create the geomtric rank–abundance distribution, where we let the BCI data tell us d, the density of the most abundant species; therefore we can multiply these by N_T to get expected abundances.

```
> d <- N[1]/Nt
> Ngeo.f <- d * (1 - d)^(0:(R - 1))
> Ngeo <- Nt * Ngeo.f
```

Last, we generate a log-series relevant to the BCI data. First, we use the **optimal.theta** function in the **untb** package to find a relevant value for Fisher's α. (See more and θ and α below under neutral theory).

```
> library(untb)
> alpha <- optimal.theta(N)
```

To calculate the rank abundance distribution for the log-series, we first need a function for the "standard exponential integral" which we then integrate for each population size.

```
> sei <- function(t = 1) exp(-t)/t
> alpha <- optimal.theta(N)
> ranks.logseries <- sapply(N, function(x) {
+     n <- x * log(1 + alpha/Nt)
+     f <- integrate(sei, n, Inf)
+     fv <- f[["value"]]
+     alpha * fv
+ })
```

Plotting other rank–abundance distributions (Fig. 10.8)

Now we can plot the BCI data, and all the distributions, which we generated above. Note that for the log-normal and the log-series, we calculated ranks, based on the species–abundance distributions, whereas in the standard form of the geometric and broken stick distributions, the expected abundances are calculated, in part, from the ranks.

```
> plot(1:R, N, ylim = c(0.1, 2000), xlim = c(1,
+     301), axes = FALSE, log = "y")
> axis(2)
> axis(1, 1 + seq(0, 300, by = 50))
> box()
> lines(1:R, NMac, lty = 4)
> lines(1:R, Ngeo, lty = 3)
> lines(ranks.logseries, N, lty = 2)
> lines(ranks.lognormal, N, lty = 1)
> legend("topright", c("Broken Stick", "Log-series",
+     "Log-normal", "Geometric"), lty = c(4, 2,
+     1, 3), bty = "n")
```

Note that we have not *fit* the the log-series or geometric distributions to the data, but rather, placed them in for comparison. Properly fitting curves to distributions is a picky business [124, 151], especially when it comes to species abundance distributions.

10.3.3 Pattern *vs.* process

Note the resemblance between stick-breaking and niche evolution — if we envision the whole stick as all of the available niche space, or energy, or limiting resources, then each fragment represents a portion of the total occupied by each species. Thus, various patterns of breakage act as models for niche partitioning and relative abundance patterns. Other biological and stochastic processes create specific distributions. For instance, completely random births, deaths, migration, and speciation will create the log-series distribution and the log-normal-like distributions (see *neutral theory* below). We noted above that independent, multiplicatively interacting factors can create the log-normal distribution.

On the other hand, all of these abundance distributions should probably be used primarily to describe *patterns* of commonness, rarity, and not to infer anything about the processes creating the patterns. These graphical descriptions are merely attractive and transparent ways to illustrate the abundances of the species in your sample/site/experiment.

The crux of the issue is that different *processes* can cause the same abundance distribution, and so, sadly, *we cannot usually infer the underlying processes from the patterns we observe.* That is, correlation is different than causation. Abundances of real species, in nature and in the lab, are the result of *mechanistic processes*, including those described in models of abundance distributions. However, we cannot say that a particular pattern was the result of a particular process, based on the pattern alone. Nonetheless, they are good summaries of ecological communities, and they show us, in a glance, a lot about *diversity*.

Nonetheless, interesting questions remain about the relations between patterns and processes. Many ecologists would argue that describing patterns and predicting them are the most important goals of ecology [156], while others argue that process is all important. However, both of these camps would agree about the fallacy of drawing conclusions about processes based on pattern — nowhere in ecology has this fallacy been more prevalent than with abundance distributions [66].

10.4 Neutral Theory of Biodiversity and Biogeography

One model of abundance distributions is particularly important, and we elaborate on it here. It is referred to variously as the *unified neutral theory of biodiversity and biogeography* [81], or often "neutral theory" for short. Neutral theory is important because it does much more than most other models of abundance distributions. It is a testable theory that makes quantitative predictions across several levels of organization, for both evolutionary and ecological processes.

Just as evolutionary biology has neutral theories of gene and allele frequencies, ecology has neutral theories of population and community structure and dynamics [9, 22, 81]. Neutral communities of species are computationally and mathematically related and often identical to models of genes and alleles [53] (Table 10.2). Thus, lessons you have learned about genetic drift often apply to neutral models of communities. Indeed, neutral ecological dynamics at the local scale are often referred to as *ecological drift*, and populations change via demographic stochasticity.[9]

Stephen Hubbell proposed his "unified neutral theory of biodiversity and biogeography" (hereafter NTBB, [81]) as a null model of the dynamics of individuals, to help explain relative abundances of tropical trees. Hubbell describes it as a direct descendant of MacArthur and Wilson's theory of island biogeography [121, 122] (see below, species–area relations). Hubbell proposed it both as a null hypothesis and also — and this is the controversial part — as a model of community dynamics that closely approximates reality.

The relevant "world" of the NTBB is a metacommunity (Fig. 10.9), that is, a collection of similar local communities connected by dispersal [102].[10] The metacommunity is populated entirely by individuals that are functionally identical. The NTBB is a theory of the *dynamics of individuals*, modeling individual

[9] Some have called neutral theory a *null model*, but others disagree [61], describing distinctions between dynamical, process-based neutral models with fitted parameters, *vs.* static data-based null models [60]. Both can be used as benchmarks against which to measure other phenomena. Under those circumstances, I suppose they might both be null hypotheses.

[10] The metacommunity concept is quite general, and a neutral metacommunity is but one caricature [102].

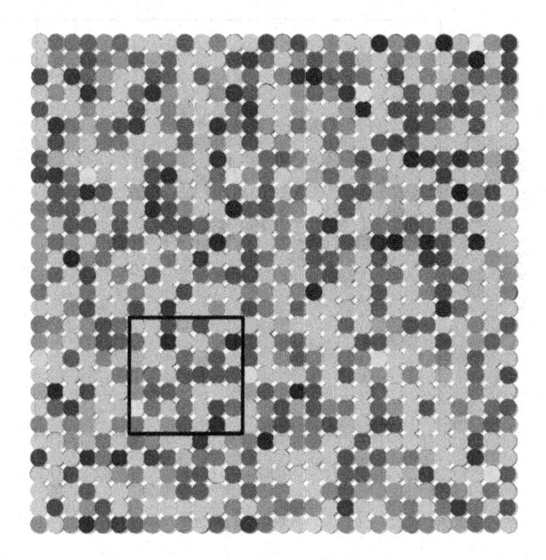

Fig. 10.9: A cartoon of a local community of forest canopy trees (small box) nested inside part of the metacommunity of a tropical forest. The true metaccommunity would extend far beyond the boundaries of this figure to include the true potential source of propagules. Shades of grey indicate different species. The local community is a sample of the larger community (such as the 50 ha forest dynamics plot on BCI) and receives migrants from the metacommunity. Mutation gives rise to new species in the metacommunity. For a particular local community, such as a 50 ha plot on an island in the Panama canal, the metacommunity will include not only the surrounding forest on the island, but also Panama, and perhaps much of the neotropics [86].

births, deaths, migration and mutation. It assumes that within a guild, such as late successional tropical trees, species are essentially neutral with respect to their fitness, that is, they exhibit *fitness equivalence.* This means that the probabilities of birth, death, mutation and migration are *identical for all individuals.* Therefore, changes in population sizes occur via *random walks*, that is, via stochastic increases and decreases with time step (Fig. 10.10). Random walks do not imply an absence of competition or other indirect enemy mediated negative density dependence. Rather, competition is thought to be diffuse, and equal among individuals. We discuss details of this in the context of simulation. Negative density dependence arises either through a specific constraint on the total number of individuals in a community [81], or as traits of individuals related to the probabilities of births, deaths, and speciation [214].

A basic paradox of the NTBB, is that in the absence of migration or mutation, diversity gradually declines to zero, or monodominance. A random walk due to fitness equivalence will eventually result in the loss of all species except one.[11] However, the loss of diversity in any single local community is predicted to be very, very slow, and is countered by immigration and speciation (we dis-

[11] This is identical to allele frequency in genetic drift.

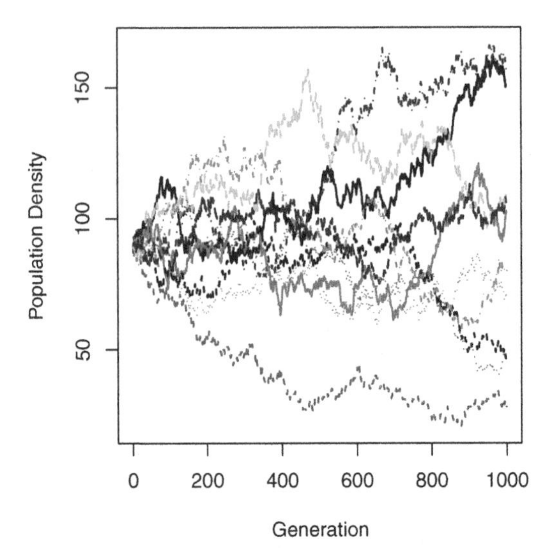

Fig. 10.10: Neutral ecological drift. Here we start with 10 species, each with 90 individuals, and let their abundances undergo random walks within a finite local community, with no immigration. Here, one generation is equal to nine deaths and nine births. Note the slow decline in unevennes — after 1000 deaths, no species has become extinct.

cuss more details below). Thus, species do not increase deterministically when rare — this makes the concept of coexistence different than the stable coexistance criteria discussed in previous chapters. Coexistance here *is not stable* but rather only a stochastic event with a limited time horizon which is balanced by the emergence of new species.

If all communities are thought to undergo random walks toward monodominance, how is diversity maintained in any particular region? Two factors maintain species in any local community. First, immigration into the local community from the metacommunity can bring in new species. Even though each local community is undergoing a random walk toward monodominance, each local community may become dominated by any one of the species in the pool because all species have equal fitness. Thus separate local communities are predicted to become dominated by different species, and these differences among local communities help maintain diversity in the metacommunity landscape.[12] Second, at longer time scales and larger spatial scales, speciation (i.e., mutation and lineage-splitting) within the entire metacommunity maintains biodiversity. Mutation and the consequent speciation provide the ultimate source of variation. Random walks toward extinction in large communities are *so* lengthy that the extinctions are balanced by speciation.

[12] This is the same as how genetic drift operates across subpopulations connected by low rates of gene exchange.

Introducing new symbols and jargon for ecological neutral theory, we state that the diversity of the metacommunity, θ, is a function of the number of individuals in the metacommunity, J_M, and the per capita rate at which new species arise (via mutation) ν ($\theta = 2J_M\nu$; Table 10.2). A local community undergoes ecological drift; drift causes the slow loss of diversity, which is balanced by a per capita (J_L) immigration rate m.

Table 10.2: Comparison of properties and jargon used in ecological and population genetic neutral theory (after Alonso et al. [1]). Here x is a continuous variable for relative abundance of a species or allele ($0 \le x \ge 1$, $x = n/J$). Note this is potentially confused with Fisher's log-series $\langle \phi_n \rangle = \theta x^n/n$, which is a discrete species abundance distribution in terms of numbers of individuals, n (not relative abundance), and where $x = b/d$.

Property	**Ecology**	**Population Genetics**
Entire System (size)	Metacommunity (J_M)	Population (N)
Nested subsystem (size)	Local community (J_L)	Subpop. or Deme (N)
Smallest neutral system unit	Individual organism	Individual gene
Diversity unit	Species	Allele
Stochastic process	Ecological drift	Genetic drift
Generator of diversity (rate symbol)	Speciation (ν)	Mutation (μ)
Fundamental diversity number	$\theta \approx 2J_M\nu$	$\theta \approx 4N\mu$
Fundamental dispersal number	$I \approx 2J_L m$	$\theta \approx 4Nm$
Relative abundance distribution ($\Phi(x)$)	$\frac{\theta}{x}(1-x)^{\theta-1}$	$\frac{\theta}{x}(1-x)^{\theta-1}$
Time to common ancestor	$\frac{-J_M x}{1-x}\log x$	$\frac{-Nx}{1-x}\log x$

It turns out that the abundance distribution of an infinitely large metacommunity is Fisher's log-series distribution, and that θ of neutral theory is α of Fisher's log-series [2, 105, 215]. However, in any one local community, random walks of rare species are likely to include zero, and thus become extinct in the local community by chance alone. This causes a deviation from Fisher's log-series in any *local community* by reducing the number of the the rarest species below that predicted by Fisher's log-series. In particular, it tends to create a log-normal–like distribution, much as we often see in real data (Fig. 10.8). These theoretical findings and their match with observed data are thus consistent with the hypothesis that communities may be governed, in some substantive part, by neutral drift and migration.

Both theory and empirical data show that species which coexist may be more similar than predicted by chance alone [103], and that similarity (i.e., fitness equality) among species helps maintain higher diversity than would otherwise be possible [27]. Chesson makes an important distinction between *stabilizing mechanisms*, which create attractors, and *equalizing mechanisms*, which reduce differences among species, slow competitive exclusion and facilitate stabilization [27].

The NTBB spurred tremendous debate about the roles of chance and determinism, of dispersal-assembly and niche-assembly, of evolutionary processes in ecology, and how we go about "doing community ecology" (see, e.g., articles in *Functional Ecology*, 19(1), 2005; *Ecology*, 87(6), 2006). This theory in its narrowest sense has been falsified with a few field experiments and observation studies [32,223]. However, the *degree* to which stochasticity and dispersal *versus* niche partitioning structure communities remains generally unknown. Continued refinements and elaboration (e.g., [2, 52, 64, 86, 164, 216]) seem promising, continuing to intrigue scientists with different perspectives on natural communities [86,99]. Even if communities turn out to be completely non-neutral, NTBB provides a baseline for community dynamics and patterns that has increased the rigor of evidence required to demonstrate mechanisms controlling coexistence and diversity. As Alonso *et al.* state, "...good theory has more predictions per free parameter than bad theory. By this yeardstick, neutral theory fares fairly well" [1].

10.4.1 Different flavors of neutral communities

Neutral dynamics in a local community can be simulated in slightly different ways, but they are all envisioned as some type of *random walk*. A random walk occurs when individuals reproduce or die at random, with the consequence that each population increases or decreases by chance.

The simplest version of a random walk assumes that births and deaths and consequent increases and decreases in population size are equally likely and equal in magnitude. A problem with this type of random walk is that a community can increase in size (number of individuals) without upper bound, or can disappear entirely, by chance. We know this doesn't happen, but it is a critically important first step in conceptualizing a neutral dynamic [22].

Hubbell added another level of biological reality by fixing the total number of individuals in a local community, J_L, as constant. When an individual dies, it is replaced with another individual, thus keeping the population size constant. Replacements come from within the local community with probability $1-m$, and replacements come from the greater metacommunity with probability m. The dynamics of the metacommunity are so slow compared to the local community that we can safely pretend that it is fixed, unchanging.

Volkov *et al.* [215] took a different approach by assuming that each species undergoes independent *biased random walks*. We imagine that each species undergoes its own completely independent random walk, *as if* it is not interacting with any other species. The key is that the birth rate, b, is slightly *less* than the death rate, d — this bias toward death gives us the name *biased random walk*. In a deterministic model with no immigration or speciation, this would result in a slow population decline to zero. In a stochastic model, however, some populations will increase in abundance by chance alone. Slow random walks toward extinctions are balanced by speciation in the metacommunity (with probability ν).

If species all undergo independent biased random walks, does this mean species don't compete and otherwise struggle for existance? No. The *reason* that

$b < d$ is precisely because all species struggle for existance, and only those with sufficiently high fitness, *and which are lucky*, survive. Neutral theory predicts that it is these species that we observe in nature — those which are lucky, and also have sufficiently high fitness.

In the metacommunity, the average number of species, $\langle \phi_n^M \rangle$, with population size n is

$$\langle \phi_n^M \rangle = \theta \frac{x^n}{n} \tag{10.10}$$

where $x = b/d$, and $\theta = 2J_M \nu$ [215]. The M superscript refers to the metacommunity, and the angle brackets indicate merely that this is the average. Here b/d is barely less than one, because it a biased random walk which is then offset by speciation, ν. Now we see that this is exactly Fisher's log-series distribution (eq. 10.7), where that $x = b/d$ and $\theta = \alpha$. Volkov *et al.* thus show that in a neutral metacommunity, x has a biological interpretation.

The expected size of the entire metacommunity is simply the sum of all of the average species' n [215].

$$J_M = \sum_{n=1}^{\infty} n \langle \phi_n^M \rangle = \theta \frac{x}{1-x} \tag{10.11}$$

Thus the size of the metacommunity is an emergent property of the dynamics, rather than an external constraint. To my mind, it seems that the number of individuals in the metacommunity must result from responses of individuals *to* their external environment.

Volkov *et al.* went on to derive expressions for births, deaths, and average relative abundances in the local community [139, 215]. Given that each individual has an equal probability of dying and reproducing, and that replacements can also come from the metacommunity with a probability proportional to their abundance in the metacommunity, one can specify rules for populations in local communities of a fixed size. These are the probability of increase, $b_{n,k}$, or decrease, $d_{n,k}$, in a species, k, of population size n.

$$b_{n,k} = (1-m)\frac{n}{J_L}\frac{J_L - n}{J_L - 1} + m\frac{\mu_k}{J_M}\left(1 - \frac{n}{J_L}\right) \tag{10.12}$$

$$d_{n,k} = (1-m)\frac{n}{J_L}\frac{J_L - n}{J_L - 1} + m\left(1 - \frac{mu_k}{J_M}\right)\frac{n}{J_L} \tag{10.13}$$

These expressions are the sum of two joint probabilities, each of which is comprised of several independent events. These events are immigration, and birth and death of individuals of different species. Here we describe these probabilities. For a population of size n of species k, we can indicate per capita, per death probabilities including

- m, the probability that a replacement is immigrant from the metacommunity, and $1 - m$, the probability that the replacement is from the local community.

- n/J_L, the probability that an individual selected randomly from the local community belongs to species k, and $1-n/J_L$ or $(J-n)/(J_L)$, the probability that an individual selected randomly from the local community belongs to any species other than k.
- $(J-n)/(J_L-1)$, the conditional probability that, given that an individual of species k has already been drawn from the population, an individual selected randomly from the local community belongs to any species other than to species k.
- μ_k/J_M, the probability that an individual randomly selected from the metacommunity belongs to species k, and $1-n/J_M$, the probability that an individual randomly selected from the metacommunity belongs to any species other than k.

Each of these probabilities is the probability of some unspecified event — that event might be birth, death, or immigration.

Before we translate eqs. 10.12, 10.13 literally, we note that b and d each have two terms. The first term is for dynamics related to the local community, which happen with probability $1-m$. The second is related to immigration from the metacommunity which occurs with probability m. Consider also that if a death is replaced by a birth of the same species, or a birth is countered by a death of the same species, they cancel each other out, as if nothing ever happened. Therefore each term requires a probability related to species k and to non-k.

Eq. 10.12, $b_{n,k}$, is the probability that an individual will be added to the population of species k. The first term is the joint probability that an addition to the population comes from within the local community $(1-m)$ *and* the birth comes from species k (n/J_L) *and* there is a death of an individual of any other species $((J_L-n)/(J_L-1))$.[13] The second term is the joint probability that the addition to the population comes from the metacommunity via immigration (m) *and* that the immigrant is of species k (μ_k/J_M) *and* is not accompanied by a death of an individual of its species (n/J_L).

An individual may be subtracted from the population following similar logic. Eq. 10.13, $d_{n.k}$, is the probability that a death will remove an individual from the population of species k. The first term is the joint probability that the death occurs in species k (n/J_L) and the replacement comes from the local community $(1-m)$ and is some species other than k $((J_L-n)/(J_L-1))$. The second term is the joint probability that the death occurs in species k (n/J_L), and that it is replaced by an immigrant (m) *and* the immigrant is any species in the metacommunity other than k $(1-\mu_k/J_M)$.

10.4.2 Investigating neutral communities

Here we explore netural communities using the `untb` package, which contains a variety of functions for teaching and research on neutral theory.

[13] The denominator of the death probability is J_L-1 instead of J_L because we have already selected the first individual who will do the reproduction, so the total number of remaining individuals is J_L-1 rather than J_L; the number of non-k individuals remains J_L-n

Pure drift

After loading the package and setting graphical parameters, let's run a simulation of drift. Recall that drift results in the slow extinction of all but one species (Fig. 10.10). We start with a local community with 20 species, each with 25 individuals[14] The simulation then runs for 1000 generations (where 9/900 individuals die per generation).[15]

```
> library(untb)
> a <- untb(start = rep(1:20, 25), prob = 0, gens = 2500,
+      D = 9, keep = TRUE)
```

We keep, in a matrix, all 450 trees from each of the 1000 time steps so that we can investigate the properties of the community. The output is a matrix where each element is an integer whose value represents a species ID. Rows are time steps and we can think of columns as spots on the ground occupied by trees. Thus a particular spot of ground may be occupied by an individual of one species for a long time, and suddenly switch identity, because it dies and is replaced by an individual of another species. Thus the community always has 450 individuals (columns), but the identities of those 450 change through time, according to the rules laid out in eqs. 10.12, 10.13. Each different species is represented by a different integer; here we show the identitites of ten individuals (columns) for generations 901–3.

```
> (a2 <- a[901:903, 1:10])
```

```
     [,1] [,2] [,3] [,4] [,5] [,6] [,7] [,8] [,9] [,10]
[1,]    2   19    5   19    9    4   16    9    1    19
[2,]    2   19    5   19    9    4   16    9    1    19
[3,]    2   19    5   19    9    4   16    9    1    19
```

Thus, in generation 901, tree no. 1 is species 2 and in generation 902, tree no. 3 is species 5.

We can make pretty pictures of the communities at time steps 1, 100, and 2500, by having a single point for each tree, and coding species identity by shades of grey.[16]

```
> times <- c(1, 50, 2500)
> sppcolors <- sapply(times, function(i) grey((a[i,
+     ] - 1)/20))
```

This function applies to the community, at each time step, the grey function. Recall that species identity is an integer; we use that integer to characterize each species' shade of grey.

Next we create the three graphs at three time points, with the appropriate data, colors, and titles.

[14] This happens to be the approximate tree density (450 trees ha^{-1}, for trees > 10 cm DBH) on BCI.

[15] Note that display.untb is great for pretty pictures, whereas untb is better for more serious simulations.

[16] We could use a nice color palette, hcl, based on hue, chroma, and luminance, for instance hcl(a[i,]*30+50)

```
> layout(matrix(1:3, nr = 1))
> par(mar = c(1, 1, 3, 1))
> for (j in 1:3) {
+     plot(c(1, 20), c(1, 25), type = "n", axes = FALSE)
+     points(rep(1:20, 25), rep(1:25, each = 20),
+         pch = 19, cex = 2, col = sppcolors[, j])
+     title(paste("Time = ", times[j], sep = ""))
+ }
```

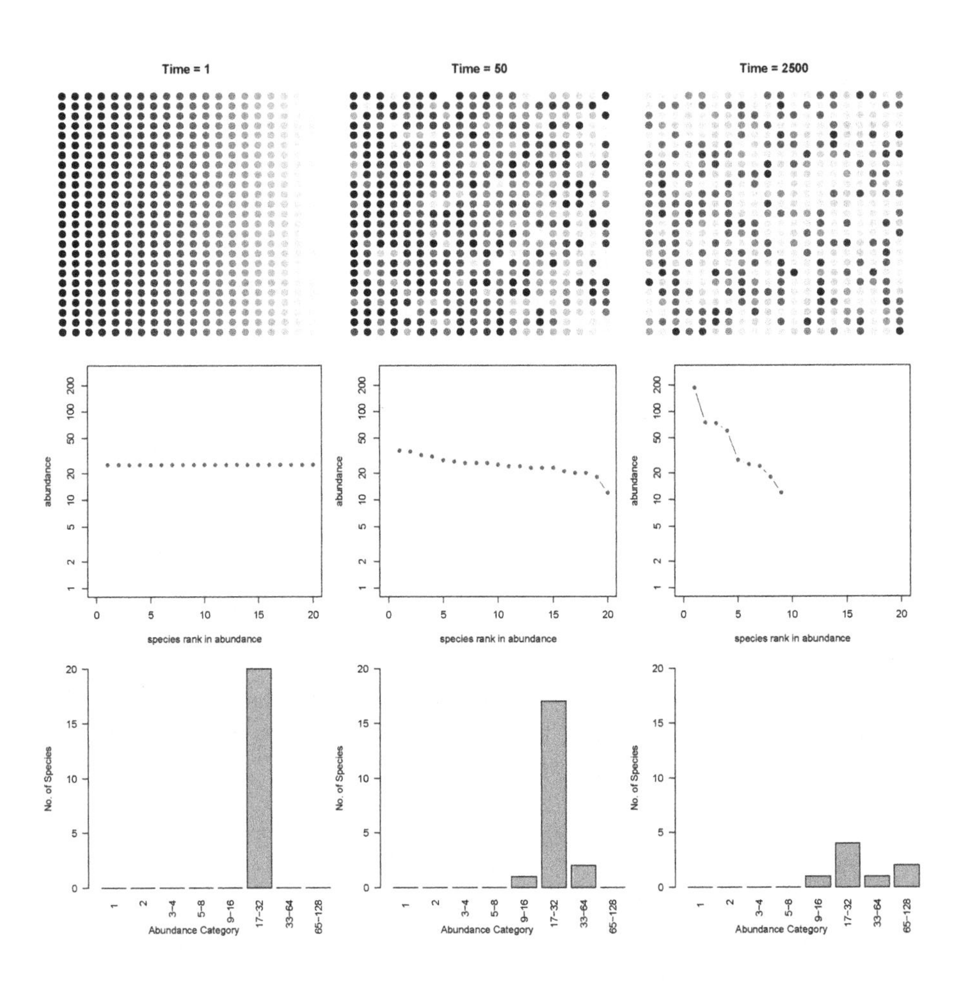

Fig. 10.11: Three snapshots of one community, drifting through time. Shades of grey represent different species. Second row contains rank abundance distributions; third row contains species abundance distributions. Drift results in the slow loss of diversity.

From these graphs (Fig. 10.11), we see that, indeed, the species identity of the 450 trees changes through time. Remember that this is *not spatially explicit* — we are not stating that individual *A* is next, or far away from, individual *B*.

Rather, this is a spatially implicit representation — all individuals are characterized by the same probabilities.

Next let's graph the rank abundance distribution of these communities (Fig. 10.11). For each time point of interest, we first coerce each "ecosystem" into a `count` object, and then plot it. Note that we plot them all on a common scale to facilitate comparison.

```
> layout(matrix(1:3, nr = 1))
> for (i in times) {
+     plot(as.count(a[i, ]), ylim = c(1, max(as.count(a[times[3],
+         ]))), xlim = c(0, 20))
+     title(paste("Time = ", i, sep = ""))
+ }
```

Next we create species abundance distributions, which are histograms of species' abundances (Fig. 10.11). If we want to plot them on a common scale, it takes a tad bit more effort. We first create a matrix of zeroes, with enough columns for the last community, use `preston` to create the counts of species whose abundances fall into logarithmic bins, plug those into the matrix, and label the matrix columns with the logarithmic bins.

```
> out <- lapply(times, function(i) preston(a[i,
+     ]))
> bins <- matrix(0, nrow = 3, ncol = length(out[[3]]))
> for (i in 1:3) bins[i, 1:length(out[[i]])] <- out[[i]]
> bins

     [,1] [,2] [,3] [,4] [,5] [,6] [,7] [,8] [,9]
[1,]    0    0    0    0    0   20    0    0    0
[2,]    0    0    0    0    1   17    2    0    0
[3,]    0    0    0    0    1    4    1    2    1

> colnames(bins) <- names(preston(a[times[3], ]))
```

Finally, we plot the species–abundance distributions.

```
> layout(matrix(1:3, nr = 1))
> for (i in 1:3) {
+     par(las = 2)
+     barplot(bins[i, ], ylim = c(0, 20), xlim = c(0,
+         8), ylab = "No. of Species", xlab = "Abundance Category")
+ }
```

Bottom line: *drift causes the slow loss of species from local communities* (Fig. 10.11). What is not illustrated here is that without dispersal, drift will cause different species to become abundant in different places because the variation is random. In that way, drift maintains diversity at large scales, in the metacommunity. Last, low rates of dispersal among local communities maintains some diversity in local communities without changing the entire metacommunity into a single large local community. Thus dispersal limitiation, but not its absence, maintains diversity.

Next, let's examine the dynamics through time. We will plot individual species trajectories (Fig. 10.12a).

```
> sppmat <- species.table(a)
> matplot(1:times[3], sppmat[1:times[3], ], type = "l",
+     ylab = "Population Density")
```

The trajectories all start at the same abundance, but they need not have. The trajectories would still have the same drifting quality.

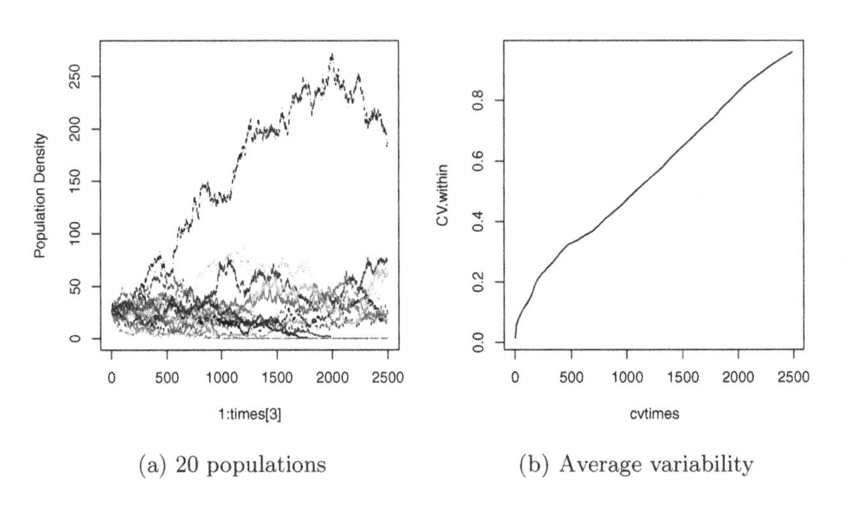

(a) 20 populations (b) Average variability

Fig. 10.12: Dynamics and average variation within populations. In random walks, average variation (measured with the coefficient of variation) increases with time.

For random walks in general, the observed variance and coefficient of variation ($CV = \hat{\sigma}/\bar{x}$) of a population will grow over time [32]. Here we calculate the average population CV of cumulative observations (note the convenient use of nested `(s)apply` functions). Let's calculate the CV's for every tenth time step.

```
> cvtimes <- seq(2, 2500, by = 10)
> CV.within <- sapply(cvtimes, function(i) {
+     cvs <- apply(sppmat[1:i, ], 2, function(x) sd(x)/mean(x))
+     mean(cvs)
+ })
```

Now plot the average CV through time. The average observed CV should increase (Fig. 10.12b).

```
> plot(cvtimes, CV.within, type = "l")
```

This shows us that the populations never approach an equilibrium, but wander aimlessly.

Real data

Last, we examine a BCI data set [36]. We load the data (data from 50 1 ha plots × 225 species, from the **vegan** package), and sum species abundances to get each species total for the entire 50 h plot (Fig. 10.13a).

```
> library(vegan)
> data(BCI)
> n <- colSums(BCI)
> par(las = 2, mar = c(7, 4, 1, 1))
> xs <- plot(preston(rep(1:225, n), n = 12, original = TRUE))
```

We would like to estimate θ and m from these data, but that requires specialized software for any data set with a realistic number of individuals. Specialized software would provide maximum likelihood estimates (in a reasonable amount of computing time) for m and θ for large communities [51,69,86]. The BCI data have been used repeatedly, so we rely on estimates from the literature ($\theta \approx 48$, $m \approx 0.1$) [51,215]. We use the approach of Volkov et al. [215] to generate expected species abundances (Fig. 10.13a).

```
> v1 <- volkov(sum(n), c(48, 0.1), bins = TRUE)
> points(xs, v1[1:12], type = "b")
```

More recently, Jabot and Chave [86] arrived at estimates that differed from previous estimates by orders of magnitude (Fig. 10.13b). Their novel approach estimated θ and m were derived from *both* species abundance data and from phylogenetic data ($\theta \approx 571$, $m \approx 0.002$). This is possible because neutral theory makes a rich array of predictions, based on both ecological and evolutionary processes. Their data were only a little bit different (due to a later census), but their analyses revealed radically different estimates, with a much greater diversity and larger metacommunity (greater θ), and much lower immigration rates (smaller m).

Here we derive expected species abundance distributions. The first is based soley on census data, and is similar to previous expections (Fig. 10.13b).

```
> v2 <- volkov(sum(n), c(48, 0.14), bins = TRUE)
> xs <- plot(preston(rep(1:225, n), n = 12, original = TRUE),
+     col = 0, border = 0)
> axis(1, at = xs, labels = FALSE)
> points(xs, v2[1:12], type = "b", lty = 2, col = 2)
```

However, when they also included phylogenetic data, they found very different expected species abundance distributions (Fig. 10.13b).

```
> v4 <- volkov(sum(n), c(571, 0.002), bins = TRUE)
> points(xs, v4[1:12], type = "b", lty = 4, col = 2)
```

If nothing else, these illustrate the effects of increasing θ and reducing m.

10.4.3 Symmetry and the rare species advantage

An important advance in neutral theory is the quantification of a *symmetric rare species advantage.* The symmetry hypothesis posits that all species have a *symmetrical rare species advantage* [214, 216]. That is, all species increase when rare to the same degree (equal negative density dependence). In a strict sense, all individuals remain the same in that their birth and death probabilities change with population size in the same manner for all species. This obviously

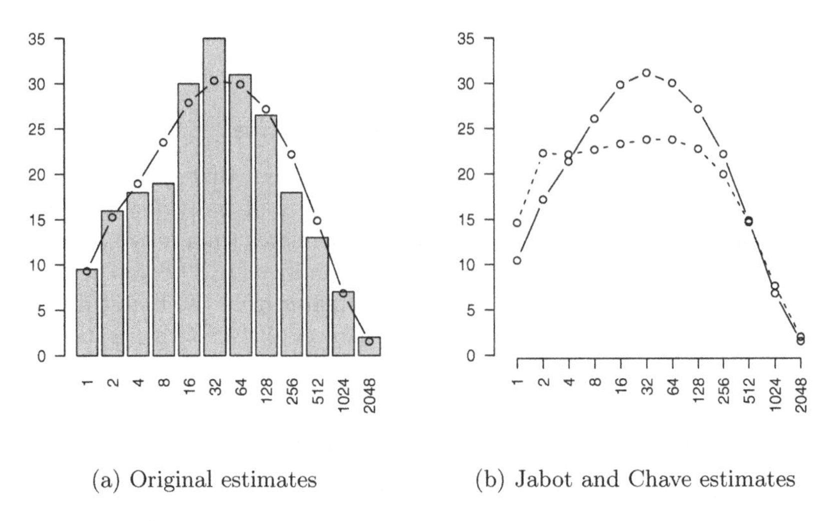

(a) Original estimates (b) Jabot and Chave estimates

Fig. 10.13: Species abundance distributions for BCI trees. (a) Data for histogram from [36], overlain with expected abundances with θ and m values fitted to the data [51,215]. (b) Jabot and Chave found that when they used only species abundances (as did previous investigators) their pattern was similar to previous findings (solid line). However, adding phylogenetic information led to very different expectations (dashed line).

reduces the chance of random walks to extinction, but is nonetheless the same among all species. Estimation of the magnitude of the rare species advantage is interesting addition to stepwise increasing complexity.

To sum up: it is safe to say that neutral theory has already made our thinking about community structure and dynamics more sophisticated and subtle, by extending island biogeography to individuals. The theory is providing quantitative, pattern-generating models, that are analogous to null hypotheses. With the advent of specialized software, theory is now becoming more useful in our analysis of data [51,69,86].

10.5 Diversity Partitioning

We frequently refer to biodiversity (i.e., richness, Simpson's, and Shannon-Wiener diversity) at different spatial scales as α, β, and γ diversity (Fig. 10.14).

- Alpha diversity, α, is the diversity of a point location or of a single sample.
- Beta diversity, β, is the diversity due to multiple localities; β diversity is sometimes thought of as *turnover* in species composition among sites, or alternatively as the number of species in a region that are *not* observed in a sample.
- Gamma diversity, γ, is the diversity of a region, or at least the diversity of all the species in a set of samples collected over a large area (with large extent relatve to a single sample).

Diversity across spatial scales can be further be *partitioned* in one of two ways, either using *additive* or *multiplicative* partitioning.

Additive partitioning [42,43,97] is represented as

$$\bar{\alpha} + \beta = \gamma \tag{10.14}$$

where $\bar{\alpha}$ is the average diversity of a sample, γ is typically the diversity of the pooled samples, and β is found by difference ($\beta = \gamma - \bar{\alpha}$). We can think of β as the *average* number of species *not* found in a sample, but which we know to be in the region. Additive partitioning allows direct comparison of average richness among samples at any hierarchical level of organization because all three measures of diversity (α, β, and γ) are *expressed in the same units.* This makes it analogous to partitioning variance in ANOVA. This is not the case for multiplicative partitioning diversity.

Partitioning can also be *multiplicative* [219],

$$\bar{\alpha}\beta = \gamma \tag{10.15}$$

where β is a conversion factor that describes the relative change in species composition among samples. Sometimes this type of β diversity is thought of as the number of different community types in a set of samples. However, one must use this interpretation with great caution, as either meaning of β diversity depends completely on the sizes or extent of the samples used for α diversity.

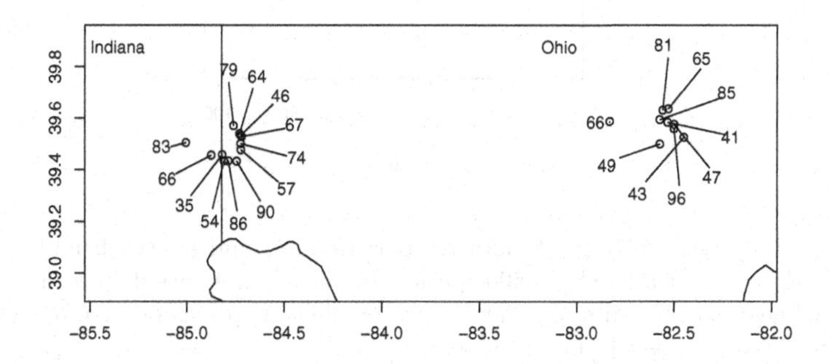

Fig. 10.14: Hierarchical sampling of moth species richness in forest patches in Indiana and Ohio, USA [198]. α-diversity is the diversity of a single site (richness indicated by numbers). γ-diversity is the total number of species found in any of the samples (here $\gamma = 230$ spp.). Additive β-diversity is the difference, $\gamma - \bar{\alpha}$, or the average number of species *not* observed in a single sample. Diversity partitioning can be done at two levels, sites within ecoregions and ecoregions within the geographic region (see example in text for details).

Let us examine the limits of β diversity in extremely small and extremely large samples. Imagine that our sample units each contain, on average, one individual (and therefore one species) and we have 10^6 samples. If richness is our measure of diversity, then $\bar{\alpha} = 1$. Now imagine that in all of our samples we find a total of 100 species, or $\gamma = 100$. Our additive and multiplicative partitions would then be $\beta_A = 99$, and $\beta_M = 100$, respectively. If the size of the sample unit increases, each sample will include more and more individuals and therefore more of the diversity, and by definition, β will decline. If each sample gets large enough, then each sample will capture more and more of the species until a sample gets so large that it includes all of the species (i.e., $\bar{\alpha} \rightarrow \gamma$). At this point, $\beta_A \rightarrow 0$ and $\beta_M \rightarrow 1$.

Note that β_A and β_M do not change at the same rates (Fig. 10.15). When we increase sample size so that each sample includes an average of two species ($\bar{\alpha} = 2$), then $\beta_A = 98$ and $\beta_M = 50$. If each sample were big enough to have on average 50 species ($\bar{\alpha} = 50$), then $\beta_A = 50$ and $\beta_M = 2$. So, the β's do not change at the same rate (Fig. 10.15).

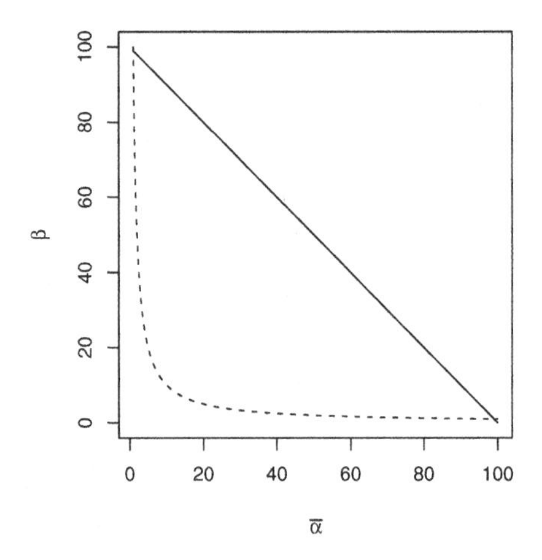

Fig. 10.15: Relations of β_A (with additive partitioning) and β_M (with multiplicative partitioning) to $\bar{\alpha}$, for a fixed $\gamma = 500$ species. In our example, we defined diversity as species richness, so the units of β_A and α are number of species per sample, and $\bar{\alpha}$ is the mean number of species in a sample.

Multiplicative β_M is sometimes thought of as the number of independent "communities" in a set of samples. This would make sense if our sampling regime were designed to capture representative parts of different communities. For example, if we sampled an elevational gradient, or a productivity gradient, and our smallest sample was sufficiently large so as to be representative of that point along the gradient[17] then β_M could provide a measure of the relative turnover in

[17] One type of sample that attempts to do this is a relevé.

composition or "number of different communities." However, we know that composition is also predicted to vary randomly across the landscape [36]. Therefore, if each sample is small, and not really representative of a "community," then the small size of the samples will inflate β_M and change the interpretation.

As an example, consider the BCI data, which consists of 50 contiguous 1 ha plots. First, we find γ (all species in the data set, or the number of columns), and $\bar{\alpha}$ (mean species number per 1 ha plot).

```
> (gamma <- dim(BCI)[2])

[1] 225

> (alpha.bar <- mean(specnumber(BCI)))

[1] 90.78
```

Next we find additive β-diversity and multiplicative β-diversity.

```
> (beta.A <- gamma - alpha.bar)

[1] 134.2

> (beta.M <- gamma/alpha.bar)

[1] 2.479
```

Now we interpret them. These plots are located in a relatively uniform tropical rainforest. Therefore, they each are samples drawn from a single community type. However, the samples are small. Therefore, each 1 ha plot (10^4 m^2 in size) misses more species than it finds, on average ($\beta_A > \bar{\alpha}$). In addition, $\beta_M = 2.48$, indicating a great deal of turnover in species composition. We could mistakenly interpret this as indicating something like ~ 2.5 independent community types in our samples. Here, however, we have a single community type — additive partitioning is a little simpler and transparent in its interpretation.

For other meanings of β-diversity, linked to neutral theory, see [36, 144].

10.5.1 An example of diversity partitioning

Let us consider a study of moth diversity by Keith Summerville and Thomas Crist [197, 198]. The subset of their data presented here consists of woody plant feeding moths collected in southwest Ohio, USA. Thousands of individuals were trapped in 21 forest patches, distributed in two adjacent ecoregions (12 sites - North Central Tillplain [NCT], and 9 sites - Western Allegheny Plateau [WAP], Fig. 10.14). This data set includes a total of 230 species, with 179 species present in the NCT ecoregion and 173 species present in the WAP ecoregion. From these subtotals, we can already see that each ecoregion had most of the combined total species (γ).

We will partition richness at three spatial scales: sites within ecoregions ($\bar{\alpha}_1$), ecoregions ($\bar{\alpha}_2$), and overall (γ). This will result in two β-diversities: β_1 among sites within each ecoregion, and β_2 between ecoregions. The relations among these are straightforward.

$$\bar{\alpha}_2 = \bar{\alpha}_1 + \beta_1 \tag{10.16}$$

$$\gamma = \bar{\alpha}_2 + \beta_2 \tag{10.17}$$

$$\gamma = \bar{\alpha}_1 + \beta_1 + \beta_2 \tag{10.18}$$

To do this in R, we merely implement the above equations using the data in Fig. 10.14 [198]. First, we get the average site richness, $\bar{\alpha}_1$. Because we have different numbers of individuals from different site, and richness depends strongly on the number of individuals in our sample, we may want to weight the sites by the number of individuals. However, I will make the perhaps questionable argument for the moment that because trapping effort was similar at all sites, we will not adjust for numbers of individuals. We will assume that different numbers of individuals reflect different population sizes, and let number of individuals be one of the local determinants of richness.

```
> data(moths)
```

```
> a1 <- mean(moths[["spp"]])
```

Next we calculate average richness richness for the ecoregions. Because we had 12 sites in NCT, and only nine sites in WAP for what might be argued are landscape constraints, we will use the weighted average richness, adjusted for the number of sites.[18] We also create an object for $\gamma = 230$.

```
> a2 <- sum(c(NCT = 179, WAP = 173) * c(12, 9)/21)
> g <- 230
```

Next, we get the remaining quantities of interest, and show that the partition is consistent.

```
> b1 <- a2 - a1
> b2 <- g - a2
> abg <- c(a1 = a1, b1 = b1, a2 = a2, b2 = b2, g = g)
> abg

    a1     b1     a2     b2      g
 65.43 111.00 176.43  53.57 230.00

> a1 + b1 + b2 == g

[1] TRUE
```

The partitioning reveals that β_1 is the largest fraction of overall γ-richness (Fig. 10.16). This indicates that in spite of the large distance between sampling areas located in different ecoregions, and the different soil types and associated flora, most of the variation occurs *among sites within regions.* If there had been a greater difference in moth community composition among ecoregions, then β_2-richness would have made up a greater proportion of the total.

[18] The arithmetic mean is $\sum a_i Y_i$, where all $a_i = 1/n$, and n is the total number of observations. A weighted average is the case where the a_i represent unequal *weights*, often the fraction of n on which each Y_i is based. In both cases, $\sum a = 1$.

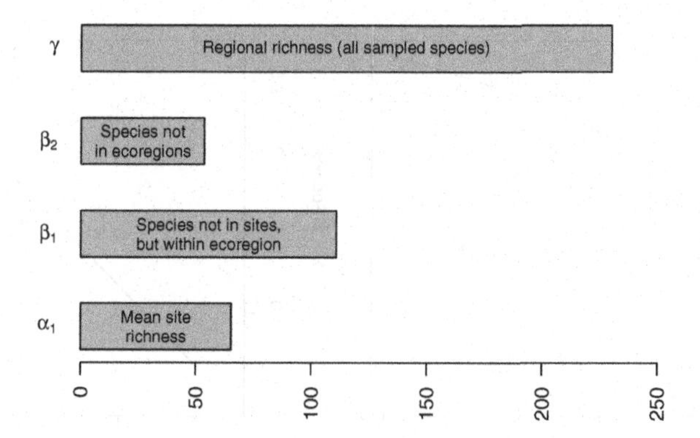

Fig. 10.16: Hierarchical partitioning of moth species richness in forest patches [198]. See Fig. 10.14 for geographical locations.

These calculations show us how simple this additive partition can be, although more complicated analyses are certainly possible. It can be very important to weight appropriately the various measures of diversity (e.g., the number of individuals in each sample, or number of samples per hierarchical level). The number of individuals in particular has a tremendous influence on richness, but has less influence on Simpson's diversity partitioning. The freely available PARTITION software will perform this additive partitioning (with sample sizes weights) and perform statistical tests [212].

10.5.2 Species–area relations

The relation between the number of species found in samples of different area has a long tradition [5, 10, 120–122, 169, 178], and is now an important part of the metastasizing subdiscipline of *macroecology* [15, 71].

Most generally, the species–area relation (SAR) is simply an empirical pattern of the number of species found in patches of different size, plotted as a function of the sizes of the respective patches (Fig. 10.17). These patches may be isolated from each other, as islands in the South Pacific [121], or mountaintops covered in coniferous forest surrounded by a sea of desert [16], or calcareous grasslands in an agricultural landscape [68]. On the other hand, these patches might be nested sets, where each larger patch contains all others [41, 163].

Quantitatively, the relation is most often proposed as a simple power law,

$$R = cA^z \tag{10.19}$$

where R is the number of species in a patch of size A, and c and z are fitted constants. This is most often plotted as a log–log relation, which makes it linear.

$$\log(R) = b + zA \tag{10.20}$$

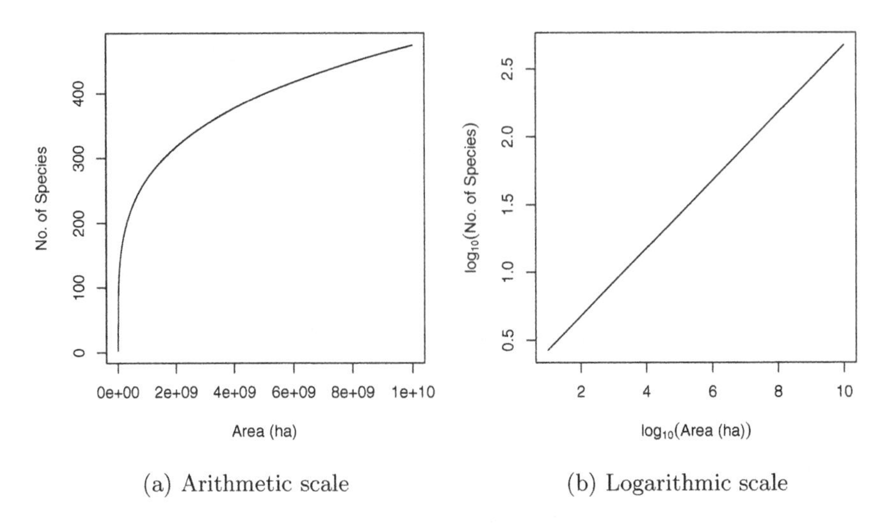

(a) Arithmetic scale (b) Logarithmic scale

Fig. 10.17: Power law species–area relations.

where b is the intercept (equal to $\log c$) and z is the slope.

Drawing power law species–area relations (Fig. 10.17)

Here we simply draw some species area curves.

```
> A <- 10^1:10
> c <- 1.5
> z <- 0.25
> curve(c * x^z, 10, 10^10, n = 500,
+     ylab = "No. of Species", xlab = "Area (ha)")

> A <- 10^1:10
> c <- 1.5
> z <- 0.25
> curve(log(c, 10) + z * x, 1, 10, ylab = quote(log[10]("No. of Species")),
+     xlab = quote(log[10]("Area (ha)")))
```

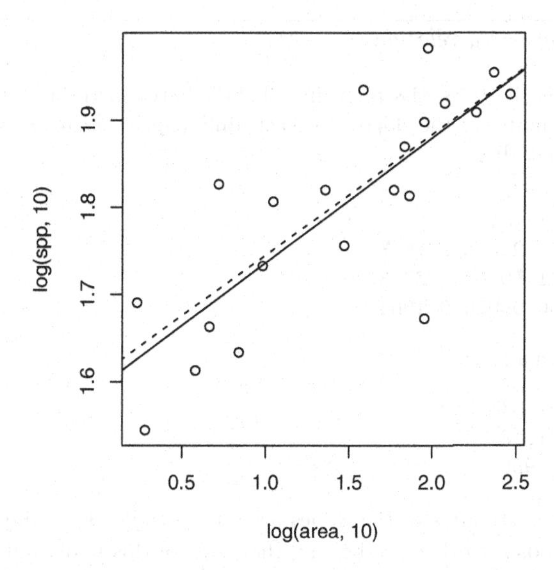

Fig. 10.18: Fitted power law species–area relations.

Fitting a species–area relation (Fig. 10.18)

Here we fit a species–area curve to data, and examine the slope. We could fit a nonlinear power relation ($y = cA^z$); this would be appropriate, for instance, if the residual noise around the line were of the same magnitude for all patch areas. We could use reduced major axis regression, which is appropriate when there is equivalent uncertainty or error on both x and y. Last (and most often), we could use a simple linear regression on the log-transformed data, which is appropriate when we know x to a high degree of accuracy, but measure y with some error, and the transformation causes the residual errors to be of similar magnitude at all areas. We start with the last (log-transformed). Here we plot the data, and fit a linear model to the common log-transformed data.

```
> plot(log(spp, 10) ~ log(area, 10), moths)
> mod <- lm(log(spp, 10) ~ log(area, 10), data = moths)
> abline(mod)
```

Next we fit the nonlinear curve to the raw data, and overlay that fit (on the log scale).

```
> mod.nonlin <- nls(spp ~ a * area^z, start = list(a = 1,
+     z = 0.2), data = moths)
> curve(log(coef(mod.nonlin)[1], 10) + x * coef(mod.nonlin)[2],
+     0, 3, add = TRUE, lty = 2)
```

Assessing species–area relations

Note that in Figure 10.18, the fits differ slightly between the two methods. Let's compare the estimates of the slope — we certainly expect them to be similar, given the picture we just drew.

```
> confint(mod)

                  2.5 % 97.5 %
(Intercept)     1.50843 1.6773
log(area, 10) 0.09026 0.1964

> confint(mod.nonlin)

       2.5%    97.5%
a 31.61609 50.1918
z  0.08492  0.1958
```

We note that the estimates of the slopes are quite similar. Determining the better of the two methods (or others) is beyond the scope of this book, but be aware that methods can matter.

The major impetus for the species–area relation came from (i) Preston's work on connections between the species–area relation and the log-normal species abundance distribution [169,170], and (ii) MacArthur and Wilson's theory of island biogeography[19] [122].

Preston posited that, given the log-normal species abundance distributions (see above), then increasingly large samples should accumulate species at particular rates. Direct extensions of this work, linking neutral theory and maximum entropy theory to species abundances and species–area relations continues today [10,71,81]

Island biogeography

MacArthur and Wilson proposed a simple theory wherein the number of species on an oceanic island was a function of the immigration rate of new species, and extinction rate of existing species (Fig. 10.19). The number of species at any one time was *a dynamic equilibrium*, resulting from both slow inevitable extinction and slow continual arrival of replacements. Thus species composition on the island was predicted to change over time, that is, to undergo turnover.

Let us imagine immigration rate, y, as a function of the number of species already on an island, x (Fig. 10.19). This relation will have a negative slope, because as the number of species rises, that chance that a new individual actually represents a new species will decline. The immigration rate will be highest when there are no species on the island, $x = 0$, and will fall to zero when every conceivable species is already there. In addition, the slope should be decelerating

[19] (Originally proposed in a paper entitled "An Equilibrium Theory of Insular Zoogeography" [121])

(concave up) because some species will be much more likely than others to immigrate. This means that the immigration rate drops steeply as the most likely immigrants arrive, and only the unlikely immigrants are missing. Immigrants may colonize quickly for two reasons. First, as Preston noted, some species are simply much more common than others. Second, some species are much better dispersers than others.

Now let us imagine extinction rate, y, as a function of the number of species, x (Fig. 10.19). This relation will have a positive slope, such that the probability of extinction increases with the number of species. This is predicted to have an accelerating slope (concave-up), for essentially the same sorts of reasons governing the shape of the immigration curve: Some species are less common than others, and therefore more likely to become extinct due to demographic and environmental stochasticity, and second, some species will have lower fitness for any number of reasons. As the number of species accumulates, the more likely it will become that these extinction-prone species (rare and/or lower fitness) will be present, and therefore able to become extinct.

The *rate of change* of the number of species on the island, ΔR, will be the difference between immimigration, I, and extinction, E, or

$$\Delta R = I - E. \tag{10.21}$$

When $\Delta R = 0$, we have an equilibrium. If we knew the quantitative form of immigration and extinction, we could solve for the equilibrium. That equilibrium would be the point on the x axis, R, where the two rates cross (Fig. 10.19).

In MacArthur and Wilson's theory of island biogeography, these rates could be driven by the *sizes* of the islands, where

- larger islands had lower extinction rates because of larger average population sizes, and
- larger islands had higher colonization rates because they were larger targets for dispersing species.

The distance between an island and sources of propagules was also predicted to influence these rates, where

- islands closer to mainlands had higher colonization rates of new species because more propagules would be more likely to arrive there, and
- the effect of area would be more important for islands far from mainlands than for islands close to mainlands.

Like much good theory, these were simple ideas, but had profound effects on the way ecologists thought about communities. Now these ideas, of dispersal mediated coexistence and landscape structure, continue to influence community ecologists [15, 81, 102].

Drawing immigration and extinction curves

It would be fun to derive a model of immigration and extinction rates from first principles [121], but here we can illustrate these relations with some simple

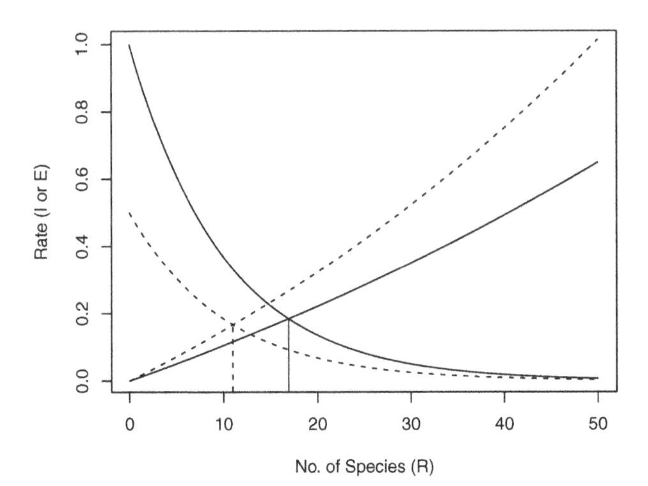

Fig. 10.19: Immigration and extinction curves for the theory of island biogeography. The declining curves represent immigration rates as functions of the number of species present on an island. The increasing curves represent extinction rates, also as functions of island richness. See text for discussion of the heights of these curves, i.e., controls on these rates. Here the dashed lines represent an island that is shrinking in size.

phenomenological graphs. We will assume that immmigration rate, I, can be represented as a simple negative exponential function $\exp(I_0 - iR)$, where I_0 is the rate of immigration to an empty island, and $-i$ is the per species negative effect on immigration rate.

```
> I0 <- log(1)
> b <- 0.1
> curve(exp(I0 - b * x), 0, 50, xlab = "No. of Species (R)",
+     ylab = "Rate (I or E)")
```

Note that extinction rate, E, must be zero if there are no species present. Imagine that extinction rate is a function of density and that average density declines as the number of species increases, or $\bar{N} = 1/R$.[20]

```
> d <- 0.01
> curve(exp(d * x) - 1, 0, 50, add = TRUE)
```

We subtract 1 merely to make sure that $E = 0$ when $R = 0$.

The number of species, R, will result from $\Delta R = 0 = I - E$, the point at which the lines cross.

$$I = e^{I_0 - bR} \tag{10.22}$$

$$E = e^{dR} - 1 \tag{10.23}$$

$$\delta R = 0 = I - E \tag{10.24}$$

[20] Why would this make sense ecologically?

Here we find this empricially by creating a function of R to minimize — we will minimize $(I-E)^2$; squaring the difference gives the quantity the convenient property that the minimum will be approached from either positive or negative values.

```
> deltaR <- function(R) {
+     (exp(I0 - b * R) - (exp(d * R) - 1))^2
+ }
```

We feed this into an optimizer for one-parameter functions, and specify that we know the optimum will be achieved somewhere in the interval between 1 and 50.

```
> optimize(f = deltaR, interval = c(1, 50))[["minimum"]]

[1] 16.91
```

The output tells us that the minimum was achieved when $R \approx 16.9$.

Now imagine that rising sea level causes island area to shrink. What is this predicted to do? It could

1. reduce the base immigration rate because the island is a smaller target,
2. increase extinction rate because of reduced densities.[21]

Let us represent reduced immigration rate by reducing I_0.

```
> I0 <- log(1/2)
> curve(exp(I0 - b * x), 0, 50, add = TRUE, lty = 2)
```

Next we increase extinction rate by increasing the per species rate.

```
> d <- 0.014
> curve(exp(d * x) - 1, 0, 50, add = TRUE, lty = 2)
```

If we note where the rates cross, we find that the number of predicted species has declined. With these new immigration and extinciton rates the predicted number of species is

```
> optimize(f = deltaR, interval = c(1, 50))[["minimum"]]

[1] 11.00
```

or 11 species, roughly a 35% decline $((17-11)/17 = 0.35)$.

The beauty of this theory is that it focuses our attention on *landscape level processes*, often outside the spatial and temporal limits of our sampling regimes. It specifies that any factor which helps determine the immigration rate or extinction rate, including island area or proximity to a source of propagules, is predicted to alter the equilibrium number of species at any point in time. We should further emphasize that the *identity* of species should change over time, that is, undergo turnover, because new species arrive and old species become extinct. The rate of turnover, however, is likely to be slow, because the species that are most likely to immigrate and least likely to become extinct will be the same species from year to year.

[21] We might also predict an increased extinction rate because of reduced rescue effect (Chapter 4).

10.5.3 Partitioning species–area relations

You may already be wondering if there is a link between island biogeography and β-diversity. After all, as we move from island to island, and as we move from small islands to large islands, we typically encounter additional species, and that is what we mean by β-diversity. Sure enough, there are connections [42].

Let us consider the moth data we used above (Fig. 10.14). The total number of species in all of the patches is, as before, γ. The average richness of these patches is $\bar{\alpha}$, and also note that *part of what determines that average is the area of the patch.* That is, when a species is missing from a patch, part of the reason might be that the patch is smaller than it could be. We will therefore partition β into yet one more category: species missing due to patch size, β_{area}. This new quantity is the average difference between $\bar{\alpha}$ and the diversity *predicted* for the largest patch (Fig. 10.20). In general then,

$$\beta = \beta_{area} + \beta_{replace} \tag{10.25}$$

where $\beta_{replace}$ is the average number of species missing that are *not* explained by patch size.

In the context of these data (Fig. 10.20), we now realize that $\beta_1 = \beta_{area} + \beta_{ecoregion}$, so the full partition becomes

$$\gamma = \bar{\alpha}_1 + \beta_{area} + \beta_{ecoregion} + \beta_{geogr.region} \tag{10.26}$$

where $\beta_{replace} = \beta_{ecoregion} + \beta_{geogr.region}$. Note that earlier in the chapter, we did not explore the effect of area. In that case, $\beta_{ecoregion}$ included both the effect of area and the effect of ecoregion; here we have further partitioned this variation into variation due to patch size, as well as variation due to ecoregion. This reduces the amount of unexplained variation among sites within each ecoregion.

Let's calculate those differences now. We will use quantities we calculated above for $\bar{\alpha}_1$, $\bar{\alpha}_2$, γ, and a nonlinear species–area model from above. We can start to create a graph similar to Fig. 10.20.

```
> plot(spp ~ area, data = moths, ylim = c(30, 230),
+      xlab = "Area (ha)", ylab = "No. of Species (R)")
> curve(coef(mod.nonlin)[1] * x^coef(mod.nonlin)[2],
+      0, max(moths[["area"]]), add = TRUE, lty = 2,
+      lwd = 2)
> abline(h = g, lty = 3)
> text(275, g, quote(gamma), adj = c(0.5, 1.5),
+      cex = 1.5)
```

Next we need to find the *predicted richness* for the maximum area. We use our statistical model to find that.

```
> (MaxR <- predict(mod.nonlin, list(area = max(moths[["area"]]))))
```

```
[1] 88.62
```

We can now find β_{area}, β_{eco} and β_{geo}.

```
> b.area <- MaxR - a1
> b.eco <- a2 - (b.area + a1)
> b.geo <- g - a2
```

Now we have partitioned γ a little bit more finely with a beastiary of β's, where

- $\bar{\alpha}_1$ is the average site richness.
- β_{area} is the average number of species not observed, due to different patch sizes.
- β_{eco} is the average number of species not observed at a site, is not missing due to patch size, but is in the ecoregion.
- β_{geo} is the average number of species not found in the samples from different ecoregions.

Finally, we add lines to our graph to show the partitions.

```
> abline(h = g, lty = 3)
> abline(h = b.eco + b.area + a1, lty = 3)
> abline(h = b.area + a1, lty = 3)
> abline(h = a1, lty = 3)
```

Now we have further quantified how forest fragment area explains moth species richness. Such understanding of the spatial distribution of biodiversity provides a way to better quantify patterns governed by both dispersal and habitat preference, and allows us to better describe and manage biodiversity in human-dominated landscapes.

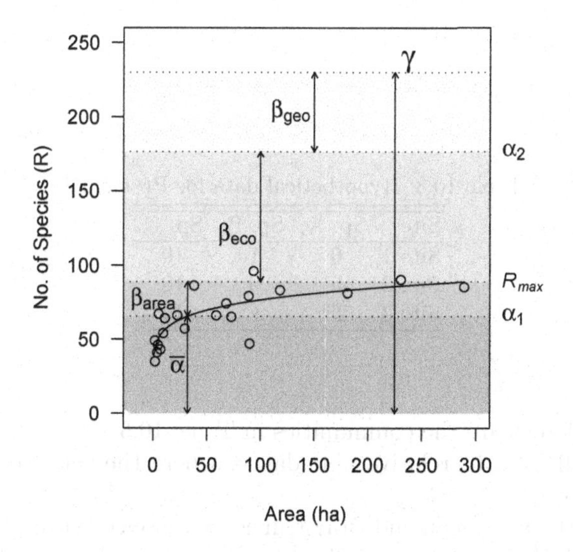

Fig. 10.20: Combining species–area relations with additive diversity partitioning. Forest fragment area explains relatively little of the diversity which accumulates in isolated patches distributed in space. However, it is likely that area associated with the collection of samples (i.e., the distances among fragments) contributes to β_{eco} and β_{geo}.

10.6 Summary

We have examined communities as multivariate entities which we can describe and compare in a variety of ways.

- Composition includes all species (multivariate data), whereas species diversity is a univariate description of the variety of species present.
- There are many ways to quantify species diversity, and they tend to be correlated. The simplest of these is richness (the number of species present), whereas other statistics take species' relative abundances into account.
- Species abundance distributions and rank abundance distributions are analogous to probability distributions, and provide more thorough ways to describe the patterns of abundance and variety in communities. These all illustrate a basic law of community ecology: most species are rare. Null models of community structure and processes make predictions about the shape of these distributions.
- Ecological neutral theory provides a dynamical model, not unlike a null model, which allows quantitative predictions relating demographic, immigration, and speciation rates, species abundance distributions, and patterns of variation in space and time.
- Another law of community ecology is that the number of species increases with sample area and appears to be influenced by immigration and extinction rates.
- We can partition diversity at different spatial scales to understand the structure of communities in landscapes.

Problems

Table 10.3: Hypothetical data for Problem 1.

Site	Sp. A	Sp. B	Sp. C
Site 1	0	1	10
Site 2	5	9	10
Site 3	25	20	10

10.1. How different are the communities in Table 10.3?
(a) Convert all data to relative abundance, where the relative abundance of each site sum to 1.
(b) Calculate the Euclidean and Bray-Curtis (Sørensen) distances between each pair of sites for both relative and absolute abundances.
(c) Calculate richness, Simpson's and Shannon-Wiener diversity for each site.

10.2. Use rarefaction to compare the tree richness in two 1 ha plots from the BCI data in the `vegan` package. Provide code, and a single graph of the expectations for different numbers of individuals; include in the graph some indication of the uncertainty.

10.3. Select one of the 1 ha BCI plots (from the `vegan` package), and fit three different rank abundance distributions to the data. Compare and contrast their fits.

10.4. Simulate a neutral community of 1000 individuals, selecting the various criteria on yur own. Describe the change through time. Relate the species abundance distributions that you observe through time to the parameters you choose for the simulation.

10.5. Using the `dune` species data (`vegan` package), partition species richness into $\bar{\alpha}$, β, and γ richness, where rows are separate sites. Do the same thing using Simpson's diversity.

A

A Brief Introduction to R

R is a language. Use it every day, and you will learn it quickly.

S, the precursor to R, is a quantitative programming environment developed at AT&T Bell Labs in the 1970s. S-Plus is a commercial, "value-added" version and was begun in the 1980s, and R was begun in the 1990s by Robert Gentleman and Ross Ihaka of the Statistics Department of the University of Auckland. Nearly 20 senior statisticians provide the core development group of the R language, including the primary developer of the original S language, John Chambers, of Bell Labs.

R is an official part of the Free Software Foundation's GNU project[1] (http://www.fsf.org/). It is free to all, and licensed to stay that way.

R is a language and environment for dynamical and statistical computing and graphics. R is similar to the S language, different implementation of S. Technically speaking, R is a "dialect" of S. R provides a very wide variety of statistical (linear and nonlinear modelling, classical statistical tests, time-series analysis, classification, clustering, ...) and graphical techniques, and is highly extensible. R has become the lingua franca of academic statistics, and is very useful for a wide variety of computational fields, such as theoretical ecology.

A.1 Strengths of R/S

- Simple and compact syntax of the language. You can learn R quickly, and can accomplish a lot with very little code.
- Extensibility. Anyone can extend the functionality of R by writing code. This may be a simple function for personal use, or a whole new family of statistical procedures in a new package.
- A huge variety of statistical and computing procedures. This derives from the ease with which R/S can be extended and shared by users around the world.
- Rapid updates.

[1] Pronounced "g-noo" — it is a recursive acronym standing for "GNU's Not Unix."

- Replicability and validation. All data analyses should be well documented, and this only happens reliably when the analyses are performed with scripts or programs, as in R or SAS. Point-and-click procedures cannot be validated by supervisors, reviewers, or auditors.
- Getting help from others is easy. Any scripted language can be quickly and easily shared with someone who can help you. I cannot help someone who says "first I clicked on this, and then I clicked on that ...."
- Repetitive tasks simplified. Writing code allows you to do anything you want a huge number of times. It also allows very simple updates with new data.
- High quality graphics. Well-designed publication-quality plots can be produced with ease, including mathematical symbols and formulae where needed. Great care has been taken over the defaults for the minor design choices in graphics, but the user retains full control.
- R is available as Free Software under the terms of the Free Software Foundation's GNU General Public License in source code form. It compiles and runs out of the box on a wide variety of UNIX platforms and similar systems (including FreeBSD and Linux). It also compiles and runs on Windows 9x/NT/2000 and Mac OS.
- Accessibility. Go now to www.r-project.org. Type "R" into Google. The R Project page is typically the first hit.

There is a tremendous amount of free documentation for R. Rcomes with a selection of manuals under the "Help" menu — explore these first. At the main R project web site, see the "Documentation" on the left sidebar. The FAQ's are very helpful. The "Other" category includes a huge variety of items; search in particular for "Contributed documentation."[2] There you will find long (100+ pages) and short tutorials. You will also find two different "RReference Cards," which are useful lists of commonly used functions.[3]

A.2 The R Graphical User Interface (GUI)

R has a very simple but useful graphical user interface (GUI; Fig. A.1). A few points regarding the GUI:

- "You call this a graphical user interface?" Just kidding — the GUI is *not* designed for point-and-click modeling.
- The R GUI is designed for package management and updates.
- The R GUI is designed for use with scripts.

The R GUI does not provide a "statistics package." R is a language and programming environment. You can download an R package called `Rcmdr` that provides a point-and-click interface for introductory statistics, if you really, really want to. In my experience, students who plan to use statistics in their

[2] I find this when I google "r 'contributed documentation'."

[3] Try googling 'R Reference Card' in quotes.

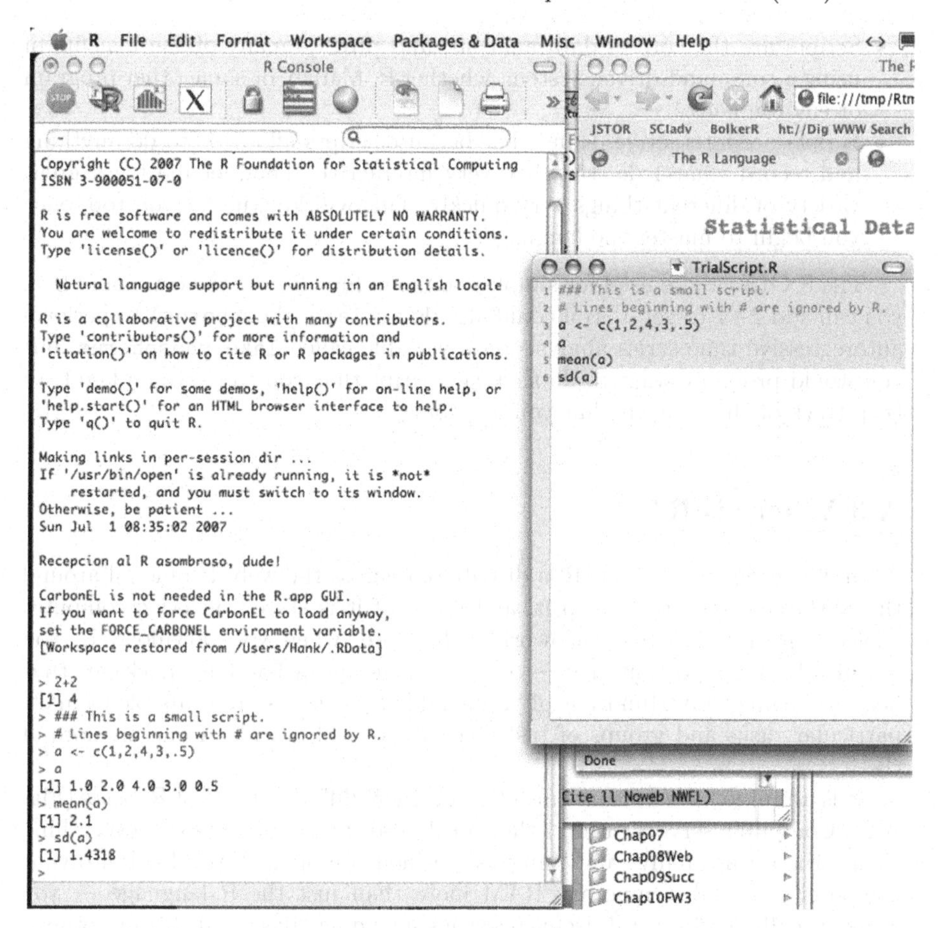

Fig. A.1: The Mac OS X R GUI. Color coded syntax not visible in this figure.

research find it more frustrating to learn this interface than to learn to take advantage of the language.

The R GUI *is* designed for maintenance. With the R GUI you can check for updates, and download any of the hundreds of small packages that extend R in hundreds of ways. (A package is not unlike a "PROC," for SAS users — first you tell call it, then you use it).

The R GUI *is* designed for using scripts. Scripts are text files that contain your analysis; that is, they contain both code to do stuff, and comments about what you are doing and why. These are opened within R and allow you to do work and save the code.

- Scripts are NOT Microsoft Word documents that require Microsoft Word to open them, but rather, simple text files, such as one could open in Notepad, or SimpleText.
- Scripts are a written record of everything you do along the way to achieving your results.

- Scripts are the core of data analysis, and provide many of the benefits of using a command-driven system, whether R, Matlab, or some other program or environment.
- Scripts are interactive. I find that because scripts allow me to do anything and record what I do, they are very interactive. They let me try a great variety of different things very quickly. This will be true for you too, once you begin to master the language.

The R GUI can be used for simple command line use. At the command line, you can add 2+2 to get 4. You could also do `ar(lake.phosphorus)` to perform autoregressive time series analysis on a variable called `lake.phosphorus`, but you would probably want to do that in a script that you can save and edit, to keep track of the analysis that you are doing.

A.3 Where is R?

As an Open Source project, R is distributed across the web. People all around the world continue to develop it, and much of it is stored on large computer servers ("mirrors") across the world. Therefore, when you download R, you download only a portion of it — the language and a few base packages that help everything run. Hundreds of "value-added" packages are available to make particular tasks and groups of tasks easier. We will download one or more of these.

It is useful to have a clear conception of where different parts of R reside (Fig. A.2). Computer servers around the world store identical copies of everything (hence the terms archive and "mirrors"). When you open R, you load into your computer's virtual, temporary RAM more than just the R language — you automatically load several useful packages including "base," "stat," and others. Many more packages exist (about a dozen come with the normal download) and hundreds are available at each mirror. These are easily downloaded through the R GUI.

A.4 Starting at the Very Beginning

To begin with, we will go through a series of steps that will get you up and running using a script file with which to interact with R, and using the proper working directory. Start here.

1. Create a new directory (i.e., a folder) in "Documents" (Mac) or "My Documents" (Windows). *and call it* "Rwork." For now, calling it the same thing as everyone else will just simplify your life. If you put "Rwork" elsewhere, adjust the relevant code as you go. For now, keep all of your script files and output files into that directory.
2. Open the R GUI in a manner that is appropriate for your operating system and setup (e.g., double-click the desktop icon).

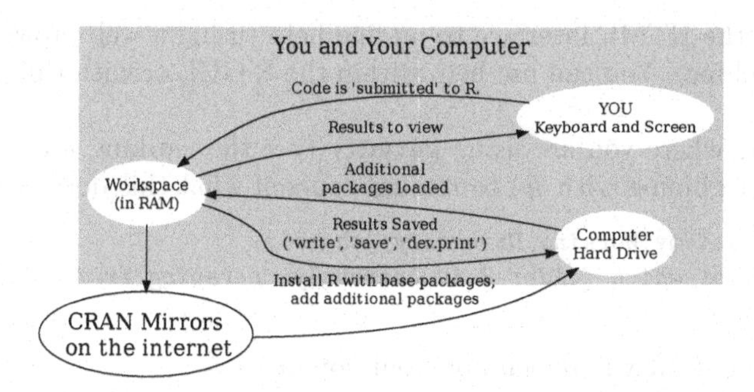

Fig. A.2: A conceptual representation of where R exists. "CRAN" stands for "Comprehensive R Archive Network." "RAM" (random access memory) is your computer's active brain; R keeps some stuff floating in this active memory and this "stuff" is the *workspace*.

3. Set the *working directory*. You can do this via Misc directory in Mac OS X or in the File menu in Windows using "Change dir...." Set the working directory to "Rwork." (If you have not already made an `Rwork` directory, do so now — put it in "Documents" or "My Documents.")
4. Open a new R script ("New Document") by using the File menu in the R GUI. On the first line, type `# My first script` with the pound sign. On the next line, type `setwd('~/Documents/Rwork')` if you are on a Mac, or `setwd('C:/Documents and Settings/Users/Jane/My Documents/Rwork')` on Windows, assuming you are named "Jane;" if not, use the appropriate pathname. Save this file in "Rwork;" save it as "RIntro.R." Windows may hide the fact that it is saving it as a ".txt" file. I will assume you can prevent this.

You should submit code to R directly from the script. *Use the script to store your ideas as comments (beginning with #) and your code, and submit code directly from the script file within R (see below for how to do that).* You do *not* need to cut-and-paste. There are slight differences between operating systems in how to submit code from the script file to the command line.

- In Microsoft Windows, place the cursor on a line in the script file *or* highlight a section of code, and then hit `Ctrl-R` to submit the code.
- In Apple Mac OS X, highlight a section of code and then hit `Command-return` to submit the code (see the Edit menu).

From this point on, enter all of the code (indicated in typewriter font, and beginning with ">") in your script file, save the file with Command-S (Mac) or Ctrl-S (Windows), and then submit the code as described above. Where a line begins with "+," ignore the plus sign, because this is just used by R to indicate continued lines of code. You may enter a single line of code on more than one line. R will simply continue as if you wrote it all on one line.

You can start the help interface with this command.

```
> help.start()
```

This starts the HTML interface to on-line help (using a web browser available at your machine). You can use help within the R GUI, or with this HTML help system.

Find out where you are using getwd() (*get* the *w*orking *d*irectory). Use a comment (beginning with #) to remind yourself what this does.

```
> # Here I Get my Working Directory; that is,
> # I find out which folder R is currently operating from.
> getwd()
```

The precise output will depend on your computer.

You can also *set* the *w*orking *d*irectory using setwd(); if you created a directory called Rwork as specified above, one of the following these should work, depending on your operating system. If these both fail, keep trying, or use the menu in the R GUI to set the working directory.

```
> setwd("~/Documents/Rwork")
```

or

```
> setwd("C:/Documents and Settings/Users/Jane/My Documents/Rwork")
```

On the Mac, a UNIX environment, the tilde-slash (~ /) represents your home directory.

I urge you to use setwd *at the beginning of each script file you write so that this script always sets you up to work in a particular, specified directory.* As you write your script file, remember,

- Text that begins with "#" will be ignored by R.
- Text that does not make sense to R will cause R to return an error message, but will not otherwise mess things up.

Remember that a strength of R is that you can provide, to yourself and others, comments that explain every step and every object. To add a comment, simply type one or more "#," followed by your comment.

Finally, have fun, be amused. Now you are ready for programming in R.

R is a language. Use it every day, and you will learn it quickly.

B

Programming in R

This material assumes you have completed Appendix A, the overview of R. Do the rest of this Appendix in a script. Make comments of your own throughout your script.

Open and save a new file (or *script*) in R. Think of your script as a pad of paper, on which you program and *also on which you write notes to yourself.* See Appendix A for instructions.

You will want to take copious notes about what you do. Make these notes in the script file, using the pound sign #. Here is an example:

```
> # This will calculate the mean of 10 random standard normal variables.
> mean( rnorm( 10 ) )
```

```
[1] 0.1053
```

You submit this (as described above) from the script directly to the Console with (select) Command-r on a Mac, or Ctrl-r on Windows.

B.1 Help

You cannot possibly use R without using its help facilities. R comes with a lot of documentation (see the relevant menu items), and people are writing more all the time (this document is an example!).

After working through this tutorial, you could go through the document "An Introduction to R" that comes with R. You can also browse "Keywords by Topic" which is found under "Search Engine & Keywords" in the Help menu.

To access help for a specific function, try

```
> `?`(mean)
```

```
Help for 'mean' is shown in browser /usr/bin/open ...
Use
        help("mean", htmlhelp = FALSE)
or
        options(htmlhelp = FALSE)
to revert.
```

or

```
> help(mean)
```

```
Help for 'mean' is shown in browser /usr/bin/open ...
Use
        help("mean", htmlhelp = FALSE)
or
        options(htmlhelp = FALSE)
to revert.
```

The help pages provide a very regular structure. There is a *name*, a brief *description*, its *usage*, its *arguments*, *details* about particular aspects of its use, the *value* (what you get when you use the function), *references*, *links* to other functions, and last, *examples*.

If you don't know the exact name of the R function you want help with, you can try

```
> help.search("mean")
> apropos("mean")
```

These will provide lists of places you can look for functions related to this keyword.

Last, a great deal of R resides in *packages* online (on duplicate servers around the world). If you are on line, help for functions that you have not yet downloaded can be retrieved with

```
> RSiteSearch("violin")
> RSiteSearch("violin", restrict = c("functions"))
```

To learn more about help functions, combine them!

```
> help(RSiteSearch)
```

B.2 Assignment

In general in R, we perform an action, and take the results of that action and *assign* the results to a new object, thereby creating a new object. Here I add two numbers and assign the result to an new object I call a.

```
> a <- 2 + 3
> a
```

```
[1] 5
```

Note that I use an arrow to make the assignment — I make the arrow with a less-than sign, <, and a dash. Note also that to reveal the contents of the object, I can type the name of the object.

I can then use the new object to perform another action, and assign

```
> b <- a + a
```

I can perform two actions on one line by separating them with a semicolon.

```
> a + a; a + b
```

```
[1] 10
[2] 15
```

Sometimes the semicolon is referred to as an "end of line" operator.

B.3 Data Structures

We refer to a single number as a scalar; a scalar is a single real number. Most objects in R are more complex. Here we describe some of these other objects: vectors, matrices, data frames, lists, and functions.

B.3.1 Vectors

Perhaps the fundamental unit of R is the *vector*, and most operations in R are performed on vectors. A vector may be just a column of scalars, for instance; this would be a column vector.

Here we create a vector called `Y`.

To enter data directly into R, we will use `c()` and create an R object, in particular a vector. A vector is, in this case, simply a group of numbers arranged in a row or column. Type into your script

```
> Y <- c(8.3, 8.6, 10.7, 10.8, 11, 11, 11.1, 11.2,
+      11.3, 11.4)
```

where the arrow is a less-than sign, <, and a dash, -. Similarly, you could use

```
> Y = c(8.3, 8.6, 10.7, 10.8, 11, 11, 11.1, 11.2, 11.3,
+      11.4)
```

These are equivalent.

R operates (does stuff) to *objects*. Those objects may be *vectors*, *matrices*, *lists*, or some other class of object.

Sequences

I frequently want to create ordered sequences of numbers. R has a shortcut for sequences of integers, and a slightly longer method that is completely flexible. First, integers:

```
> 1:4
```

```
[1] 1 2 3 4
```

```
> 4:1
```

```
[1] 4 3 2 1
```

```
> -1:3
```

```
[1] -1  0  1  2  3
```

```
> -(1:3)
```

```
[1] -1 -2 -3
```

Now more complex stuff, specifying either the units of the sequence, or the total length of the sequence.

```
> seq(from = 1, to = 3, by = 0.2)
```

```
 [1] 1.0 1.2 1.4 1.6 1.8 2.0 2.2 2.4 2.6 2.8 3.0
```

```
> seq(1, 3, by = 0.2)
```

```
 [1] 1.0 1.2 1.4 1.6 1.8 2.0 2.2 2.4 2.6 2.8 3.0
```

```
> seq(1, 3, length = 7)
```

```
[1] 1.000 1.333 1.667 2.000 2.333 2.667 3.000
```

I can also fill in with repetitive sequences. Compare carefully these examples.

```
> rep(1, 3)
```

```
[1] 1 1 1
```

```
> rep(1:3, 2)
```

```
[1] 1 2 3 1 2 3
```

```
> rep(1:3, each = 2)
```

```
[1] 1 1 2 2 3 3
```

B.3.2 Getting information about vectors

Here we can ask R to tell us about `Y`, getting the length (the number of elements), the mean, the maximum, and a six number summary.

```
> sum(Y)
```

```
[1] 105.4
```

```
> mean(Y)
```

```
[1] 10.54
```

```
> max(Y)
```

```
[1] 11.4
```

```
> length(Y)
```

```
[1] 10
```

```
> summary(Y)
```

```
   Min. 1st Qu.  Median    Mean 3rd Qu.    Max.
    8.3    10.7    11.0    10.5    11.2    11.4
```

A vector could be character, or logical as well, for instance

```
> Names <- c("Sarah", "Yunluan")
> Names

[1] "Sarah"   "Yunluan"

> b <- c(TRUE, FALSE)
> b

[1]  TRUE FALSE
```

Vectors can also be dates, complex numbers, real numbers, integers, or factors. For factors, such as experimental treatments, see section B.3.5. We can also ask R what classes of data these belong to.

```
> class(Y)

[1] "numeric"

> class(b)

[1] "logical"
```

Here we test whether each element of a vector is greater than a particular value or greater than its mean. *When we test an object, we get a logical vector back that tells us, for each element, whether the condition was true or false.*

```
> Y > 10

 [1] FALSE FALSE  TRUE  TRUE  TRUE  TRUE  TRUE  TRUE  TRUE  TRUE

> Y > mean(Y)

 [1] FALSE FALSE  TRUE  TRUE  TRUE  TRUE  TRUE  TRUE  TRUE  TRUE
```

We test using >, <, >=, <=, ==, ! = and other conditions. Here we test whether a vector is equal to a number.

```
> Y == 11

 [1] FALSE FALSE FALSE FALSE  TRUE  TRUE FALSE FALSE FALSE FALSE
```

A test of "not equal to"

```
> Y != 11

 [1]  TRUE  TRUE  TRUE  TRUE FALSE FALSE  TRUE  TRUE  TRUE  TRUE
```

This result turns out to be quite useful, including when we want to *extract* subsets of data.

Algebra with vectors

In R, we can add, subtract, multiply and divide vectors. When we do this, we are really *operating on the elements in the vectors*. Here we add vectors a and b.

```
> a <- 1:3
> b <- 4:6
> a + b
```

```
[1] 5 7 9
```

Similarly, when we multiply or divide, we also operate on each pair of elements in the pair of vectors.

```
> a * b
```

```
[1]  4 10 18
```

```
> a/b
```

```
[1] 0.25 0.40 0.50
```

We can also use scalars to operate on vectors.

```
> a + 1
```

```
[1] 2 3 4
```

```
> a * 2
```

```
[1] 2 4 6
```

```
> 1/a
```

```
[1] 1.0000 0.5000 0.3333
```

What R is doing is *recycling* the scalar (the 1 or 2) as many times as it needs to in order to match the length of the vector. Note that if we try to multiply vectors of unequal length, R performs the operation but may or may not give a warning. Above, we got no warningmessage. However, if we multiply a vector of length 3 by a vector of length 2, R returns a warning.

```
> a * 1:2
```

```
[1] 1 4 3
```

R *recycles the shorter vector* just enough to match the length of the longer vector. The above is the same as

```
> a * c(1, 2, 1)
```

```
[1] 1 4 3
```

On the other hand, if we multiply vectors of length 4 and 2, we get no error, because four is a multple of 2.

```
> 1:4 * 1:2

[1] 1 4 3 8
```

Recycling makes the above the same as the following.

```
> 1:4 * c(1, 2, 1, 2)

[1] 1 4 3 8
```

B.3.3 Extraction and missing values

We can extract or subset elements of the vector.

I extract subsets of data in two basic ways, by

- identifying which rows (or columns) I want (i.e. the first row), or
- providing a *logical* vector (of TRUE's and FALSE's) of the same length as the vector I am subsetting.

Here I use the first method, using a single integer, and a sequence of integers.

```
> Y[1]

[1] 8.3

> Y[1:3]

[1]  8.3  8.6 10.7
```

Now I want to extract all girths greater than the average girth. Although I don't have to, I remind myself what the logical vector looks like, and then I use it.

```
> Y > mean(Y)

 [1] FALSE FALSE  TRUE  TRUE  TRUE  TRUE  TRUE  TRUE  TRUE  TRUE

> Y[Y > mean(Y)]

[1] 10.7 10.8 11.0 11.0 11.1 11.2 11.3 11.4
```

Note that I get back all the values of the vector where the condition was TRUE.

In R, *missing data* are of the type "NA." This means "not available," and R takes this appellation seriously. thus is you try to calculate the mean of a vector with missing data, R resists doing it, because if there are numbers missing from the set, how could it possibly calculate a mean? If you ask it to do something with missing data, the answer will be missing too.

Given that R treats missing data as missing data (and not something to be casually tossed aside), there are special methods to deal with such data. For instance, we can test which elements are missing with a special function, `is.na`.

```
> a <- c(5, 3, 6, NA)
> a

[1]  5  3  6 NA

> is.na(a)
```

```
[1] FALSE FALSE FALSE  TRUE
```

```
> !is.na(a)
```

```
[1]  TRUE  TRUE  TRUE FALSE
```

```
> a[!is.na(a)]
```

```
[1] 5 3 6
```

```
> na.exclude(a)
```

```
[1] 5 3 6
attr(,"na.action")
[1] 4
attr(,"class")
[1] "exclude"
```

Some functions allow you to remove missing elements on the fly. Here we let a function fail with missing data, and then provide three different ways to get the same thing.

```
> mean(a)
```

```
[1] NA
```

```
> mean(a, na.rm = TRUE)
```

```
[1] 4.667
```

```
> d <- na.exclude(a)
> mean(d)
```

```
[1] 4.667
```

Note that R takes missing data seriously. If the fourth element of the set really is missing, I cannot calculate a mean because I don't know what the vector is.

B.3.4 Matrices

A matrix is a two dimensional set of elements, for which *all elements are of the same type*. Here is a character matrix.

```
> matrix(letters[1:4], ncol = 2)
```

```
     [,1] [,2]
[1,] "a"  "c"
[2,] "b"  "d"
```

Here we make a numeric matrix.

```
> M <- matrix(1:4, nrow = 2)
> M
```

```
     [,1] [,2]
[1,]    1    3
[2,]    2    4
```

Note that the matrix is filled in by columns, or *column major order*. We could also do it by rows.

```
> M2 <- matrix(1:4, nrow = 2, byrow = TRUE)
> M2
```

```
     [,1] [,2]
[1,]    1    2
[2,]    3    4
```

Here is a matrix with 1s on the diagonal.

```
> I <- diag(1, nrow = 2)
> I
```

```
     [,1] [,2]
[1,]    1    0
[2,]    0    1
```

The identity matrix plays a special role in matrix algebra; in many ways it is equivalent to the scalar 1. For instance, the inverse of a matrix, $\mathbf{M}$, is $\mathbf{M}^{-1}$, which is the matrix which satisfies the equality $\mathbf{M}\mathbf{M}^{-1} = \mathbf{I}$, where $\mathbf{I}$ is the identity matrix. We solve for the inverse using a few different methods, including

```
> Minv <- solve(M)
> M %*% Minv
```

```
     [,1] [,2]
[1,]    1    0
[2,]    0    1
```

QR decomposition is available (e.g., `qr.solve()`).

Note that R recycles the "1" until the specified number of rows and columns are filled. If we do not specify the number of rows and columns, R fills in the matrix with what you give it (as it did above).

Extraction in matrices

I extract elements of matrices in the same fashion as vectors, but specify both rows and columns.

```
> M[1, 2]
```

```
[1] 3
```

```
> M[1, 1:2]
```

```
[1] 1 3
```

If I leave either rows or columns blank, R returns all rows (or columns).

```
> M[, 2]
```

```
[1] 3 4
```

```
> M[, ]
```

```
     [,1] [,2]
[1,]    1    3
[2,]    2    4
```

Simple matrix algebra

Basic matrix algebra is similar to algebra with scalars, but with a few very important differences. Let us define another matrix.

```
> N <- matrix(0:3, nrow = 2)
> N
```

```
     [,1] [,2]
[1,]    0    2
[2,]    1    3
```

To perform scalar, or element-wsie operations, we have

$$\mathbf{A} = \begin{pmatrix} a\ b \\ c\ d \end{pmatrix}; \ \mathbf{B} = \begin{pmatrix} m\ o \\ n\ \ p \end{pmatrix} \tag{B.1}$$

$$\mathbf{AB} = \begin{pmatrix} am\ bo \\ cn\ dp \end{pmatrix} \tag{B.2}$$

The element-wise operation on these two is the default in R,

```
> M * N
```

```
     [,1] [,2]
[1,]    0    6
[2,]    2   12
```

where the element in row 1, column 1 in M is multiplied by the element in the same position in N.

To perform *matrix mulitplication*, recall from Chap. 2 that,

$$\mathbf{A} = \begin{pmatrix} a\ b \\ c\ d \end{pmatrix}; \ \mathbf{B} = \begin{pmatrix} m\ o \\ n\ \ p \end{pmatrix} \tag{B.3}$$

$$\mathbf{AB} = \begin{pmatrix} (am + bn)\ (ao + bp) \\ (cm + dn)\ (co + dp) \end{pmatrix} \tag{B.4}$$

To perform *matrix mulitplication* in R, we use %*%,

```
> M %*% N
```

```
     [,1] [,2]
[1,]    3   11
[2,]    4   16
```

Refer to Chapter 2 (or "matrix algebra" at Wikipedia) for why this is so.

Note that matrix multiplication is not commutative, that is, $\mathbf{NM} \neq \mathbf{MN}$. Compare the previous result to

```
> N %*% M
```

```
     [,1] [,2]
[1,]    4    8
[2,]    7   15
```

Note that a *vector* in R is not defined *a priori* as a column matrix or a row matrix. Rather, it is used as either depending on the circumstances. Thus, we can either left multiply or right multiply a vector of length 2 and M.

```
> 1:2 %*% M

     [,1] [,2]
[1,]    5   11

> M %*% 1:2

     [,1]
[1,]    7
[2,]   10
```

If you want to be very, very clear that your vector is really a matrix with one column (a column vector), you can make it thus.

```
> V <- matrix(1:2, ncol = 1)
```

Now when you multiply M by V, you will get the expected sucesses and failure, according to the rules of matrix algebra.

```
> M %*% V

     [,1]
[1,]    7
[2,]   10

> try(V %*% M)
```

R has formal rules about how it converts vectors to matrices on-the-fly, but it is good to be clear on your own.

Other matrix operations are available. Whenever we add or subtract matrices together, or add a matrix and a scalar, it is always element-wise.

```
> M + N

     [,1] [,2]
[1,]    1    5
[2,]    3    7

> M + 2

     [,1] [,2]
[1,]    3    5
[2,]    4    6
```

The transpose of a matrix is the matrix we get when we substitute rows for columns, and columns for rows. To transpose matrices, we use `t()`.

```
> t(M)

     [,1] [,2]
[1,]    1    2
[2,]    3    4
```

More advanced matrix operations are available as well, for singular value decomposition (`svd`), eigenanalysis (`eigen`), finding determinants (`det`), QR decomposition (`qr`), Choleski factorization (`chol`), and related functions. The `Matrix` package was designed to handle with aplomb large sparse matrices.

B.3.5 Data frames

Data frames are two dimensional, a little like spreadsheets and matrices. All columns having exactly the same number of rows. Unlike matrices, each column can be a different data type (e.g., numeric, integer, charactor, complex, imaginary). For instance, the columns of a data frame could contain the names of species, the experimental treatment used, and the dimensions of species traits, as character, factor, and numeric variables, respectively.

```
> dat <- data.frame(species = c("S.altissima", "S.rugosa",
+     "E.graminifolia", "A. pilosus"), treatment = factor(c("Control",
+     "Water", "Control", "Water")), height = c(1.1,
+     0.8, 0.9, 1), width = c(1, 1.7, 0.6, 0.2))
> dat
```

```
         species treatment height width
1    S.altissima   Control    1.1   1.0
2       S.rugosa     Water    0.8   1.7
3 E.graminifolia   Control    0.9   0.6
4     A. pilosus     Water    1.0   0.2
```

We can extract data from data frames just the way we can with matrices.

```
> dat[2, ]
```

```
   species treatment height width
2 S.rugosa     Water    0.8   1.7
```

```
> dat[3, 4]
```

```
[1] 0.6
```

We can test elements in data frames, as here where I test whether each element column 2 is "Water." I then use that to extract rows of data that are associated with this criterion.

```
> dat[, 2] == "Water"
```

```
[1] FALSE  TRUE FALSE  TRUE
```

```
> dat[dat[, 2] == "Water", ]
```

```
     species treatment height width
2   S.rugosa     Water    0.8   1.7
4 A. pilosus     Water    1.0   0.2
```

I could also use the subset function

```
> subset(dat, treatment == "Water")
```

```
     species treatment height width
2   S.rugosa     Water    0.8   1.7
4 A. pilosus     Water    1.0   0.2
```

There are advantages to using data frames which will become apparent.

Factors

Factors are a class of data; as such they could belong above with our discussion of character and logical and numeric vectors. I tend, however, to use them in data frames almost exclusively, because I have a data set that includes a bunch of response variables, and *the factors imposed by my experiment.*

When defining a factor, R by default orders the factor levels in alphabetic order — we can reorder them as we like. Here I demonstrate each piece of code and then use the pieces to make a factor in one line of code.

```
> c("Control", "Medium", "High")

[1] "Control" "Medium"  "High"

> rep(c("Control", "Medium", "High"), each = 3)

[1] "Control" "Control" "Control" "Medium"  "Medium"  "Medium"
[7] "High"    "High"    "High"

> Treatment <- factor(rep(c("Control", "Medium", "High"),
+     each = 3))
> Treatment

[1] Control Control Control Medium  Medium  Medium  High
[8] High    High
Levels: Control High Medium
```

Note that R orders the factor alphabetically. This may be relevant if we do something with the factor, such as when we plot it (Fig. B.1a).

```
> levels(Treatment)

[1] "Control" "High"    "Medium"

> stripchart(1:9 ~ Treatment)
```

Now we can re-specify the factor, telling R the order of the levels we want, taking care to remember that R can tell the difference between upper and lower case (Fig. B.1b). See also the function `relevel`.

```
> Treatment <- factor(rep(c("Control", "Medium", "High"),
+     each = 3), levels = c("Control", "Medium", "High"))
> levels(Treatment)

[1] "Control" "Medium"  "High"

> stripchart(1:9 ~ Treatment)
```

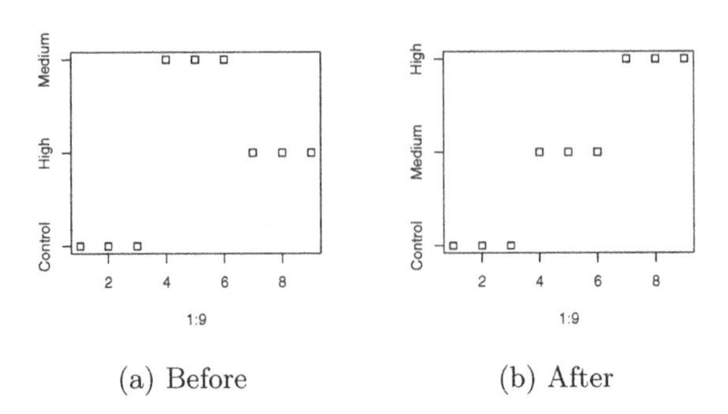

(a) Before (b) After

Fig. B.1: Graphics before and after the factor was releveled to place the factor levels in a logical order.

B.3.6 Lists

An amazing data structure that R boasts is the *list*. A *list* is simply a collection of other objects kept together in a hierarchical structure. Each component of the list can be a complete different class of object. Let's build one.

```
> my.list <- list(My.Y = Y, b = b, Names, Weed.data = dat,
+      My.matrix = M2, my.no = 4)
> my.list

$My.Y
 [1]  8.3  8.6 10.7 10.8 11.0 11.0 11.1 11.2 11.3 11.4

$b
[1] 4 5 6

[[3]]
[1] "Sarah"   "Yunluan"

$Weed.data
         species treatment height width
1    S.altissima   Control    1.1   1.0
2       S.rugosa     Water    0.8   1.7
3 E.graminifolia   Control    0.9   0.6
4     A. pilosus     Water    1.0   0.2

$My.matrix
     [,1] [,2]
[1,]    1    2
[2,]    3    4

$my.no
[1] 4
```

We see that this list is a set of objects: a numeric vector, a logical vector, a character vector, a data frame, a matrix, and a scalar (a number). Lists can be nested within other lists.

Note that if we do not specify a name for a component, we can still extract it using the number of the component.

I extract list components in several ways, including by name, and by number (see ?'[' for more information).

```
> my.list[["b"]]

[1] 4 5 6

> my.list[[2]]

[1] 4 5 6
```

If I use a name, there are a few ways, including

```
> my.list[["b"]]

[1] 4 5 6

> my.list$b

[1] 4 5 6
```

If by number, that are two ways, with one or two brackets. In addition to two brackets, as above, we can use one bracket. This allows for extraction of more than one component of the list.

```
> my.list[1:2]

$My.Y
 [1]  8.3  8.6 10.7 10.8 11.0 11.0 11.1 11.2 11.3 11.4

$b
[1] 4 5 6
```

Note that I can extract a subset of one component.

```
> my.list[["b"]][1]

[1] 4
```

If one way of extraction is working for you, experiment with others.

B.3.7 Data frames are also lists

You can also think of a data frame as a list of columns of identical length. I like to extract columns the same way — by name.

```
> mean(dat$height)

[1] 0.95
```

B.4 Functions

A function is a command that does something. You have already been using functions, throughout this document. Let's examine functions more closely.

Among other things, a function has a name, arguments, and values. For instance,

```
> help(mean)
```

This will open the help page (again), showing us the *arguments*. The first argument `x` is the object for which a mean will be calculated. The second argument is `trim=0`. If we read about this argument, we find that it will "trim" a specified fraction of the most extreme observations of `x`. *The fact that the argument `trim` is already set equal to zero means that is the default.* If you do not use `trim`, then the function will use `trim=0`. Thus, these two are equivalent.

```
> mean(1:4)

[1] 2.5

> mean(1:4, trim = 0)

[1] 2.5
```

R is an "object-oriented" language. A consequence of this is that the same function name will perform different actions, depending on the *class* of the object.[1]

```
> class(1:10)

[1] "integer"

> class(warpbreaks)

[1] "data.frame"

> summary(1:10)

   Min. 1st Qu.  Median    Mean 3rd Qu.    Max. 
   1.00    3.25    5.50    5.50    7.75   10.00 

> summary(warpbreaks)

     breaks      wool   tension
 Min.   :10.0   A:27   L:18   
 1st Qu.:18.2   B:27   M:18   
 Median :26.0          H:18   
 Mean   :28.1                 
 3rd Qu.:34.0                 
 Max.   :70.0                 
```

In the `warpbreaks` data frame, summary provides the six number summary for each numeric or integer column, but provides "tables" of the factors, that is, it counts the occurrences of each level of a factor and sorts the levels. When we use summary on a linear model, we get output of the regression,

[1] R has hundreds of built-in data sets for demonstrating things. We use one here called 'warpbreaks.' You can find some others by typing `data()`.

```
> summary(lm(breaks ~ wool, data = warpbreaks))

Call:
lm(formula = breaks ~ wool, data = warpbreaks)

Residuals:
   Min     1Q Median     3Q    Max
-21.04  -9.26  -3.65   4.71  38.96

Coefficients:
            Estimate Std. Error t value Pr(>|t|)
(Intercept)    31.04       2.50   12.41   <2e-16
woolB          -5.78       3.54   -1.63     0.11

Residual standard error: 13 on 52 degrees of freedom
Multiple R-squared: 0.0488,     Adjusted R-squared: 0.0305
F-statistic: 2.67 on 1 and 52 DF,  p-value: 0.108
```

B.4.1 Writing your own functions

One very cool thing in R is that you can write your own functions. Indeed it is the extensibility of R that makes it the home of cutting edge working, because edge cutters (i.e., leading scientists) can write code that we all can use. People actually write entire *packages*, which are integrated collections of functions, and R has been extended with hundreds of such packages available for download at all the R mirrors.

Let's make our own function to calculate a mean. Let's further pretend you work for an unethical boss who wants you to show that average sales are higher than they really are. Therefore your function should provide a mean plus 5%.

```
> MyBogusMean <- function(x, cheat = 0.05) {
+     SumOfX <- sum(x)
+     n <- length(x)
+     trueMean <- SumOfX/n
+     (1 + cheat) * trueMean
+ }
> RealSales <- c(100, 200, 300)
> MyBogusMean(RealSales)

[1] 210
```

Thus a function can take any input, do stuff, including produce graphics, or interact with the operating system, or manipulated numbers. You decide on the arguments of the function, in this case, `x` and `cheat`. Note that we supplied a number for `cheat`; this results in the `cheat` argument having a *default* value, and we do not have to supply it. If an argument does not have a default, we have to supply it. If there is a default value, we can change it. Now try these.

```
> MyBogusMean(RealSales, cheat = 0.1)

[1] 220
```

```
> MyBogusMean(RealSales, cheat = 0)

[1] 200
```

B.5 Sorting

We often like to sort our numbers and our data sets; a single vector is easy. To do something else is only a little more difficult.

```
> e <- c(5, 4, 2, 1, 3)
> e

[1] 5 4 2 1 3

> sort(e)

[1] 1 2 3 4 5

> sort(e, decreasing = TRUE)

[1] 5 4 3 2 1
```

If we want to sort all the rows of a data frame, keeping records (rows) intact, we can use `order`. This function is a little tricky, so we explore its use in a vector.

```
> e

[1] 5 4 2 1 3

> order(e)

[1] 4 3 5 2 1

> e[order(e)]

[1] 1 2 3 4 5
```

Here `order` generates an *index* to properly *order* something. Above, this index is used to tell R to select the 4th element of `e` first — `order` puts the number '4' into the first spot, indicating that R should put the 4th element of `e` first. Next, it places '3' in the second spot because the 3rd element of `e` belongs in the 2nd spot of an ordered vector, and so on.

We can use `order` to sort the rows of a data frame. Here I order the rows of the data frame according to increasing order of plant heights.

```
> dat

         species treatment height width
1    S.altissima   Control    1.1   1.0
2       S.rugosa     Water    0.8   1.7
3 E.graminifolia   Control    0.9   0.6
4     A. pilosus     Water    1.0   0.2

> order.nos <- order(dat$height)
> order.nos
```

```
[1] 2 3 4 1
```

This tells us that to order the rows, we have to use the 2nd row of the original data frame as the first row in the ordered data frame, the 3rd row as the new second row, etc. Now we use this index to select the rows of the original data frame in the correct order to sort the whole data frame.

```
> dat[order.nos, ]

          species treatment height width
2        S.rugosa     Water    0.8   1.7
3 E.graminifolia   Control    0.9   0.6
4      A. pilosus     Water    1.0   0.2
1     S.altissima   Control    1.1   1.0
```

We can reverse this too, of course.

```
> dat[rev(order.nos), ]

          species treatment height width
1     S.altissima   Control    1.1   1.0
4      A. pilosus     Water    1.0   0.2
3 E.graminifolia   Control    0.9   0.6
2        S.rugosa     Water    0.8   1.7
```

B.6 Iterated Actions: the apply Family and Loops

We often want to perform an action again and again and again..., perhaps thousands or millions of times. In some cases, each action is independent — we just want to do it a lot. In these cases, we have a choice of methods. Other times, each action depends on the previous action. In this case, I always use for-loops.[2] Here I discuss first methods that work only for independent actions.

B.6.1 Iterations of independent actions

Imagine that we have a matrix or data frame and we want to do the same thing to each column (or row). For this we use `apply`, to "apply" a function to each column (or row). We tell `apply` what data we want to use, we tell it the "margin" we want to focus on, and then we tell it the function. The *margin* is the *side* of the matrix. We describe matrices by their number of rows, then columns, as in "a 2 by 5 matrix," so rows constitute the first margin, and columns constitute the second margin. Here we create a 2×5 matrix, and take the mean of rows, for the first margin. Then we sum the columns for the second margin.

```
> m <- matrix(1:10, nrow = 2)
> m
```

[2] There are other methods we could use. These are discussed by others, under various topics, including "flow control." We use ODE solvers for continuous ordinary differential equations.

```
     [,1] [,2] [,3] [,4] [,5]
[1,]    1    3    5    7    9
[2,]    2    4    6    8   10
```

```
> apply(m, MARGIN = 1, mean)
```

```
[1] 5 6
```

```
> apply(m, MARGIN = 2, sum)
```

```
[1]  3  7 11 15 19
```

See ?rowMeans for simple, and even faster, operations.

Similarly, lapply will "apply" a function to each element of a list, or each column of a data frame, and always returns a list. sapply does something similar, but will simplify the result, to a less complex data structure if possible.

Here we do an independent operation 10 times using sapply, defining a function on-the-fly to calculate the mean of a random draw of five observations from the standard normal distribution.

```
> sapply(1:10, function(i) mean(rnorm(5)))
```

```
 [1] -0.5612 -0.4815 -0.4646  0.7636  0.1416 -0.5003 -0.1171
 [8]  0.2647  0.6404 -0.1563
```

B.6.2 Dependent iterations

Often the repeated actions depend on previous outcomes, as with population growth. Here we provide a couple of examples where we accomplish this with *for loops*.

One thing to keep in mind for *for loops* in R: the computation of this is fastest if we first make a holder for the output. Here I simulate a random walk, where, for instance, we start with 25 individuals at time = 0, and increase or decrease by some amount that is drawn randomly from a normal distribution, with a mean of zero and a standard deviation 2. We will round the "amount" to the nearest integer (the zero-th decimal place). Your output will differ because it is a random process.

```
> gens <- 10
> output <- numeric(gens + 1)
> output[1] <- 25
> for (t in 1:gens) output[t + 1] <- output[t] + round(rnorm(n = 1,
+     mean = 0, sd = 2), 0)
> output
```

```
 [1] 25 29 25 26 28 29 30 32 33 29 30
```

B.7 Rearranging and Aggregating Data Frames

B.7.1 Rearranging or reshaping data

We often need to rearrange our data. A common example in ecology is to collect repeated measurements of an experimental unit and enter the data into multiple columns of a spreadsheet, creating a *wide* format. R prefers to analyze data in a single column, in a *long* format. Here we use **reshape** to rearrange this.

These data are carbon dioxide uptake in 12 individual plants. They are currently structured as longitudinal data; here we rearrange them in the wide format, as if we record uptake seven sequential observations on each plant in different columns. See **?reshape** for details. Here **v.names** refers to the column name of the response variable, **idvar** refers to the column name for the variable that identifies an individual on which we have repeated measurements, and **timevar** refers to the column name which identifies different observations of *the same individual plant.*

```
> summary(CO2)
```

```
     Plant             Type         Treatment           conc     
 Qn1    : 7   Quebec     :42   nonchilled:42   Min.   :  95  
 Qn2    : 7   Mississippi:42   chilled   :42   1st Qu.: 175  
 Qn3    : 7                                    Median : 350  
 Qc1    : 7                                    Mean   : 435  
 Qc3    : 7                                    3rd Qu.: 675  
 Qc2    : 7                                    Max.   :1000  
 (Other):42                                                  
     uptake     
 Min.   : 7.7  
 1st Qu.:17.9  
 Median :28.3  
 Mean   :27.2  
 3rd Qu.:37.1  
 Max.   :45.5  
```

```
> CO2.wide <- reshape(CO2, v.names = "uptake", idvar = "Plant",
+     timevar = "conc", direction = "wide")
> names(CO2.wide)
```

```
 [1] "Plant"       "Type"        "Treatment"   "uptake.95"  
 [5] "uptake.175"  "uptake.250"  "uptake.350"  "uptake.500" 
 [9] "uptake.675"  "uptake.1000"
```

This is often how we might record data, with an experimental unit (individual, or plot) occupying a single row. If we import the data in this format, we would typically like to reorganize it in the long format, because most analyses we want to do may require this. Here, **v.names** and **timevar** are the names we want to use for some new columns, for the response variable and the identifier of the repeated measurement (typically the latter may be a time interval, but here it is a CO_2 concentration). **times** supplies the identifier for each repeated observation.

```
> CO2.long <- reshape(CO2.wide, v.names = "Uptake",
+     varying = list(4:10), timevar = "Concentration",
+     times = c(95, 175, 250, 350, 500, 675, 1000))
> head(CO2.long)
```

```
     Plant   Type  Treatment Concentration Uptake id
1.95   Qn1 Quebec nonchilled            95   16.0  1
2.95   Qn2 Quebec nonchilled            95   13.6  2
3.95   Qn3 Quebec nonchilled            95   16.2  3
4.95   Qc1 Quebec    chilled            95   14.2  4
5.95   Qc2 Quebec    chilled            95    9.3  5
6.95   Qc3 Quebec    chilled            95   15.1  6
```

If we wanted to, we could use `order()` to re-sort the data frame, for instance to match the original.

```
> CO2.long2 <- with(CO2.long, CO2.long[order(Plant,
+     Concentration), ])
> head(CO2.long2)
```

```
      Plant   Type  Treatment Concentration Uptake id
1.95    Qn1 Quebec nonchilled            95   16.0  1
1.175   Qn1 Quebec nonchilled           175   30.4  1
1.250   Qn1 Quebec nonchilled           250   34.8  1
1.350   Qn1 Quebec nonchilled           350   37.2  1
1.500   Qn1 Quebec nonchilled           500   35.3  1
1.675   Qn1 Quebec nonchilled           675   39.2  1
```

See also the very simple functions `stack` and `unstack`.

B.7.2 Summarizing by groups

We often want to summarize a column of data *by groups* identified in another column. Here I summarize CO_2 uptake by the means of each experimental treatment, chilling. The code below provides the column to be summarized (`uptake`), a vector (or list of vectors) containing the group id's, and the function to use to summarize each subset (means). We calculate the mean CO_2 uptake for each group.

```
> tapply(CO2[["uptake"]], list(CO2[["Treatment"]]),
+     mean)
```

```
nonchilled    chilled
     30.64      23.78
```

We can get fancier, as well, with combinations of groups, for each combination of Type and Treatment.

```
> tapply(CO2[["uptake"]], list(CO2[["Treatment"]],
+     CO2[["Type"]]), sd)
```

```
           Quebec Mississippi
nonchilled  9.596       7.402
chilled     9.645       4.059
```

We can also define a function on-the-fly to calculate both mean and standard deviation of Type and Treatment combination. We will need, however, to define groups differently, by creating the interaction of the two factors.

```
> tapply(CO2[["uptake"]], list(CO2[["Treatment"]],
+     CO2[["Type"]]), function(x) c(mean(x), sd(x)))

           Quebec    Mississippi
nonchilled Numeric,2 Numeric,2
chilled    Numeric,2 Numeric,2
```

See also by that actually uses tapply to operate on data frames.

When we summarize data, as in tapply, we often want the result in a nice neat data frame. The function aggregate does this. Its use is a bit like tapply — you provide (i) the numeric columns of a data frame, or a matrix, (ii) a list of named factors by which to organize the responses, and then (iii) the function to summarize (or aggregate) the data. Here we summarize both concentration and uptake.

```
> aggregate(CO2[, 4:5], list(Plant = CO2[["Plant"]]),
+     mean)

   Plant conc uptake
1    Qn1  435  33.23
2    Qn2  435  35.16
3    Qn3  435  37.61
4    Qc1  435  29.97
5    Qc3  435  32.59
6    Qc2  435  32.70
7    Mn3  435  24.11
8    Mn2  435  27.34
9    Mn1  435  26.40
10   Mc2  435  12.14
11   Mc3  435  17.30
12   Mc1  435  18.00
```

A separate package entitled reshape supplies some very elegant and intuitive approaches to the sorting, reshaping and aggregating of data frames. I typically use the reshape *package* (with functions melt and cast), rather than the reshape function supplied in the stat. I do so merely because I find it a little more intuitive. R also has strong connections to relational database systems such as MySQL.

B.8 Getting Data out of and into the Workspace

We often want to get data into R, and we sometimes want to get it out, as well. Here we start with the latter (referred to as *writing* data), and finish with the former (referred to as *reading* data).

Here I create a data frame of numbers, and write it to a text file in two different formats. The first is a file where the observations in each row are separated by tabs, and the second separates them by commas.

```
> dat <- data.frame(Name = rep(c("Control", "Treatment"),
+     each = 5), First = runif(10), Second = rnorm(1))
> write.table(dat, file = "dat.txt")
> write.csv(dat, file = "dat.csv")
```

Open these in a spreadsheet such as Calc (in OpenOffice and NeoOffice). We can then read these into R using the `read.*` family of functions.

```
> dat.new <- read.csv("dat.csv")
> dat.new2 <- read.table("dat.txt", header = TRUE)
```

These objects will both be data frames.

Now let's get a statistical summary and export that.

```
> mod.out <- summary(aov(First ~ Name, data = dat))
> mod.out[[1]]
```

```
            Df Sum Sq Mean Sq F value Pr(>F)
Name         1 0.1562  0.1562    4.44  0.068
Residuals    8 0.2814  0.0352
```

```
> write.csv(mod.out[[1]], "ModelANOVA.csv")
```

Open this in a spreadsheet, such as Calc, in OpenOffice, or in any other application.

See also the `xtable` package for making tables in LaTeX or HTML formats.

B.9 Probability Distributions and Randomization

R has a variety of probability distributions built-in. For the normal distribution, for instance, there are four functions:

`dnorm` The probability density function, that creates the widely observed bell-shaped curve.
`pnorm` The cumulative probability function that we usually use to describe the probability that a test statistic is greater than or equal to a critical value.
`qnorm` The quantile function that takes probabilities as input.
`rnorm` A random number generator which draws values (quantiles) from a distribution with a specified mean and standard deviation.

For each of these, default parameter values return the standard normal distribution ($\mu = 0$, $\sigma = 1$), but these parameters can be changed.

Here we have the 95% confidence intervals.

```
> qnorm(p = c(0.025, 0.975))
```

```
[1] -1.96  1.96
```

Next we create a histogram using 20 random draws from a normal distribution with a mean of 11 and a standard deviation of 6; we overlay this with the probability density function (Fig. B.2).

```
> myplot <- hist(rnorm(20, m = 11, sd = 6), probability = TRUE)
> myplot

$breaks
[1]  0  5 10 15 20 25

$counts
[1] 1 8 6 4 1

$intensities
[1] 0.01 0.08 0.06 0.04 0.01

$density
[1] 0.01 0.08 0.06 0.04 0.01

$mids
[1]  2.5  7.5 12.5 17.5 22.5

$xname
[1] "rnorm(20, m = 11, sd = 6)"

$equidist
[1] TRUE

attr(,"class")
[1] "histogram"

> lines(myplot$mids, dnorm(myplot$mids, m = 11, sd = 6))
```

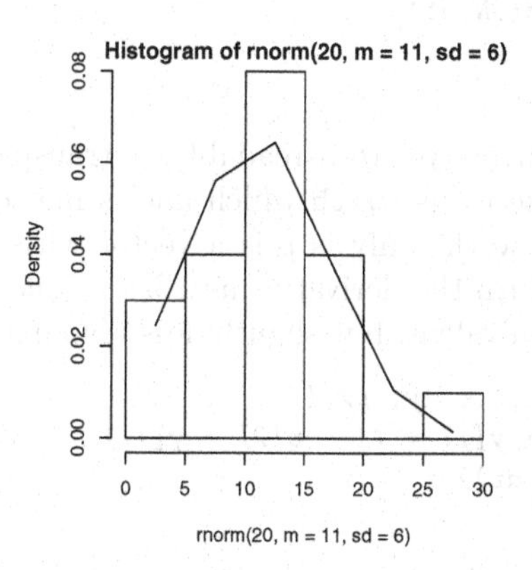

Fig. B.2: Histogram of random numbers drawn from a normal distribution with $\mu = 11$ and $\sigma = 6$. The normal probability density function is drawn as well.

B.10 Numerical integration of ordinary differential equations

In order to study continuous population dynamics, we often would like to integrate complex nonlinear functions of population dynamics. To do this, we need to use numerical techniques that turn the infinitely small steps of calculus, $\mathrm{d}x$, into very small, but finite steps, in order to approximate the change in y, given the change in x, or $\mathrm{d}y/\mathrm{d}x$. Mathematicians and computer scientists have devised very clever ways of doing this very accurately and precisely. In R, the best package for this is `deSolve`, which contains several *solvers* for differential equations that perform numerical integration. We will access these solvers (i.e. numerical integraters) using the `ode` function in the `deSolve` package. This function, `ode`, is a "wrapper" for the underlying suite of functions that do the work. That is, it provides a simple way to use any one of the small suite of functions.

When we have an ordinary differential equation (ODE) such as logistic growth,[3] we say that we "solve" the equation for a particular time interval given a set of parameters and initial conditions or initial population size. For instance, we say that we solve the logistic growth model for time at $t = 0, 1 \ldots 20$, with parameters $r = 1$, $\alpha = 0.001$, and $N_0 = 10$.

Let's do an example with `ode`, using logistic growth. We first have to define a function in a particular way. The arguments for the function must be time, a vector of populations, and a vector or list of model parameters.

```
> logGrowth <- function(t, y, p) {
+     N <- y[1]
+     with(as.list(p), {
+         dN.dt <- r * N * (1 - a * N)
+         return(list(dN.dt))
+     })
+ }
```

Note that I like to convert y into a readable or transparent state variable (N in this case). I also like to use `with` which allows me to use the names of my parameters [157]; this works only is `p` is a vector with named paramters (see below). Finally, we return the derivative as a list of one component.

The following is equivalent, but slightly less readable or transparent.

```
> logGrowth <- function(t, y, p) {
+     dN.dt <- p[1] * y[1] * (1 - p[2] * y[1])
+     return(list(dN.dt))
+ }
```

To solve the ODE, we will need to specify parameters, and initial conditions. Because we are using a vector of named parameters, we need to make sure we name them! We also need to supply the time steps we want.

```
> p <- c(r = 1, a = 0.001)
> y0 <- c(N = 10)
> t <- 1:20
```

[3] *e.g.* $dN/dt = rN(1 - \alpha N)$

Now you put it all into ode, with the correct arguments. The output is a matrix, with the first column being the time steps, and the remaining being your state variables. First we load the deSolve package.

```
> library(deSolve)
> out <- ode(y = y0, times = t, func = logGrowth, parms = p)
> out[1:5, ]

     time      N
[1,]    1  10.00
[2,]    2  26.72
[3,]    3  69.45
[4,]    4 168.66
[5,]    5 355.46
```

If you are going to model more than two species, y becomes a vector of length 2. Here we create a function for Lotka-Volterra competition, where

$$\frac{dN_1}{dt} = r_1 N_1 (1 - \alpha_{11} N_1 - \alpha_{12} N_2) \tag{B.5}$$

$$\frac{dN_2}{dt} = r_2 N_2 (1 - \alpha_{22} N_2 - \alpha_{21} N_1) \tag{B.6}$$

(B.7)

```
> LVComp <- function(t, y, p) {
+     N <- y
+     with(as.list(p), {
+         dN1.dt <- r[1] * N[1] * (1 - a[1, 1] * N[1] -
+             a[1, 2] * N[2])
+         dN2.dt <- r[2] * N[2] * (1 - a[2, 1] * N[1] -
+             a[2, 2] * N[2])
+         return(list(c(dN1.dt, dN2.dt)))
+     })
+ }
```

Note that LVComp assumes that N and r are vectors, and the competition coefficients are in a matrix. For instance, the function extracts the the first element of r for the first species (r[1]); for the intraspecific competition coefficient for species 1, it uses the element of a that is in the first column and first row (a[1,1]). The vector of population sizes, N, contains one value for each population *at one time point.* Thus here, the vector contains only two elements (one for each of the two species); it holds only these values, but will do so repeatedly, at each time point. Only the output will contain all of the population sizxes through time.

To integrate these populations, we need to specify new initial conditions, and new parameters for the two-species model.

```
> a <- matrix(c(0.02, 0.01, 0.01, 0.03), nrow = 2)
> r <- c(1, 1)
> p2 <- list(r, a)
> N0 <- c(10, 10)
```

```
> t2 <- c(1, 5, 10, 20)
> out <- ode(y = N0, times = t2, func = LVComp, parms = p2)
> out[1:4, ]

     time     1     2
[1,]    1 10.00 10.00
[2,]    5 35.54 21.80
[3,]   10 39.61 20.36
[4,]   20 39.99 20.01
```

The ode function uses a superb ODE solver, lsoda, which is a very powerful, well tested tool, superior to many other such solvers. In addition, it has several bells and whistles that we will not need to take advantage of here, although I will mention one, hmax. This tells lsoda the largest step it can take. Once in a great while, with a very *stiff* ODE (a very wiggly complex dynamic), ODE assumes it can take a bigger step than it should. Setting hmax to a smallish number will limit the size of the step to ensure that the integration proceeds as it should.

One of the other solvers in the deSolve, lsodar, will also return roots (or equilibria), for a system of ODEs, if they exist. Here we find the roots (i.e. the solutions, or equilibria) for a two species enemy-victim model.

```
> EV <- function(t, y, p) {
+     with(as.list(p), {
+         dv.dt <- b * y[1] * (1 - 0.005 * y[1]) -
+             a * y[1] * y[2]
+         de.dt <- a * e * y[1] * y[2] - s * y[2]
+         return(list(c(dv.dt, de.dt)))
+     })
+ }
```

To use lsodar to find equilibria, we need to specify a root finding function whose inputs are are the sme of the ODE function, and which returns a scalar (a single number) that determines whether the rate of change (dy/dx) is sufficiently close to zero that we can say that the system has stopped changed, that is, has reached a steady state or equilibrium. Here we sum the absolute rates of change of each species, and then subtract 10^{-10}; if that difference is zero, we decide that, for all pratcial purposes, the system has stopped changing.

```
> rootfun <- function(t, y, p) {
+     dstate <- unlist(EV(t, y, p))
+     return(sum(abs(dstate)) - 1e-10)
+ }
```

Note that unlist changes the list returned by EV into a simple vector, which can then be summed.

Next we specify parameters, and time. Here all we want is the root, so we specify that we want the value of y after a really long time ($t = 10^{10}$). The lsodar function will stop sooner than that, and return the equilibrium it finds, and the time step at which it occurred.

```
> p <- c(b = 0.5, a = 0.02, e = 0.1, s = 0.2)
> t <- c(0, 1e+10)
```

Now we run the function.

```
> out <- ode(y = c(45, 200), t, EV, parms = p, rootfun = rootfun,
+     method = "lsodar")
> out[, ]

      time   1     2
[1,]   0.0  45 200.0
[2,] 500.8 100  12.5
```

Here we see that the steady state population sizes are $V = 100$ and $E = 12.5$, and that given our starting point, this steady state was achieved at $t = 500.8$. Other information is available; see `?lsodar` after loading the `deSolve package`.

B.11 Numerical Optimization

We frequently have a function or a model that we think can describe a pattern or process, but we need to "play around with" the numerical values of the constants in order to make the right shape with our function/model. That is, we need to find the value of the constant (or constants) that create the "best" representation of our data. This problem is known as *optimization*.

Optimization is an entire scientific discipline (or two). It boils down to quickly and efficiently finding parameters (i.e. constants) that meet our criteria. This is what we are doing when we "do" statistics. We fit models to data by telling the computer the structure of the model, and asking it to find values of the constants that minimize the residual error.

Once you have a model of the reality you want to describe, the basic steps toward optimization we consider are (i) create an *objective function*, (ii) use a routine to *minimize* (or *maximize*) the objective function through optimal choice of parameter values, and (iii) see if the "optimal" parameters values make sense, and perhaps refine and interpret them.

An *objective function* compares the data to the predicted values from the model, and returns a quantitative measure of their difference. One widely used objective function the *least-squares criterion*, that is, the objective function is the average or the sum of the squared deviations between the model values and the data — just like a simple ANOVA might. An optimization routine then tries to find model parameters that minimize this criterion.

Another widely used objective function is the likelihood function, or *maximum likelihood*. The likelihood function uses a probability distribution of our choice (often the normal distribution). The objective function then calculates the collective probability of observing those data, given the parameters and fit of the model. In other words, we pretend that the model and the predicted values are true, measure how far off each datum is from the predicted value, and then use a probability distribution to calculate the probability of seeing each datum. It then multiplies all those probabilities to get the *likelihood* of observing those data, given the selected parameters. An optimization routine then tries to find model parameters that maximize this likelihood. In practice, it is more computationally stable to calculate the negative of the sum of the logarithms of the probabilities, and try to minimize that quantity, rather than maximize the likelihood — but in principle they are they same thing.

Once we have an objective function, an optimization routine makes educated guesses regarding good values for the parameters, until it finds the best values it can, those which minimize the objective function. There are a great variety of optimization routines, and they all have their strengths. One important tradeoff they exhibit is that the fastest and most accurate methods are sometimes the least able to handle difficult data [13]. Below, we rely on a combination to take advantage of the strengths of each type.

Here we introduce two of R's general purpose functions in the `base` package, `optimize` and `optim`, and another, in the `bbmle` package, `mle2` [13]. The function `optimize` should be used where we are in search of one parameter; use others when more than one parameter is being optimized. There are many other optimization routines in R, but we start here[4].

Here we start with one of R's general optimization functions, the one designed for finding a single parameter. Let us find the mean ($\bar{x}$) of some data through optimization. We will start with data, y, and let our conceptual model of that data be μ, the mean. We then create a *objective function* whose output will get smaller as the parameter of our model approaches the value we want. Poughly speaking, the mean is the value that minimizes the total difference between all the data and the itself. We will use the least-squares criterion, where the sum of all the squared deviations reaches a minimum when μ approaches the mean.

```
> y <- c(1, 0:10)
> f <- function(y, mu) {
+     sum((y - mu)^2)
+ }
```

Our function, `f`, subtracts μ from each value of y, squares each of these differences, and then sums these squared differences, to get the sum of squares. Our goal is to minimize this. If we guessed at it by hand, we would get this (Fig. B.3).

```
> guesses <- seq(4, 6, by = 0.05)
> LS.criterion <- sapply(guesses, function(mu) f(mu = mu,
+     y = y))
> plot(guesses, LS.criterion, type = "l")
```

Fig. B.3 shows us that the minimum of the objective function occurs when `mu` is a little over 4.5. Now let's let R minimize our least squared deviations. With `optimize`, we provide the function first, we then provide a range of possible values for the parameter of interest, and then give it the values of parameters or data used by the function, other than the parameter we want to fit.

[4] Indeed, all statistical models are fancy optimization routines.

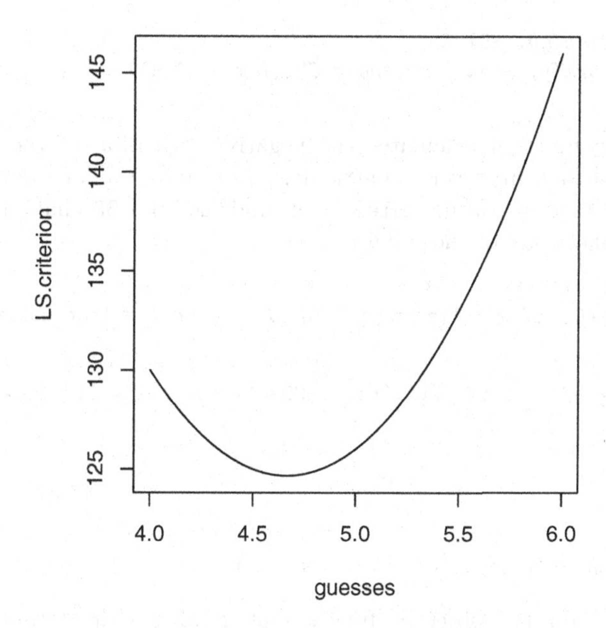

Fig. B.3: Illustration of the least squares criterion. Our objective function returns (i.e. generates) the squared deviations between the fitted model and the data. Optimization minimizes the criterion ("LS.criterion") and thereby finds the right guess (x axis).

```
> (results <- optimize(f, c(0, 10), y = y))

$minimum
[1] 4.667

$objective
[1] 124.7
```

We see that `optimize` returns two components in a list. The first is called `minimum`, which is the parameter value that causes our function `f` to be at a minimum. The second component, `objective` is the value of `f` when mu = 4.667.

Next we demonstrate `mle2`, a function for maximum likelihood estimation. Maximum likelihood relies on probability distributions to find the probability of observing a particular data set, assuming the model is correct. This class of optimization routines finds the parameters that maximize that probability.

Let us solve the same problem as above. For the same data, `y`, we create a maximum likelihood function to calculate the mean. In maximum likelihood, we actually minimize the negative logarithm of the likelihood because it is more computationally stable — the same parameters that minimize the negative log-likelihood also maximize the likelihood. We assume that the data are normally distributed, so it makes sense to assume that the probabilities derive from the normal probability density function.

```
> LL <- function(mu, SD) {
+     -sum(dnorm(y, mean = mu, sd = SD, log = TRUE))
+ }
```

This objective function calculates the negative logarithm of the probability density of each datum, given a particular mean and standard deviation, `mu`, `SD`. The optimization routine, `mle2`, then finds `mu` and `SD` that minimize the negative log-likelihood of those data.

```
> library(bbmle)
> (fit <- mle2(LL, start = list(mu = 5, SD = 1), control = list(maxit = 10^5)))

Call:
mle2(minuslogl = LL, start = list(mu = 5, SD = 1), control = list(maxit = 10^5))

Coefficients:
   mu    SD
4.667 3.223

Log-likelihood: -31.07
```

Another way to put this objective function into `mle2` is with a formula interface.

```
> mle2(y ~ dnorm(mu, sd = SD), start = list(mu = 1,
+     SD = 2))

Call:
mle2(minuslogl = y ~ dnorm(mu, sd = SD), start = list(mu = 1,
    SD = 2))

Coefficients:
   mu    SD
4.667 3.223

Log-likelihood: -31.07
```

We can examine this more closely, examing the probablities associated with the profile confidence intervals.

```
> summary(fit)

Maximum likelihood estimation

Call:
mle2(minuslogl = LL, start = list(mu = 5, SD = 1), control = list(maxit = 10^5))

Coefficients:
   Estimate Std. Error z value   Pr(z)
mu    4.667      0.930    5.02 5.3e-07
SD    3.223      0.658    4.90 9.6e-07

-2 log L: 62.14

> pr <- profile(fit)
```

```
> par(mar = c(5, 4, 3, 2))
> plot(pr)
```

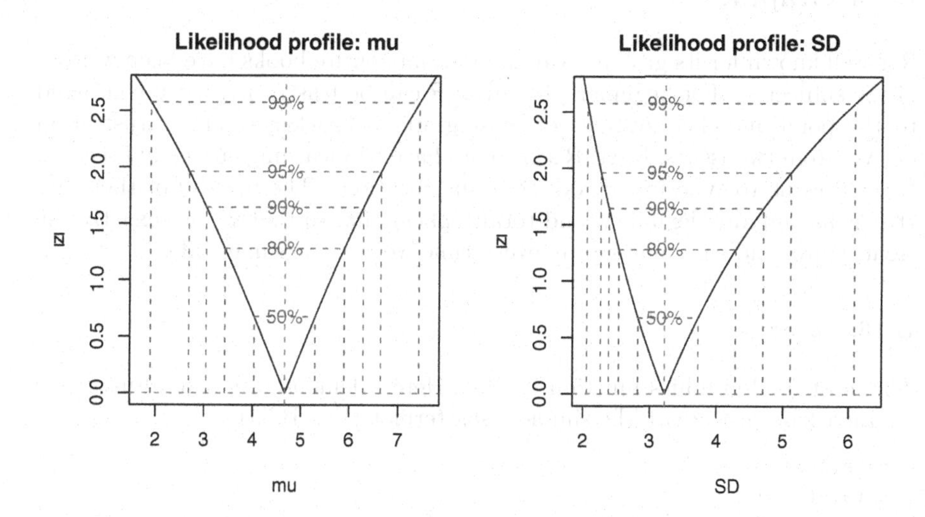

Fig. B.4: Profile confidence intervals for various limits, based on `mle2`.

Often we have reason to limit parameters to particular bounds. Most often, we may need to ensure that a parameter is greater than zero, or less than zero, or less often between zero and one. Sometimes we have a rationale based on physical or biological constraints that will limit a parameter within particular values.

To constrain parameters, we could use a routine that applies constraints directly (see particular optimization methods under `mle2`,`nlminb`, and `optim`). We could also transform the parameters, so that the optimizer uses the transformed version, while the ODE model uses the parameters in their original units. For instance, we could let the optimizer find the best value of a logarithm of our parameter that allows the original parameter to make model predictions that fit the data. Another consequence of using logarithms, rather than the original scale is that it facilitates computational procedures in estimating vary large and very small numbers. An example helps make this clear — see an extended example in Chap. 6, on disease models.

B.12 Derivatives

We can use `deriv` and `D` to have R provides derivatives. First we supply an expression, and then we get gradients.

```
> host1 <- expression(R * H * (1 + a * P)^-k)
> D(host1, "H")
```

```
R * (1 + a * P)^-k
```

B.13 Graphics

R is well known for its graphics capabilities, and entire books have been written of the subject(s). For beginners, however, R can be frustrating when compared to the point-and-click systems of most graphics "packages." This frustration derives from two issues. First, R's graphics have of a learning curve, and second, R requires us to type in, or code, our specifications. The upsides of these are that R has infinite flexibility, and total replicability, so that we get exactly the right figure, and the same figure, every time we run the same code.

B.13.1 plot

The most used graphics function is `plot`. Here I demonstrate several uses.

First let's just create the simplest scatterplot (Fig. B.5a).

```
> data(trees)
> attach(trees)
> plot(Girth, Height)
```

To this we can add a huge variety of variation, using arguments to `plot`.

B.13.2 Adding points, lines and text to a plot

After we have started a plot, we may want to add more data or information. Here set up a new graph without plotting points, add text at each point, then more points, a line and some text.

```
> par(mar = c(5, 4, 3, 2))
> plot(Girth, Volume, type = "n", main = "My Trees")
> points(Girth, Volume, type = "h", col = "lightgrey",
+     pch = 19)
```

Now we want to add points for these data, using the tree heights as the plotting symbol. We are going to use an alternate coloring system, designed with human perception in mind (`hcl`). We scale the colors so that the hue varies between 30 and 300, depending on the height of the tree; I allow the symbols to be transparent (90% opaque) overlapping. I also allow the size of the numbers to vary with height (`cex = 0.5 + hts`) Last, we add a legend (Fig. B.5b).

```
> hts <- (Height - min(Height))/max(Height - min(Height))
> my.colors <- hcl(h = 30 + 270 * hts, alpha = 0.9)
> text(Girth, Volume, Height, col = my.colors, cex = 0.5 +
+     hts)
```

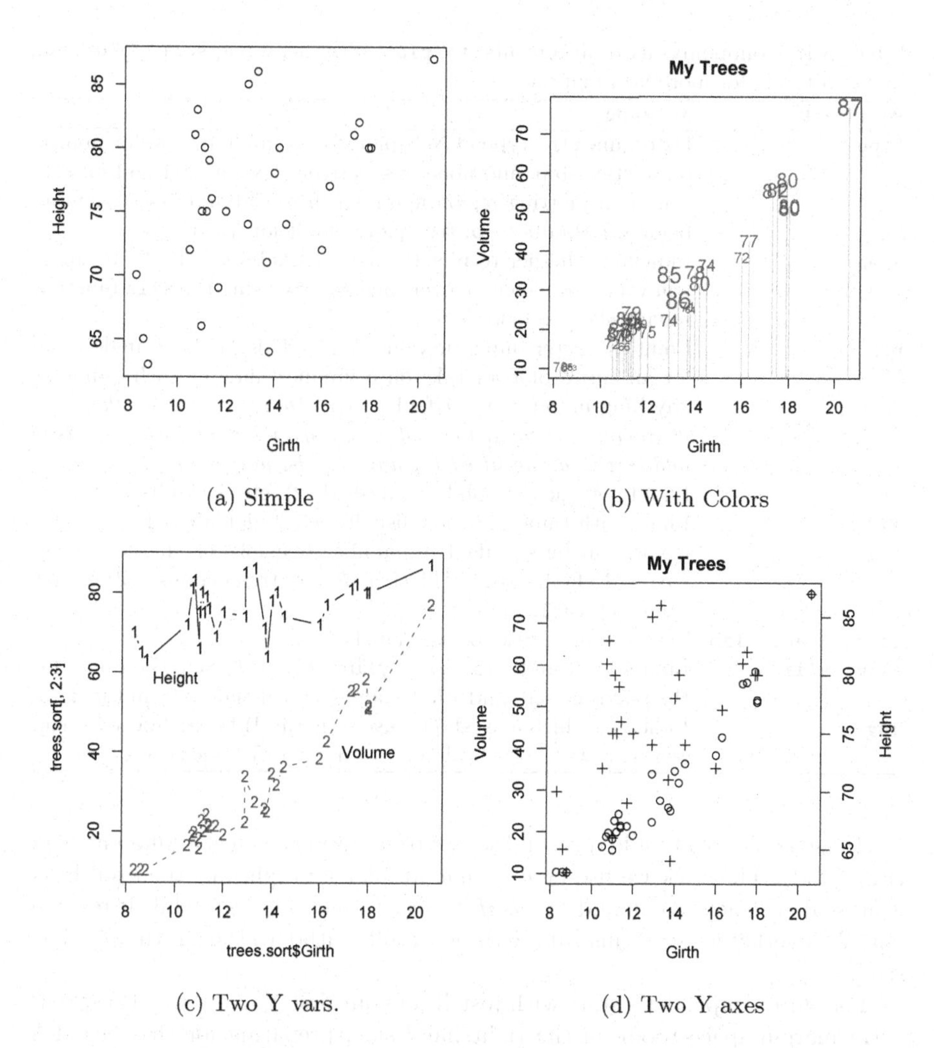

(a) Simple (b) With Colors

(c) Two Y vars. (d) Two Y axes

Fig. B.5: See code for graphics parameters used to generate these plots. Fig. (b) uses an alternate color scheme that provides human perception-adjusted hsv (hue, saturation, and value) specification.

B.13.3 More than one response variable

We often plot more than one response variable on a single axis. We could use `lines` or `points` to add each additional variable. We could also use `matplot` to plot a matrix of variables *vs.* one predictor (Fig. B.5c).

```
> trees.sort <- trees[order(trees$Girth, trees$Height),
+     ]
> matplot(trees.sort$Girth, trees.sort[, 2:3], type = "b")
> text(18, 40, "Volume", col = "darkred")
> text(10, 58, "Height")
```

Table B.1: Commonly used arguments to **plot**. See help pages at **?plot** and **?plot.default** for more information.

Argument	Meaning
`type`	Determines the type of X-Y plot, for example **p**, **l**, **s**, for points, lines, stair-step, and none, respectively. "None" is useful for setting up a plotting region upon which to elaborate (see example below). Defaults to **p**; see ?plot.default for other types.
`axes`	Indicates whether to plot the axes; defaults to TRUE. Useful if you want more control over each axis by using the **axis** function separately (see below).
`pch`	Point character (numeric value, 1–21). This can be a single value for an entire plot, or take on a unique value for each point, or anything in between. Defaults to 1. *To add text more than one character in length, for instance a species name, we can easily add text to a plot at each point (see the next section).*
`lty`	Line type, such as solid (1), dashed (2), etc. Defaults to 1.
`lwd`	Line width (numeric value usually 0.5–3; default is 1).
`col`	Color; can be specified by number (e.g., 2), or character (e.g. "red"). Defaults to 1 ("black"). R has tremendous options for color; see `?hcl`.
`main, ylab, xlab`	Text for main title, or axis labels.
`xlim, ylim`	Limits for x and y axes, e.g. `ylim=c(0, 1.5)` sets the limits for the y-axis at zero and 1.5. Defaults are calcualted from the data.
`log`	Indicates which axes should use a (natural) logarithm scale, e.g. `log = 'xy'` causes both axes to use logarithmic scales.

We frequently want to add a second y-axis to a graph that has a different scale (Fig. B.5d). The trick we use here is that we plot a graph, but then tell R we want to do the next command "*... as if it was on a new device*"[5] while it really is not. We overlay what we just did with new stuff, without clearing the previous stuff.

For our example, let's start with just X and our first Y. Note we also specify extra margin space room on the right hand side, preparing for the second Y axis.

```
> quartz(, 4, 4)
> par(mar = c(5, 4, 2, 4))
> plot(Girth, Volume, main = "My Trees")
```

Now we try our trick. We draw a new plot "as if" it were a new graph. We use the same X values, and the new Y data, and we also specify no labels. We also use a different line type, for clarity.

```
> par(new = TRUE)
> plot(Girth, Height, axes = FALSE, bty = "n", xlab = "",
+      ylab = "", pch = 3)
```

[5] From the **par** help page.

Now we put the new Y values on the fourth side, the right hand Y axis. We add a Y axis label using a function for *marginal text* (Fig. B.5d).

```
> axis(4)
> mtext("Height", side = 4, line = 3)
```

```
> par(mar = c(5, 4, 2, 4))
> plot(Girth, Volume, main = "My Trees")
> par(new = TRUE)
> plot(Girth, Height, axes = FALSE, bty = "n", xlab = "",
+      ylab = "", pch = 3)
> axis(4)
> mtext("Height", side = 4, line = 3)
```

B.13.4 Controlling Graphics Devices

When we make a graph with the `plot` function, or other function, it will typically open a graphics window on the computer screen automatically; if we desire more control, we can use several functions to be more deliberate. We create new graphics "devices" or graphs in several ways, including the functions `windows()` (Microsoft Windows OS), `quartz()` (Mac OS), `x11()` (X11 Window system). For instance, to open a "graphics device" on a Mac computer that is 5 inches wide and 3 inches tall, we write

```
> quartz(width = 5, height = 3)
```

To do the same thing on a computer running Windows, we type

```
> windows(width = 5, height = 3)
```

To control the *par*ameters of the graph, that is, what it looks like, aside from data, we use arguments to the `par` function. Many of these arguments refer to *sides* of the graph. These a numbered 1–4 for the bottom X axis, the left side Y axis, the top, and the right side Y axis. Arguments to `par` are many (see `?par`), and include the following.

`mar` controls the width of margins on each side; units are number of lines of text; defaults to c(5, 4, 4, 2) + 0.1, so the bottom has the most room, and the right hand side has the least room.

`mgp` controls the spacing of the axis title, labels and the actual line itself; units of number of lines of text, and default to c(3, 1, 0), so the axis title sits three lines away from the edge of the plotting region, the axis labels, one line away and the axis line sits at the edge of the plotting region.

`tcl` tick length, as a fraction of the height of a line of text; negative values put the tick marks outside, positive values put the tick marks inside. Defaults to -0.5.

We can build each side of the graph separately by initiating a graph but not plotting axes `plot(..., axes = FALSE)`, and then adding the axes separately. For instance, `axis(1)` adds the bottom axis.

Last, we can use `layout` to make graph with several smaller subgraphs (see also (`mfrow` and `mfcol` arguments to `par` and the function `split.screen`). The function `layout` takes a matrix as its argument, the matrix contains a sequence of numbers that tells R how to fill the regions Graphs can fit in more than one of these regions if indicated by the same number.

Here we create a compound graphic organized on top of a 4×4 grid; it will have two rows, will be be filled in by rows. The first graph will be the upper left, the second the upper right, and the third will fill the third and fourth spots in the second. We will fill each with a slightly different plot of the same data (Fig. B.6).

```
> quartz(, 5, 5)
> layout(matrix(c(1, 2, 3, 3), nrow = 2, byrow = TRUE))
> plot(Girth, Height)
```

Now we add the second and third ones but with different settings.

```
> par(mar = c(3, 3, 1, 1), mgp = c(1.6, 0.2, 0), tcl = 0.2)
> plot(Girth, Height)
> par(mar = c(3, 3, 2, 1), mgp = c(1.6, 0.2, 0), tcl = 0.2)
> plot(Girth, Height, axes = FALSE, xlim = c(8, 22))
> axis(1, tcl = -0.3)
> axis(2, tick = F)
> rug(Height, side = 2, col = 2)
> title("A Third, Very Wide, Plot")
```

B.13.5 Creating a Graphics File

Now that you have made this beautiful thing, I suppose you would like to stick it into a manuscript. One way to get graphics out of R and into something else (presentation software, a manuscript), is to create a graphics device, and then save it with `dev.print` in a format that you like, such as PDF, postscript, PNG, or JPEG.

For instance, we might do this to save a graphics file in our working directory.

```
> getwd()
> quartz(, 4, 4)
> plot(Height, Volume, main = "Tree Data")
> dev.print(pdf, "MyTree.pdf")
```

This should have saved a small PDF figure in your current working directory, returned by `getwd`.

You will have to find your own way to make graphics files that suits your operating system, your preferred applications, and your personality.

B.14 Graphical displays that show distributions

Here we take a quick look at ways to reveal distributions of data. First, two views to see in the Console, a six number summary of quantiles and the mean, and the good ol' stem and leaf plot, a favorite of computational botanists everywhere.

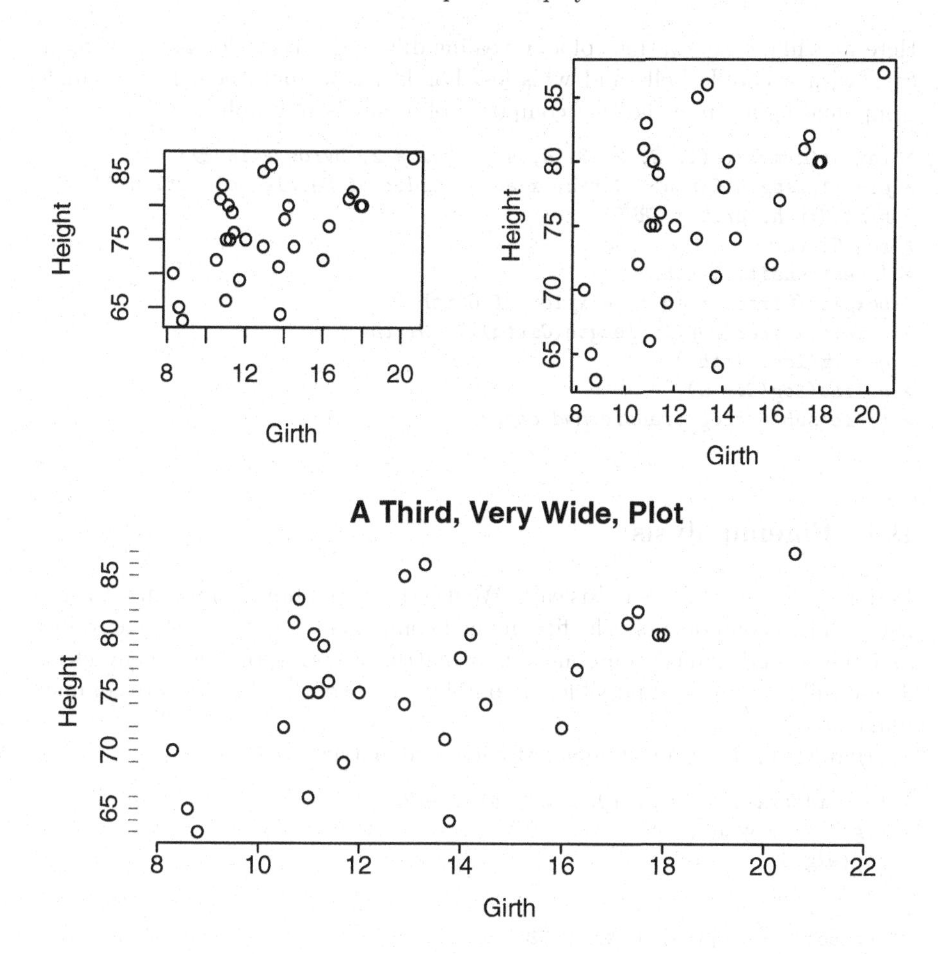

Fig. B.6: A variety of examples with different graphics *par*ameters.

```
> summary(Girth)

   Min. 1st Qu.  Median    Mean 3rd Qu.    Max.
    8.3    11.0    12.9    13.2    15.2    20.6

> stem(Girth)

  The decimal point is at the |

   8 | 368
  10 | 57800123447
  12 | 099378
  14 | 025
  16 | 03359
  18 | 00
  20 | 6
```

Here we will create 4 various plots revealing different ways to look at your data, each with a couple bells and whistles. For kicks, we put them into a single compound figure, in a "layout" composed of a matrix of graphs.

```
> layout(matrix(c(1, 2, 2, 3, 4, 4), nrow = 2, byrow = TRUE))
> plot(1:length(Girth), Girth, xlab = "Order of Sample Collection?")
> hist(Girth, prob = TRUE)
> rug(Girth)
> lines(density(Girth))
> boxplot(Girth, main = "Boxplot of Girth")
> points(jitter(rep(1, length(Girth))), Girth)
> qqnorm(log(Girth))
> qqline(log(Girth))
> title(sub = "Log transformed data")
```

B.15 Eigenanalysis

Performing eigenanalysis in Ris easy. We use the `eigen` function which returns a list with two components. The first named component is a vector of eigenvalues and the second named component is a matrix of corresponding eigenvectors. These will be numeric if possible, or complex, if any of the elements are complex numbers.

Here we have a typical demographic stage matrix.

```
> A <- matrix(c(0, 0.1, 10, 0.5), nrow = 2)
> eig.A <- eigen(A)
> str(eig.A)
```

```
List of 2
 $ values : num [1:2] 1.28 -0.78
 $ vectors: num [1:2, 1:2] -0.9919 -0.127 -0.997 0.0778
```

Singular value decomposition (SVD) is a generalization of eigenanalysis and is used in Rfor some applications where eigenanalysis was used historically, but where SVD is more numerically accurate (`prcomp` for principle components analysis).

B.16 Eigenanalysis of demographic versus Jacobian matrices

Eigenanalyses of demographic and Jacobian matrices are worth comparing. In one sense, they have similar meanings — they both describe the asymptotic (long-term) properties of a system, either population size (demographic matrix) or a perturbation at an equilibrium. The quantitative interpretation of the eigenvalues will therefore differ.

In the case of the stage (or age) structured demographic model, the elements of the demographic matrix are discrete per capita increments of change over a

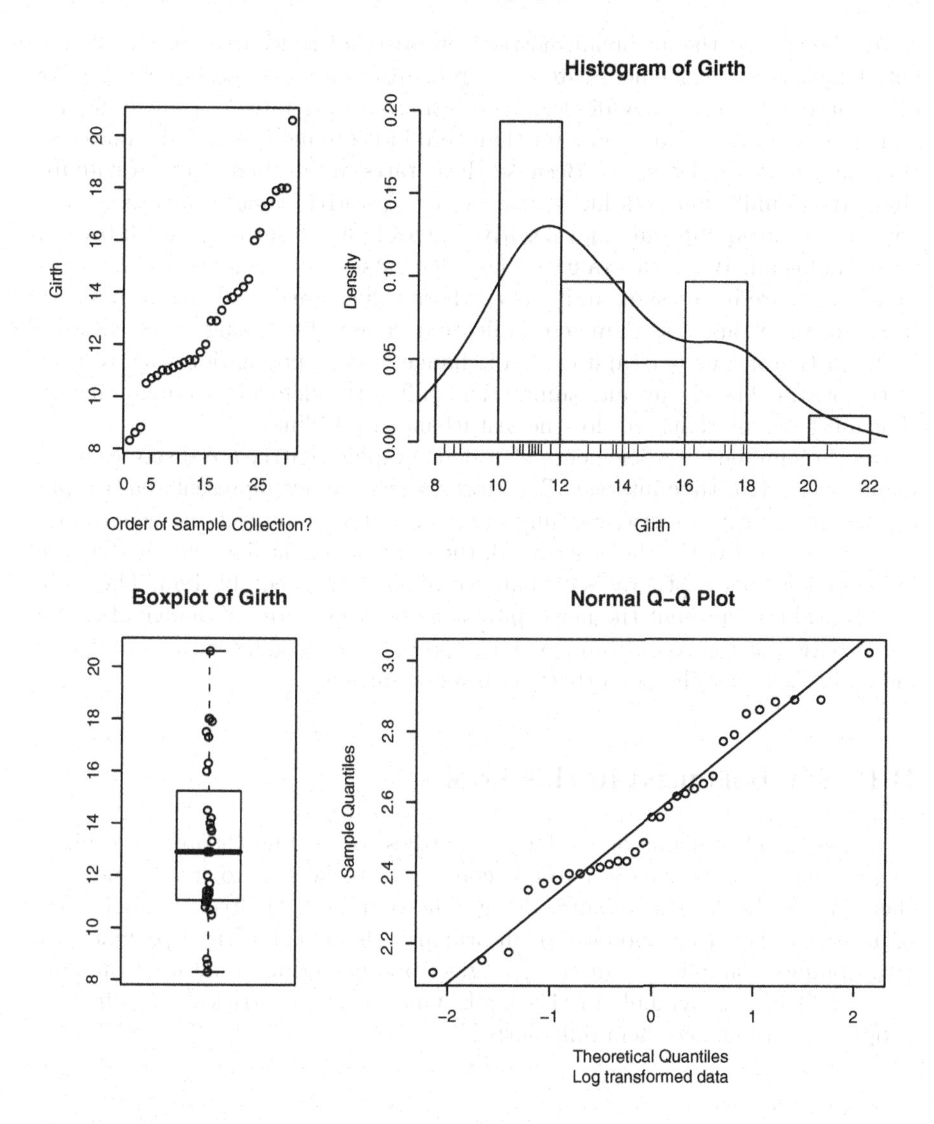

Fig. B.7: Examples of ways to look at the distribution of your data. See `?hist`, for example, for more information.

specified time interval. This is directly analogous to the finite rate of increase, λ, in discrete unstructured models. (Indeed, an unstructured discrete growth model is a stage-structured model with one stage). Therefore, the eigenvalues of a demographic matrix will have the same units — a per capita increment of change. That is why the dominant eigenvalue has to be greater than 1.0 for the population to increase, and less than 1 (not merely less than zero) for the population to decline.

In the case of the Jacobian matrix, comprised of continuous partial differential equations, the elements are per capita *instantaneous* rates of change. As differential equations, they describe the *instantanteous* rates of change, analogous to r. Therefore, values greater than zero indicate increases, and values less than zero indicate decreases. Because these rates are evaluated at an equilibrium, the equilibrium acts like a new zero — positive values indicate growth away from the equilibrium, and negative values indicate shrinkage back toward the equilibrium. When we evaluate these elements at the equilibrium, the numbers we get are in the same units as r, where values greater than zero indicate increase, and values less than zero indicate decrease. The change they describe is the instantaneous per capita rate of change of each population with respect to the others. The eigenvalues summarizing all of the elements Jacobian matrix thus must be less than zero for the disturbance to decline.

So, in summary, the elements of a demographic matrix are discrete increments over a real time interval. Therefore its eigenvalues represent relative per capita growth rates a discrete time interval, and we interpret the eigenvalues with respect to 1.0. On the other hand, the elements of the Jacobian matrix are instantaneous per captia rates of change evaluated at an equilibrium. Therefore its eigenvalues represent the per capita *instantaneous* rates of change of a tiny perturbation at the equilibrium. We interpret the eigenvalues with respect to 0 indicating whether the perturbation grows or shrinks.

B.17 Symbols used in this book

I am convinced that one of the biggest hurdles to learning theoretical ecology — and the one that is easiest to overcome — is to be able to "read" and hear them in your head. This requires being able to pronounce Greek symbols. Few of us learned how to pronounce "α" in primary school. Therefore, I provide here an incomplete simplistic American English pronunciation guide for (some of) the rest of us, for symbols in this book. Only a few are tricky, and different people will pronounce them differently.

Table B.2: Symbols and their pronunciation; occasional usage applies to lowercase, unless otherwise specified. A few symbols have common variants. Any symbol might be part of any equation; ecologists frequently ascribe other meanings which have to be defined each time they are used. See also http://en.wikipedia.org/wiki/Greek_letters

Symbol	Spelling	Pronunciation; occasional or conventional usage
A, α	alpha	al'-fa; point or local diversity (or a parameter in the logseries abundance distribution)
B, β	beta	bay'-ta; turnover diversity
Γ, γ	gamma	gam'-ma; regional diversity
Δ, δ, ∂	delta	del'-ta; change or difference
E, ϵ, ε	epsilon	ep'-si-lon; error
Θ, θ	theta	thay'-ta ("th" as in "thanks"); in neutral theory, biodiversity.
Λ, λ	lambda	lam'-da; eigenvalues, and finite rate of increase
M, μ	mu	meeoo, myou; mean
N, ν	nu	noo, nou
Π, π	pi	pie; uppercase for product (of the elements of a vector)
P, ρ	rho	row (as in "a boat"); correlation
Σ, σ, ς	sigma	sig'-ma; standard deviation (uppercase is used for summation)
T, τ	tau	(sounds like what you say when you stub your toe - "Ow!" but with a "t").
Φ, ϕ	phi	fie, figh
X, χ	chi	kie, kigh
Ψ, ψ	psi	sie, sigh
Ω, ω	omega	oh-may'-ga; degree of omnivory

References

1. D. Alonso, R. S. Etienne, and A. J. McKane. The merits of neutral theory. *Trends in Ecology & Evolution*, 21:451–457, 2006.
2. D. Alonso and A. J. McKane. Sampling hubbell's neutral theory of biodiversity. *Ecology Letters*, 7:901–910, 2004.
3. J. Antonovics and H.M. Alexander. Epidemiology of anther-smut infection of *Silene alba* (= *S. latifolia*) caused by *Ustilago violacea* – patterns of spore deposition in experimental populations. *Proceedings of the Royal Society B-Biological Sciences*, 250:157–163, 1992.
4. R. A. Armstrong. Fugitive species: Experiments with fungi and some theoretical considerations. *Ecology*, 57:953–963, 1976.
5. O. Arrhenius. Species and area. *Journal of Ecology*, 9:95–99, 1921.
6. J. P. Bakker and F. Berendse. Constraints in the restoration of ecological diversity in grassland and heathland communities. *Trends in Ecology & Evolution*, 14:63–68, 1999.
7. F. A. Bazzaz. *Plants in Changing Environments.* Cambridge University Press, Boston, 1996.
8. L. Becks, F. M. Hilker, H. Malchow, K. Jurgens, and H. Arndt. Experimental demonstration of chaos in a microbial food web. *Nature*, 435:1226–1229, 2005.
9. G. Bell. The distribution of abundance in neutral communities. *The American Naturalist*, 155:606–617, 2000.
10. G. Bell. Neutral macroecology. *Science*, 293(5539):2413–2418, 2001.
11. E. L. Berlow, A. M. Neutel, J. E. Cohen, P. C. de Ruiter, B. Ebenman, M. Emmerson, J. W. Fox, V. A. A. Jansen, J. I. Jones, G. D. Kokkoris, D. O. Logofet, A. J. McKane, J. M. Montoya, and O. Petchey. Interaction strengths in food webs: Issues and opportunities. *Journal of Animal Ecology*, 73:585–598, 2004.
12. E. J. Berry, D. L. Gorchov, B. A. Endress, and M. H. H. Stevens. Source-sink dynamics within a plant population: the impact of substrate and herbivory on palm demography. *Population Ecology*, 50:63–77, 2008.
13. B. M. Bolker. *Ecological Models and Data in R.* Princeton University Press, Princeton, NJ, 2008.
14. B. M. Bolker, S. W. Pacala, and C. Neuhauser. Spatial dynamics in model plant communities: What do we really know? *American Naturalist*, 162:135–148, 2003.
15. J.H. Brown. *Macroecology.* The University of Chicago Press, Chicago, 1995.
16. J.H. Brown and D.W. Davidson. Competition between seed-eating rodents and ants in desert ecosystems. *Science*, 196:880–882, 1977.

17. J.H. Brown and A. Kodric-Brown. Turnover rate in insular biogeography: effect of immigration on extinction. *Ecology*, 58:445–449, 1977.
18. K P. Burnham and D. R. Anderson. *Model Selection and Multimodel Inference: A Practical Information-Theoretic Approach*. Springer, New York, 2002.
19. J. E. Byers, K. Cuddington, C. G. Jones, T. S. Talley, A. Hastings, J. G. Lambrinos, J. A. Crooks, and W. G. Wilson. Using ecosystem engineers to restore ecological systems. *Trends in Ecology & Evolution*, 21:493–500, 2006.
20. C.E. Caceres. Temporal variation, dormancy, and coexistence: A field test of the storage effect. *Proceedings of the National Academy of Sciences, U.S.A.*, 94:9171–9175, 1997.
21. T. J. Case. *An Illustrated Guide to Theoretical Ecology*. Oxford University Press, Inc., New York, 2000.
22. H. Caswell. Community structure: A neutral model analysis. *Ecological Monographs*, 46:327–354, 1976.
23. H. Caswell. *Matrix Population Models: Construction, Analysis, and Interpretation*. Sinauer Associates, Inc., Sunderland, MA, USA, 2nd edition, 2001.
24. J. M. Chase and M. Leibold. *Ecological Niches: Linking Classical and Contemporary Approaches*. University of Chicago Press, Chicago, 2003.
25. X. Chen and J. E. Cohen. Global stability, local stability and permanence in model food webs. *Journal of Theoretical Biology*, 212:223–235, 2001.
26. P. L. Chesson. Multispecies competition in variable environments. *Theoretical Population Biology*, 45:227–276, 1994.
27. P. L. Chesson. General theory of competitive coexistence in spatially-varying environments. *Theoretical Population Biology*, 58:211–237, 2000.
28. P. L. Chesson. Mechanisms of maintenance of species diversity. *Annual Review of Ecology and Systematics*, 31:343–366, 2000.
29. P. L. Chesson. Quantifying and testing coexistence mechanisms arising from recruitment fluctuations. *Theoretical Population Biology*, 64:345–357, 2003.
30. P. L. Chesson and R. R. Warner. Environmental variability promotes coexistence in lottery competitive systems. *The American Naturalist*, 117:923–943, 1981.
31. J. S. Clark, S. LaDeau, and I. Ibanez. Fecundity of trees and the colonization-competition hypothesis. *Ecological Monographs*, 74:415–442, 2004.
32. J. S. Clark and J. S. McLachlan. Stability of forest biodiversity. *Nature*, 423:635–638, 2003.
33. J. E. Cohen. Human population: The next half century. *Science*, 302:1172–1175, 2003.
34. J.E. Cohen, F. Briand, and C.M. Newman. *Community Food Webs: Data and Theory*, volume 20 of *Biomathematics*. Springer-Verlag, Berlin, 1990.
35. S. L. Collins and S. M. Glenn. Importance of spatial and temporal dynamics in species regional abundance and distribution. *Ecology*, 72:654–664, 1991.
36. R. Condit, N. Pitman, E. G. Leigh, J. Chave, J. Terborgh, R. B. Foster, P. Nunez, S. Aguilar, R. Valencia, G. Villa, H. C. Muller-Landau, E. Losos, and S. P. Hubbell. Beta-diversity in tropical forest trees. *Science*, 295:666–669, 2002.
37. J.H. Connell. Diversity in tropical rain forests and coral reefs. *Science*, 199:1302–1310, 1978.
38. J.H. Connell and W.P. Sousa. On the evidence needed to judge ecological stability or persistence. *The American Naturalist*, 121:789–824, 1983.
39. R.F. Constantino, J.M. Cushing, B. Dennis, and R.A. Desharnais. Experimentally induced transitions in the dynamic behaviour of insect populations. *Nature*, 375:227–230, 1995.
40. J. M. Craine. Reconciling plant strategy theories of grime and tilman. *Journal of Ecology*, 93:1041–1052, 2005.

41. M. J. Crawley and J. E. Harral. Scale dependence in plant biodiversity. *Science*, 291:864–868, 2001.
42. T. O. Crist and J. A. Veech. Additive partitioning of rarefaction curves and species-area relationships: Unifying alpha-, beta- and gamma-diversity with sample size and habitat area. *Ecology Letters*, 9:923–932, 2006.
43. T. O. Crist, J. A. Veech, J. C. Gering, and K. S. Summerville. Partitioning species diversity across landscapes and regions: A hierarchical analysis of alpha, beta, and gamma diversity. *American Naturalist*, 162:734–743, 2003.
44. D. T. Crouse, L. B. Crowder, and H. Caswell. A stage-based population model for loggerhead sea turtles and implications for conservation. *Ecology*, 68:1412–1423, 1987.
45. J. F. Crow and M. Kimura. *An Introduction to Population Genetics Theory.* Harper & Row, New York, 1970.
46. A. C. Davison and D. V. Hinkley. *Bootstrap Methods and Their Applicaiton.* Cambridge Series on Statistical and Probabilistic Mathematics. Cambridge University Press, New York, 1997.
47. J. A. Dunne, R. J. Williams, and N. D. Martinez. Food-web structure and network theory: The role of connectance and size. *Proceedings of the National Academy of Sciences of the United States of America*, 99:12917–12922, 2002.
48. S. P. Ellner and J. Guckenheimer. *Dynamic Models in Biology.* Princeton University Press, Princeton, NJ, 2006.
49. S. P. Ellner and P. Turchin. When can noise induce chaos and why does it matter: A critique. *Oikos*, 111:620–631, 2005.
50. B. A. Endress, D. L. Gorchov, and R. B. Noble. Non-timber forest product extraction: Effects of harvest and browsing on an understory palm. *Ecological Applications*, 14:1139–1153, 2004.
51. R. S. Etienne. A new sampling formula for neutral biodiversity. *Ecology Letters*, 8:253–260, 2005.
52. R. S. Etienne. A neutral sampling formula for multiple samples and an 'exact' test of neutrality. *Ecology Letters*, 10:608–618, 2007.
53. W.J. Ewens. *Mathematical Population Genetics, I. Theoretical Introduction*, volume 27 of *Interdisciplinary Applied Mathematics*. Springer, New York, 2nd ed. edition, 2004.
54. J.M. Facelli, P. Chesson, and N. Barnes. Differences in seed biology of annual plants in arid lands: A key ingredient of the storage effect. *Ecology*, 86:2998–3006, 2005.
55. R. A. Fisher, A. S. Corbet, and C. B. Williams. The relation between the number of species and the number of individuals in a random sample of an animal population. *Jounral of Animal Ecology*, 12:42–58, 1943.
56. G. F. Fussmann, S. P. Ellner, N. G. Hairston, L. E. Jones, K. W. Shertzer, and T. Yoshida. Ecological and evolutionary dynamics of experimental plankton communities. *Advances in Ecological Research*, 37:221–243, 2005.
57. D. L. Gorchov and D. Christensen. Projecting the effect of harvest on the population biology of goldenseal, *Hydrasts canadensis* l., a medicianl herb in Ohio, USA forests. In New York Botanical Garden Society for Economic Botany, editor, *Symposium on Assessing Local Management of Wild Plant Resources: Approaches in Population Biology*, 2002.
58. N. J. Gotelli. *A Primer of Ecology.* Sinauer Associates, Inc., Sunderland, MA, 3rd ed. edition, 2001.
59. N. J. Gotelli and R. K. Colwell. Quantifying biodiversity: procedures and pitfalls in the measurement and comparison of species richness. *Ecology Letters*, 4:379–391, 2001.

60. N. J. Gotelli and G. R. Graves. *Null Models in Ecology*. Smithsonian Institution Press, Washington, DC, 1996.
61. N. J. Gotelli and B. J. McGill. Null versus neutral models: what's the difference? *Ecography*, 29:793–800, 2006.
62. N.G. Gotelli and W.G. Kelley. A general model of metapopulation dynamics. *Oikos*, 68:36–44, 1993.
63. N.J. Gotelli. Metapopulation models: the rescue effect, the propagule rain, and the core-satellite hypothesis. *The American Naturalist*, 138:768–776, 1991.
64. J. L. Green and J. B. Plotkin. A statistical theory for sampling species abundances. *Ecology Letters*, 10:1037–1045, 2007.
65. K. L. Gross and P. A. Werner. Colonizing abilities of "biennial" plant species in relation to ground cover: implications for their distriubtions in a successional sere. *Ecology*, 63:921–931, 1982.
66. N. G. Hairston, Sr. *Ecological Experiments: Purpose, Design, and Execution*. Cambridge Studies in Ecology. Cambridge University Press, Cambridge, UK, 1991.
67. J.M. Halley. Ecology, evolution and 1/f-noise. *Trends in Ecology & Evolution*, 11:33–37, 1996.
68. P. A. Hamback, K. S. Summerville, I. Steffan-Dewenter, J. Krauss, G. Englund, and T. O. Crist. Habitat specialization, body size, and family identity explain lepidopteran density-area relationships in a cross-continental comparison. *Proceeding of the National Academy of Sciences, USA*, 104:8368–8373, 2007.
69. Hankin, R.K.S. *untb: ecological drift under the UNTB*, 2007. R package version 1.3-3.
70. I. Hanski. Dynamics of regional distribution: The core and satellite species hypothesis. *Oikos*, 38:210–221, 1982.
71. J. Harte, T. Zillio, E. Conlisk, and A. B. Smith. Maximum entropy and the state-variable approach to macroecology. *Ecology*, 89:2700–2711, 2008.
72. M.P. Hassell. *The Dynamics of Arthropod Predator-Prey Systems*, volume 13 of *Monographs in Population Biology*. Princeton University Press, Princeton, 1978.
73. A. Hastings. Disturbance, coexistence, history and competition for space. *Theoretical Population Biology*, 18:363–373, 1980.
74. A. Hastings. Transients: the key to long-term ecological understanding? *Trends in Ecology and Evolution*, 19:39–45, 2004.
75. A. Helm, I. Hanski, and M. Partel. Slow response of plant species richness to habitat loss and fragmentation. *Ecology Letters*, 9:72–77, 2006.
76. M. O. Hill. Diversity and evenness: a unifying notion and its consequences. *Ecology*, 54:427–432, 1973.
77. C.S. Holling. Some characteristics of simple types of predation and parasitism. *Canadian Entomologist*, 91:385, 1959.
78. C.S. Holling. Resilience and stability of ecological systems. *Annual Review of Ecology and Systematics*, 4:1–23, 1973.
79. R.D. Holt and G.A Polis. A theoretical framework for intraguild predation. *The American Naturalist*, 149:745–764, 1997.
80. H. S. Horn and R. H. MacArthur. Competition among fugitive species in a harlequin environment. *Ecology*, 53:749–752, 1972.
81. S. P. Hubbell. *A Unified Theory of Biodiversity and Biogeography*. Monographs in Population Biology. Princeton University Press, Princeton, N.J., 2001.
82. S. H. Hurlbert. The nonconcept of species diversity: A critique and alternative parameters. *Ecology*, 52:577–586, 1971.

83. G. C. Hurtt and S. W. Pacala. The consequences of recruitment limitation: Reconciling chance, history, and competitive differences between plants. *Journal of Theoretical Biology*, 176:1–12, 1995.
84. G. E. Hutchinson. *An Introduction to Population Ecology*. Yale University Press, New Haven, CT, 1978.
85. G. R. Huxel and K. McCann. Food web stability: The influence of trophic flows across habitats. *American Naturalist*, 152:460–469, 1998.
86. F. Jabot and J. Chave. Inferring the parameters of the neutral theory of biodiversity using phylogenetic information, and implications for tropical forests. *Ecology Letters*, 12:239–248, 2009.
87. S. A. Juliano. Nonlinear curve fitting: predation and functional response curves. In S. M. Scheiner and J. Gurevitch, editors, *Design and Analysis of Ecological Experiments*. Oxford University Press, 2001.
88. P. Kareiva and U. Wennergren. Connecting landscape patterns to ecosystem and population processes. *Nature*, 373:299–302, 1995.
89. C. K. Kelly and M. G. Bowler. Coexistence and relative abundance in forest trees. *Nature*, 417:437–440, 2002.
90. B. E. Kendall, J. Prendergast, and O. N. Bjornstad. The macroecology of population dynamics: Taxonomic and biogeographic patterns in population cycles. *Ecology Letters*, 1:160–164, 1998.
91. W. O. Kermack and W. G. McCormick. A contribution to the mathematical theory of epidemics. *Proceedings of the Royal Society, Series A*, 115:700–721, 1927.
92. C. J. Keylock. Simpson diversity and the shannon-wiener index as special cases of a generalized entropy. *Oikos*, 109:203–207, 2005.
93. S. E. Kingsland. *Modeling Nature*. University of Chicago Press, Chicago, 1985.
94. C.J. Krebs, S. Boutin, R. Boonstra, A.R.E. Sinclair, J.N.M. Smith, M.R.T. Dale, K. Martin, and R. Turkington. Impact of food and predation on the snowshoe hare cycle. *Science*, 269:1112–1115, 1995.
95. R. Lande. Extinction thresholds in demographic models of territorial populations. *American Naturalist*, 130:624–635, 1987.
96. R. Lande. Demographic models of the northern spotted owl (*Strix occidentalis caurina*). *Oecologia*, 75:601–607, 1988.
97. R. Lande. Statistics and partitioning of species diversity, and similarity among multiple communities. *Oikos*, 76:5–13, 1996.
98. R. Lande, P. J. DeVries, and T. R. Walla. When species accumulation curves intersect: Implications for ranking diversity using small samples. *Oikos*, 89:601–605, 2000.
99. A. M. Latimer, J. A. Silander, and R. M. Cowling. Neutral ecological theory reveals isolation and rapid speciation in a biodiversity hot spot. *Science*, 309:1722–1725, 2005.
100. L. P. Lefkovitch. The study of population growth in organisms grouped by stages. *Biometrics*, 21:1–18, 1965.
101. C.L. Lehman and D. Tilman. Competition in spatial habitats. In D. Tilman and P. Kareiva, editors, *Spatial Ecology: The Role of Space in Population Dynamics and Interspecific Interactions*, pages 185–203. Princeton University Press, Princeton, NJ, 1997.
102. M. A. Leibold, M. Holyoak, N. Mouquet, P. Amarasekare, J. M. Chase, M. F. Hoopes, R. D. Holt, J. B. Shurin, R. Law, D. Tilman, M. Loreau, and A. Gonzalez. The metacommunity concept: A framework for multi-scale community ecology. *Ecology Letters*, 7:601–613, 2004.

103. M. A. Leibold and M. A. McPeek. Coexistence of the niche and neutral perspectives in community ecology. *Ecology*, 87:1399–1410, 2006.
104. M.A. Leibold. A graphical model of keystone predators in food webs: trophic regulation of abundance, incidence, and diversity patterns in communities. *The American Naturalist*, 147:784–812, 1996.
105. Jr. Leigh, E. G. *Tropical Forest Ecology: A View from Barro Colorado Island.* Oxford University Press, Oxford, UK, 1999.
106. F. Leisch. Sweave: Dynamic generation of statistical reports using literate data analysis. In Wolfgang Härdle and Bernd Rönz, editors, *Compstat 2002 — Proceedings in Computational Statistics*, pages 575–580. Physica Verlag, Heidelberg, 2002.
107. P. H. Leslie. On the use of matrices in certain population mathematics. *Biometrika*, 35:183–212, 1945.
108. S. A. Levin, J. E. Cohen, and A. Hastings. Dispersal strategies in a patchy environment. *Theoretical Population Biology*, 26:165–191, 1984.
109. R. Levins. The strategy of model building in population biology. *American Scientist*, 54:421–431, 1966.
110. R. Levins. Some demographic and genetic consequences of environmental heterogeneity for biological control. *Bulletin of the Entomological Society of America*, 15:237–240, 1969.
111. R. Levins and D. Culver. Regional coexistence of species and competition between rare species. *Proceeding of the National Academy of Sciences, U.S.A.*, 68:1246–1248, 1971.
112. R.C. Lewontin. The meaning of stability. In *Diversity and Stability in Ecological Systems*, volume 22 of *Brookhaven Symposia in Biology*, pages 13–24, Upton, NY, 1969. Brookhaven National Laboratory.
113. R. Lincoln, G. Boxshall, and P. Clark. *A Dictionary of Ecology, Evolution and Systematics.* Cambridge University Press, Cambridge UK, 2nd edition, 1998.
114. R. Lindborg and O. Eriksson. Historical landscape connectivity affects prresent plant species diversity. *Ecology*, 85:1840–1845, 2004.
115. Z. T. Long and I. Karel. Resource specialization determines whether history influences community structure. *Oikos*, 96:62–69, 2002.
116. M. Loreau, N. Mouquet, and R. D. Holt. Meta-ecosystems: A theoretical framework for a spatial ecosystem ecology. *Ecology Letters*, 6:673–679, 2003.
117. A. J. Lotka. *Elements of Mathematical Biology.* Dover Publications, Inc., Mineola, NY, 1956.
118. R. H. MacArthur. On the relative abundance of bird species. *Proceedings of the National Academy of Sciences of the United States of America*, 43:293–295, 1957.
119. R. H. MacArthur. Some generalized theorems of natural selection. *Proceedings of the National Academy of Sciences, USA*, 48:1893–1897, 1962.
120. R. H. MacArthur. *Geographical Ecology: Patterns in the Distribution of Species.* Harper & Row, New York, 1972.
121. R. H. MacArthur and E. O. Wilson. An equilibrium theory of insular zoogeography. *Evolution*, 17:373–387, 1963.
122. R. H. MacArthur and E. O. Wilson. *The Theory of Island Biogeography.* Monographs in Population Biology. Princeton University Press, Princeton, 1967.
123. R.H. MacArthur and E.R. Pianka. On optimal use of a patchy environment. *The American Naturalist*, 100:603–609, 1966.
124. A. E. Magurran. *Measuring Biological Diversity.* Princeton University Press, 2004.

125. T. R. Malthus. *An Essay on the Principle of Population.* J. Johnson, London, 1798.
126. B.F.J. Manly. *Randomization, Bootstrap and Monte Carlo Methods in Biology.* Chapman and Hall, London, 1997.
127. R. M. May. *Stability and Complexity in Model Ecosystems*, volume 6 of *Monographs in Population Biology.* Princeton University Press, Princeton, NJ, 1973.
128. R. M. May. Biological populations with nonoverlapping generation: stable points, stable cycles, and chaos. *Science*, 186:645–647, 1974.
129. R. M. May. *Ecology and Evolution of Communities*, chapter Patterns of species abundance and diversity, pages 81–120. Harvard University Press, Cambridge, MA, 1975.
130. R. M. May. Thresholds and multiple breakpoints in ecosystems with a multiplicity of states. *Nature*, 269:471–477, 1977.
131. R. M. May. Host-parasitoid systems in patchy environments: A phenomenological model. *Journal of Animal Ecology*, 47:833–844, 1978.
132. R. M. May. *Stability and Complexity in Model Ecosystems.* Princeton Landmarks in Biology. Princeton University Press, 2001.
133. R. M. May. Network structure and the biology of populations. *Trends in Ecology & Evolution*, 21:394–399, 2006.
134. R.M. May. Will a large complex system be stable? *Nature*, 238:413–414, 1972.
135. R.M. May. *Theoretical Ecology: Principles and Applications.* Blackwell Scientific Publications, Oxford, 1976.
136. H. McCallum, N. Barlow, and J. Hone. How should pathogen transission be modeled? *Trends in Ecology and Evolution*, 16:295–300, 2001.
137. K. McCann. Density-dependent coexistence in fish communities. *Ecology*, 79:2957–2967, 1998.
138. K. McCann and A. Hastings. Re-evaluating the omnivory - stability relationship in food webs. *Proceedings of the Royal Society of London Series B-Biological Sciences*, 264:1249–1254, 1997.
139. A. McKane, D. Alonso, and R. V. Sole. Mean-field stochastic theory for species-rich assembled communities. *Physical Review E*, 62:8466–8484, 2000.
140. M. A. McPeek and S. Kalisz. Population sampling and bootstrapping in complex designs: Demographic analysis. In S.M. Scheiner and J. Gurevitch, editors, *Design and Analysis of Ecological Experiments*, pages 232–252. Chapman & Hall, New York, 1993.
141. F. Messier. Ungulate population models with predation - a case study with the north american moose. *Ecology*, 75:478–488, 1994.
142. P. J. Morin. *Community Ecology.* Blackwell Science, Inc., Malden, MA, 1999.
143. P. J. Morin. Productivity, intraguild predation, and population dynamics in experimental food webs. *Ecology*, 80:752–760, 1999.
144. H. Morlon, G. Chuyong, R. Condit, S.P. Hubbell, D. Kenfack, D. Thomas, R. Valencia, and J. L. Green. A general framework for the distance-decay of similarity in ecological communities. *Ecology Letters*, 11:904–917, 2008.
145. I. Motomura. A statistical treatment of associations [in Japanese]. *Japanese Journal of Zoology*, 44:379–383, 1932.
146. S. Nee and R.M. May. Dynamics of metapopulations: habitat destruction and competitive coexistence. *Journal of Animal Ecology*, 61:37–40, 1992.
147. M. Nei. *Molecular Evolutionary Genetics.* Columbia University Press, New York, 1987.
148. M.G. Neubert and H. Caswell. Alternatives to resilience for measuring the responses of ecological systems to perturbations. *Ecology*, 78:653–665, 1997.

149. A.J. Nicholson and V.A. Bailey. The balance of animal populations - Part I. *Proceedings of the Zoological Society of London*, pages 551–598, 1935.
150. I. Noy-Meir. Stability of grazing systems: An application of predator-prey graphs. *The Journal of Ecology*, 63:459–481, 1975.
151. J. Oksanen, R. Kindt, P. Legendre, R. O'Hara, G. L. Simpson, P. Solymos, M. H. H. Stevens, and H. Wagner. *vegan: Community Ecology Package*, 2008. R package version 1.15-1.
152. S. W. Pacala and M. Rees. Models suggesting field experiments to test two hypotheses explaining successional diversity. *The American Naturalist*, 152:729–737, 1998.
153. S.W. Pacala. Dynamics of plant communities. In M.J. Crawley, editor, *Plant Ecology*, pages 532–555. Blackwel Science, Ltd., Oxford, UK, 1997.
154. S.W. Pacala, M.P. Hassell, and R.M. May. Host-parasitoid associations in patchy environments. *Nature*, 344:150–153, 1990.
155. O.L. Petchey. Environmental colour affects the dynamics of single species populations. *Proceedings of the Royal Society of London B*, 267:747–754, 2001.
156. R. H. Peters. Some general problems in ecology illustrated by food web theory. *Ecology*, 69:1673–1676, 1988.
157. T. Petzoldt. R as a simulation platform in ecological modelling. *R News*, 3:8–16, 2003.
158. T. Petzoldt and K. Rinke. simecol: An object-oriented framework for ecological modeling in r. *Journal of Statistical Software*, 22:1–31, 2007.
159. S.L. Pimm and J.H. Lawton. Number of trophic levels in ecological communities. *Nature*, 268:329 331, 1977.
160. S.L. Pimm and J.H. Lawton. On feeding on more than one trophic level. *Nature*, 275:542–544, 1978.
161. J. Pinheiro and D. Bates. *Mixed-effects Models in S and S-PLUS*. Springer, New York, 2000.
162. W. Platt and I. Weis. Resource partitioning and competition within a guild of fugitive prairie plants. *The American Naturalist*, 111:479–513, 1977.
163. J. B. Plotkin, M. D. Potts, D. W. Yu, S. Bunyavejchewin, R. Condit, R. Foster, S. Hubbell, J. LaFrankie, N. Manokaran, L.H. Seng, R. Sukumar, M. A. Nowak, and P. S. Ashton. Predicting species diversity in tropical forests. *Proceedings of the National Academy of Sciences, U.S.A.*, 97(20):10850–10854, 2000.
164. J.B. Plotkin and H.C. Muller-Landau. Sampling the species composition of a landscape. *Ecology*, 83:3344–3356, 2002.
165. G. A. Polis, C. A. Myers, and R. D. Holt. The ecology and evolution of intraguild predation: potential competitors that eat each other. *Annual Review of Ecology and Systematics*, 20:297–330, 1989.
166. G.A. Polis. Complex trophic interactions in deserts: an empirical critique of food web ecology. *The American Naturalist*, 138:123–155, 1991.
167. D. M. Post. The long and short of food-chain length. *Trends in Ecology & Evolution*, 17:269–277, 2002.
168. F. W. Preston. The commonness, and rarity, of species. *Ecology*, 29:254–283, 1948.
169. F. W. Preston. Time and space and variation of variation. *Ecology*, 41:611–627, 1960.
170. F. W. Preston. The canonical distribution of commonness and rarity: part I. *Ecology*, 43:185–215., 1962.
171. S. Pueyo, F. He, and T. Zillio. The maximum entropy formalism and the idiosyncratic theory of biodiversity. *Ecology Letters*, 10:1017–1028, 2007.

172. H. R. Pulliam. Sources, sinks, and population regulation. *The American Naturalist*, 132:652–661, 1988.
173. R Development Core Team. *R: A language and environment for statistical computing*. R Foundation for Statistical Computing, Vienna, Austria, version 2.8.1 edition, 2008.
174. D. Rabinowitz and J. K. Rapp. Dispersal abilities of seven sparse and common grasses from a Missouri prairie. *American Journal of Botany*, 68:616–624, 1981.
175. H. L. Reynolds and S. W. Pacala. An analytical treatment of root-to-shoot ratio and plant competition for soil nutrient and light. *American Naturalist*, 141:51–70, 1993.
176. J. F. Riebesell. Paradox of enrichment in competitive systems. *Ecology*, 55:183–187, 1974.
177. M. L. Rosenzweig. Why the prey curve has a hump. *American Naturalist*, 103:81–87, 1969.
178. M. L. Rosenzweig. *Species Diversity in Space and Time*. Cambridge University Press, Cambridge, 1995.
179. M.L. Rosenzweig. Paradox of enrichment: destabilization of exploitation ecosystems in ecological time. *Science*, 171:385–387, 1971.
180. M.L. Rosenzweig and R.H. MacArthur. Graphical representation and stability conditions of predator-prey interactions. *American Naturalist*, 97:209–223, 1963.
181. J. Roughgarden. *Primer of Ecological Theory*. Prentice-Hall, Inc., Upper Saddle River, NJ, USA, 1998.
182. J. Roughgarden and F. Smith. Why fisheries collapse and what to do about it. *Proceedings of the National Academy of Sciences, U.S.A.*, 93:5078–5083, May 1996.
183. P. F. Sale. The maintenance of high diversity in coral reef fish communities. *The American Naturalist*, 111:337–359, 1977.
184. M. Scheffer, S. Szabo, A. Gragnani, E. H. van Nes, S. Rinaldi, N. Kautsky, J. Norberg, R. M. M. Roijackers, and R. J. M. Franken. Floating plant dominance as a stable state. *Proceeding of the National Academy of Sciences, U.S.A.*, 100:4040–4045, 2003.
185. O. J. Schmitz. Perturbation and abrupt shift in trophic control of biodiversity and productivity. *Ecology Letters*, 7:403–409, 2004.
186. A. Schröder, L. Persson, and A. M. De Roos. Direct experimental evidence for alternative stable states: A review. *Oikos*, 110:3–19, 2005.
187. A. Shmida and S. P. Ellner. Coexistance of plant species with similar niches. *Vegetatio*, 58:29–55, 1984.
188. R. M. Sibly, D. Barker, M. C. Denham, J. Hone, and M. Pagel. On the regulation of populations of mammals, birds, fish, and insects. *Science*, 309:607–610, 2005.
189. J. G. Skellam. Random dispersal in theoretical populations. *Biometrika*, 38:196–218, 1951.
190. K. Soetaert, T. Petzoldt, and R. W. Setzer. *deSolve: General solvers for ordinary differential equations (ODE) and for differential algebraic equations (DAE)*. R package version 1.2-3.
191. N. C. Stenseth, W. Falck, O. N. Bjornstad, and C. J. Krebs. Population regulation in snowshoe hare and Canadian lynx: Asymmetric food web configurations between hare and lynx. *Proceedings of the National Academy of Sciences of the United States of America*, 94:5147–5152, 1997.
192. P. A. Stephens and W. J. Sutherland. Consequences of the allee effect for behaviour, ecology and conservation. *Trends in Ecology & Evolution*, 14:401–405, 1999.

193. R. W. Sterner and J. J. Elser. *Ecological Stoichiometry: The Biology of Elements From Molecules to the Biosphere.* Princeton University Press, Princeton, N.J., 2002.
194. M. H. H. Stevens, O. L. Petchey, and P. E. Smouse. Stochastic relations between species richness and the variability of species composition. *Oikos*, 103:479–488, 2003.
195. M. H. H. Stevens and C. E. Steiner. Effects of predation and nutrient enrichment on a food web with edible and inedible prey. *Freshwater Biology*, 51:666–671, 2006.
196. G. Sugihara. Minimal community structure: an explanation of species abundance patterns. *The American Naturalist*, 116:770–787, 1980.
197. K. S. Summerville and T. O. Crist. Determinants of lepidopteran community composition and species diversity in eastern deciduous forests: roles of season, eco- region and patch size. *Oikos*, 100:134–148, 2003.
198. K. S. Summerville and T. O. Crist. Contrasting effects of habitat quantity and quality on moth communities in fragmented landscapes. *Ecography*, 27:3–12, 2004.
199. R. M. Thompson, M. Hemberg, B. M. Starzomski, and J. B. Shurin. Trophic levels and trophic tangles: The prevalence of omnivory in real food webs. *Ecology*, 88:612–617, 2007.
200. D. Tilman. *Resource Competition and Community Structure.* Monographs in Population Biology. Princeton University Press, Princeton, NJ, 1982.
201. D. Tilman. *Plant Strategies and the dynamics and structure of plant communities*, volume 26 of *Monographs in Population Biology.* Princeton University Press, Princeton, NJ, 1988.
202. D. Tilman. Competition and biodiversity in spatially structured habitats. *Ecology*, 75:2–16, 1994.
203. D. Tilman, R. M. May, C. L. Lehman, and M. A. Nowak. Habitat destruction and the extinction debt. *Nature*, 371:65–66, 1994.
204. David Tilman. Resource competition and plant traits: A response to Craine et al. 2005. *Journal of Ecology*, 95:231–234, 2007.
205. M. Tokeshi. Niche apportionment or random assortment: species abundance patterns revisited. *Journal of Animal Ecology*, 59:1129–1146, 1990.
206. M. Tokeshi. *Species Coexistence: Ecological and Evolutionary Perspectives.* Blackwell Science, Ltd, Oxford, UK, 1999.
207. J. Vandermeer. Omnivory and the stability of food webs. *Journal of Theoretical Biology*, 238:497–504, 2006.
208. J. Vandermeer. Oscillating populations and biodiversity maintenance. *Bioscience*, 56:967–975, 2006.
209. J. Vandermeer, I. de la Cerda, I. G.and Perfecto, D. Boucher, J. Ruiz, and A. Kaufmann. Multiple basins of attraction in a tropical forest: Evidence for nonequilibrium community structure. *Ecology*, 85:575–579, 2004.
210. J. Vandermeer and P. Yodzis. Basin boundary collision as a model of discontinuous change in ecosystems. *Ecology*, 80:1817–1827, 1999.
211. M. J. Vanni, W. H. Renwick, J. L. Headworth, J. D. Auch, and M. H. Schaus. Dissolved and particulate nutrient flux from three adjacent agricultural watersheds: A five-year study. *Biogeochemistry*, 54:85–114, 2001.
212. J.A. Veech and T.O. Crist. Partition: software for hierarchical additive partitioning of species diversity, version 2.0. http:// www.users.muohio.edu/ cristto/ partition.htm, 2007.
213. M. Vellend and M. A. Geber. Connections between species diversity and genetic diversity. *Ecology Letters*, 8:767–781, 2005.

214. I. Volkov, J. R. Banavar, F. L. He, S. P. Hubbell, and A. Maritan. Density dependence explains tree species abundance and diversity in tropical forests. *Nature*, 438:658–661, 2005.
215. I. Volkov, J. R. Banavar, S. P. Hubbell, and A. Maritan. Neutral theory and relative species abundance in ecology. *Nature*, 424:1035–1037, 2003.
216. I. Volkov, J. R. Banavar, S. P. Hubbell, and A. Maritan. Patterns of relative species abundance in rainforests and coral reefs. *Nature*, 450:45–49, 2007.
217. R. R. Warner and P. L. Chesson. Coexistence mediated by recruitment fluctuations: a field guide to the storage effect. *The American Naturalist*, 125:769–787, 1985.
218. D. Wedin and D. Tilman. Competition among grasses along a nitrogen gradient: initial conditions and mechanisms of competition. *Ecological Monographs*, 63:199–229, 1993.
219. R. H. Whittaker. Vegetation of the siskiyou mountains, oregon and california. *Ecological Monographs*, 30:279–338, 1960.
220. J. A. Wiens. Spatial scaling in ecology. *Functional Ecology*, 3:385–397, 1989.
221. R. J. Williams and N. D. Martinez. Simple rules yield complex food webs. *Nature*, 404:180–182, 2000.
222. S. D. Wilson and D. Tilman. Competitive responses of 8 old-field plant-species in 4 environments. *Ecology*, 76:1169–1180, 1995.
223. J. T. Wootton. Field parameterization and experimental test of the neutral theory of biodiversity. *Nature*, 433:309–312, 2005.

Index